This *Atlas of Galactic Neutral Hydrogen* contains maps showing the distribution of emission from atomic hydrogen, the principal component of the interstellar medium in the Milky Way, as measured over a five-year period with the 25-meter radio telescope of the Netherlands Foundation for Research in Astronomy. Each map corresponds to a particular velocity interval. The maps are displayed in several projections.

The Leiden/Dwingeloo survey covers the entire sky above declination –30°, on a half-degree grid, over a velocity range of 1000 kilometers per second at a resolution of 1 kilometer per second. The limiting brightness temperature sensitivity is less than 0.07 K.

Atlas of Galactic Neutral Hydrogen

Atlas of Galactic Neutral Hydrogen

Dap Hartmann & W. B. Burton

CAMBRIDGE
UNIVERSITY PRESS

PUBLISHED BY THE PRESS SYNDICATE OF THE UNIVERSITY OF CAMBRIDGE
The Pitt Building, Trumpington Street, Cambridge CB2 1RP, United Kingdom

CAMBRIDGE UNIVERSITY PRESS
The Edinburgh Building, Cambridge CB2 2RU, United Kingdom
40 West 20th Street, New York, NY 10011-4211, USA
10 Stamford Road, Oakleigh, Melbourne 3166, Australia

First published 1997

Printed in the United Kingdom at the University Press, Cambridge

Typeset by the author

A catalogue record for this book is available from the British Library

Library of Congress Cataloguing in Publication data available

ISBN 0 521 47111 7 hardback (and accompanying CD-ROM)

Contents

Preface

Hydrogen in its neutral atomic state is the principal component of the interstellar medium in our Galaxy. It is also probably the most easily observed interstellar constituent. But the sky is large, and the thought of surveying the entire accessible sky was a daunting one. We were fully aware that we were embarking on a half-decade voyage when the Netherlands Foundation for Research in Astronomy (NFRA) granted us the use of its Dwingeloo 25-meter radio telescope in 1988.

The Dwingeloo telescope is a venerable one. In fact, it is the oldest radio telescope in the world still operating. With this instrument the 3-kpc arm and other vagaries in the center of our Galaxy had been found, shortly after its dedication in 1956. The discovery of the still puzzling High-Velocity Clouds and subsequently of the Intermediate-Velocity Clouds followed in the early 1960s. But discussions on the future of the telescope, as it ended its 30th productive year, included the possibility of demolition. We accepted the Foundation's provision that use the telescope was granted on a do-it-yourself basis. Such use of a facility, indeed survey work in general, may not appeal to all. But we found the prospect of an all-sky survey of the fun one.

Operating out of a shed with a leaking roof, we would begin each observing session by emptying the strategically placed plastic buckets, mopping the floor, and adjusting the makeshift tents draped over the antiquated but adequate computer and pen plotters. (The punched-card readers, paper-tape devices, and Brown recorders had long since been abandoned to their watery fate). The out-dated data-acquisition software was rewritten as the survey began. The data-pipeline software had to be designed for an early-generation PC; all of the data manipulation (with the exception of the stray-radiation calculations) and most of the maps displayed in this Atlas were generated with this home-brewed software operating on a single PC.

Although we enjoyed an interlude of doing science in a way more common in an earlier era, we would be exaggerating if we implied that the NFRA interpreted their do-it-yourself terms as leaving us entirely to our own devices. The Foundation supported our efforts in a variety of ways. Most importantly, although the mechanical structure had not changed fundamentally since the late 1950s, the telescope had been outfitted with modern electronics. The receiver we used was similar to those currently in operation on the Westerbork Synthesis Telescope. Particularly important to the kinematic coverage of the survey was the Dwingeloo Autocorrelation Spectrometer (DAS), developed by Albert Bos as a prototype for the DAS now the workhorse spectrometer at the James Clerk Maxwell Telescope on Mauna Kea, Hawaii. The NFRA staff carried out all heavy maintenance on the telescope structure and ensured that the electronics remained in good order. The data-acquisition program was debugged and modified where needed, and routine checks on the vitality of the observing program were done in our absence. During the five-year observing run (the entire first year was devoted to observations which helped debugging the autocorrelator and data-acquisition software), the telescope wore out several pairs of bearings, had a cryogenic compressor replaced, was abused by vandals who placed coins, cans, and tree branches on the azimuth track, and was visited by burglars who stole and destroyed some non-essential equipment, which cluttered up the telescope cabin in greater amounts than the essential equipment.

The final quality of the survey depends crucially on the success of the correction for stray radiation. Most astronomical instruments map a point source as a blurred or somewhat distorted image. The antenna pattern, the response of the antenna to the radiation presented to it, is the radio-astronomy analogue to the point-spread function, more familiar to optical astronomers. A telescope like the Dwingeloo 25-m is sensitive not just in the direction in which it is pointed, but also to emission from all over the sky, as well as to radiation reflected off the ground. The radiation thus spuriously recorded accounts for as much as half of the total emission measured in some directions.

Not only was the telescope sensitive to 21-cm line radiation coming from all directions on the sky, but its response was furthermore different for different times of day and year, and changed as the telescope pointed to different directions on the sky, and also depended on the orientation of the telescope with respect to the ground. Because the antenna was sensitive to the entire hydrogen sky, we had to measure the complete sky first, before this spurious emission could be calculated and subtracted. This meant that we had to sail blindly during four full years of observing. During that time, diligently repeated observations of single directions differed substantially, and the extensive series of standard calibration fields showed bewildering scatter. Only after the entire northern sky had been observed could the stability of the system be verified.

Before the survey commenced, we had established contact with our colleagues Peter Kalberla and Ulrich Mebold at the University of Bonn. They are the acknowledged authorities on the problems of stray radiation in a parabolic antenna like that in Dwingeloo. After the final spectrum had been recorded in Dwingeloo, the Bonn algorithm was adjusted to the Dwingeloo telescope, and applied to the observations. The repeated spectra now matched to within the specifications; the calibration spectra showed that the system had been stable over the entire period of the survey. We are grateful to our Bonn colleagues for their efforts in this crucial matter.

We wish also to record our gratitude to our colleagues Harvey Liszt, Thomas Bania, and Gerrit Verschuur, whose early advice and subsequent interest in the progress of the survey was a source of valued encouragement.

The geographical latitude (+53°) of Dwingeloo in the astronomical far north restricted our efforts to declinations above −30°. At the Instituto Argentino de Radioastronomia near Buenos Aires, Esteban Bajaja and his group are now engaged in observing the southern sky in order to fill the gaps so obvious in the maps in this Atlas. They are using the 100-foot IAR radio telescope, equipped with a receiver and a spectrometer quite similar to what we used in Dwingeloo. When their work is completed the combined material will cover the sky fully in a quite homogeneous manner.

The Dwingeloo telescope has recently entered its fifth decade of full-time observing, operating with modest means and exploiting its principal, in fact probably only, advantage over more modern facilities, namely time. There remain a number of survey and monitoring projects for which this instrument, which we regard with great affection, may well continue to make useful contributions.

Dap Hartmann, *Center for Astrophysics, Harvard University*

W.B. Burton, *Sterrewacht, Leiden University*

1. Introduction

This Atlas presents in graphical as well as digital form observations made with the Dwingeloo radio telescope of emission from neutral atomic hydrogen, HI, lying in our own Galaxy. Observations of HI provide information on a broad range of physical and morphological characteristics of the galactic interstellar environment, including information on temperature, density, and motion. HI contributes more mass than any other observed interstellar constituent. It is furthermore so widely spread that no direction devoid of HI has ever been observed. Under most circumstances, the interstellar medium is transparent enough to emission at 21-cm wavelength that HI investigations refer to much of the entire Galaxy, well beyond the optical horizon. Under less common circumstances, the optical depth of the HI line is high, so that information on the gas temperature is revealed. Because an intrinsically narrow spectral line is involved, the observations measure the kinematics of the interstellar medium in great detail.

Several recent reviews stressing the astrophysical relevance of HI data are available. The review by Dickey & Lockman (1990) discusses HI evidence regarding temperature and density, especially pertaining to the local, vertical variation of these parameters; the review by Kulkarni & Heiles (1988) discusses the differing temperature regimes represented in 21-cm spectra; and the reviews by Burton (1988, 1992) deal with aspects of galactic morphology amenable to studies using the HI tracer.

Almost three decades after the first major northern-sky surveys of galactic neutral atomic hydrogen were made with the former Hat Creek radio telescope, new material was needed to deepen understanding of the interstellar medium. The availability of high-quality all-sky surveys at other wavelengths, in particular in the far-infrared and in the X-ray regimes, prompted the undertaking of a new HI survey using modern equipment. The observational parameters of earlier surveys of the HI 21-cm emission line which have been used for general investigations of galactic morphology and of the galactic interstellar medium have been tabulated by Burton (1988, 1992). The principal parameters are the extent and resolution of the coverage of the sky, the extent and resolution of the coverage in velocity, and the sensitivity to weak emission. The Leiden/Dwingeloo survey presented here improves on the earlier material in each of these parameters.

The principal merits of the new data compared with those of the Hat Creek 85-ft survey of Weaver & Williams (1973) lie in the improved sensitivity and sky coverage; compared with the Hat Creek survey of Heiles & Habing (1974), in the improved sensitivity and velocity coverage; compared with the Leiden/Green Bank 140-ft survey of Burton (1985), in the improved sky coverage and resolution; compared with the Bell Labs survey of Stark *et al.* (1992), in the improved velocity and spatial resolution.

The surveys of galactic HI emission from the southern sky made by Kerr *et al.* (1986) using the Parkes 60-ft telescope at $|b|<10°$, and at higher $|b|$ by Cleary, Heiles, & Haslam (1979) using the Parkes 60-ft and by Colomb, Pöppel, & Heiles (1980) using the IAR 100-ft, remain the standards at $\delta<-30°$, although efforts are currently underway to map the southern sky with modern receivers.

Without attempting a complete listing of earlier 21-cm survey work, we mention two previous Dwingeloo surveys, namely that of Tolbert (1971; see also

Wesselius & Fejes 1973), which remained a prime source of information on the class of intermediate-velocity clouds seen at high latitudes until the appearance of the Bell Labs survey, and the high-velocity-cloud survey of Hulsbosch & Wakker (1988; see also Wakker 1990) which provided the first comprehensive sky coverage of this class of objects.

An older telescope can remain competitive if equipped with state-of-the-art receivers and if the project is such that the modest size of the telescope can be compensated for by a major investment in observing time. It is the case for optical as well as for radio telescopes that technological developments regarding mirrors or antennas, and the general supporting mechanics, have changed more slowly than the detector or receiver electronics. Furthermore, telescopes of modest size are the only ones practicable for survey work attempting coverage of major portions of the sky.

The observations in this Atlas utilized the Dwingeloo 25-meter radio telescope of the Netherlands Foundation for Research in Astronomy (NFRA) for a five-year period. The telescope structure itself had been little changed since its dedication in 1956; the antenna surface had been renewed in 1969 and little changed since then. The detector electronics were, however, of competitive modern design. Improvements in the velocity coverage were made possible by the new-generation DAS spectrometer whose prototype was installed on the Dwingeloo telescope as the project began; improvements in the sensitivity were made possible by the availability of a low-noise receiver of the same sort currently in use on the telescopes of the Westerbork Synthesis Radio Telescope (WSRT).

2. Project motivation

When the Netherlands Foundation for Research in Astronomy was approached for support for the survey, including full-time use of the Dwingeloo radio telescope over a period of some years, our proposal stressed several particular uses of the data which we hoped to gather. Once completed, the new material would be valuable in addressing a range of scientific problems. Among the problems of most interest to us are the following:

1. Many interstellar cirrus dust features have gas counterparts traced by HI to the anomalous-velocity realm of intermediate-velocity clouds (IVCs, with velocities differing from the simply rotating, plane-parallel case, by some 40 to 70 $\mathrm{km\,s^{-1}}$). IVC structures, furthermore, are commonly found correlated with normal-velocity material. The all-sky topography of the IVC gas would be established and its relation to the normal-velocity HI explored. The link of HI at anomalous velocities with dust cirri had suggested the importance of exploring the possibility of some of the HI having been produced by star-formation processes, rather than being precursor to them; the IVC gas would then be material which has been processed near the galactic plane and subsequently accelerated away from the plane as a consequence of star-formation activity. The new data would reveal additional details of the high-velocity cloud (HVC) distribution. The HVC/IVC distinction still remains arbitrary; an effort would be made to establish if this distinction has a physical basis.
2. Regions of exceptionally low total N_{HI} in addition to the 'hole' studied by Lockman, Jahoda, & McCammon (1986) would be sought. Perhaps even more interesting would be regions in which the N_{HI} at *conventional* velocities (within, say, 20 $\mathrm{km\,s^{-1}}$ of $v_{\mathrm{lsr}}=0$ $\mathrm{km\,s^{-1}}$) is exceptionally low ($<7\times10^{19}$ $\mathrm{cm^{-2}}$), but where substantial densities do occur at *anomalous* velocities. Regions of low N_{HI} are par-

ticularly interesting because of the role of HI in obscuring X-rays. Indications would be pursued that X-ray shadows are cast by HI features in regions of low total $N_{\rm HI}$.

3. The extension of velocity information beyond the coverage and resolution available earlier would enable analysis of the HI data in terms of the areal filling factor; derivation of the volume filling factor would be a more difficult matter, although important constraints could be put on this parameter even using the $0\overset{\circ}{.}5$ angular resolution of the new survey. The new material was expected to show many structures with velocity-width dispersions of the order of 1 $\rm km\,s^{-1}$ and less, and which are clearly isolated in velocity; the commonly accepted picture of high-latitude HI which attributes variations in $N_{\rm HI}$ largely to gradients in a generally smooth, diffuse distribution is one which the new data might well be able to refute. Determinations of filling factors (and, in general, of characteristic kinematic- and structural- scale lengths) are important to discussions of interstellar energetics, in particular of turbulence and scale-height maintenance, of radiation penetration, and of global HI optical depth.
4. The gas-to-dust interrelationships among HI column densities, Lick galaxy counts, and reddening of galactic and extragalactic objects had been analyzed by Burstein and Heiles (1978, 1982), leading to the establishment of correlations predicting galactic reddening. Their work raised important astrophysical questions which could not be verified with the data then at hand: a zero-point offset was found in the relation between reddening and $N_{\rm HI}$ (that is, regions with essentially no reddening showed nevertheless substantial HI intensities), and the HI gas-to-dust ratio was found to vary widely from region to region, for physical reasons not identified. The Burstein and Heiles work had been based on Berkeley HI data so confined in velocity that in many directions much of the gas was ignored (in particular the crucial contribution from the warped outer layer), as were spectra of sensitivity lower than what would characterize the proposed new data; the information on the diffuse dust is also richer now than when the 1978 analysis was carried out. The gas/dust/reddening problem would be re-analyzed, exploiting the qualities of the new HI survey and of the IRAS and other complementary data. Particular attention would be paid to the evident breakdown of the correlation between HI and dust emissivities, which is tight in the inner Galaxy and locally, but evidently not in the outer Galaxy, probably because of changing environmental conditions.
5. Much HI gas is marshalled in filamentary structures which have been variously named from a subjective lexicon as shells, supershells, bubbles, worms, chimneys, etc. The form and motions of such structures reveal aspects of the macroscopic energetics of the ISM. They have not yet been fully studied, largely because of limitations in the available HI data, the most important limitation being the narrow velocity extent of the $|b|>10^\circ$ Berkeley data. A few shells have been traced to velocities well outside the range of the earlier survey data. The larger extent suggests an upward revision of the currently accepted energetics; if verified as general, this conclusion would be important to interpretation of the IVC class of objects, as well as to considerations of the energetics.

In entertaining new proposals in 1987, the NFRA Board had stipulated that future use of the Dwingeloo telescope would be granted only on a 'do-it-yourself' basis, with essential maintenance provided by the Foundation, but without routine operating or data-reduction support. Our proposal was granted, on this basis, for a 1988 start.

3. Background

During World War II, J.H. Oort, professor of astronomy at the Leiden Observatory, and H.C. van de Hulst, graduate student then at the University of Utrecht, discussed whether it would be possible to observe the general gaseous interstellar medium at radio wavelengths. Oort was fully aware of the pioneering work of Jansky and of Reber, which had demonstrated the presence of cosmic *continuum* radiation. He realized that observations of a radio *spectral line* would reveal the kinematics of the interstellar medium, and that distance measures, so notoriously difficult in astronomy, would follow from the kinematics. Van de Hulst set out to investigate which spectral lines might be observable at radio frequencies. At the 75th meeting of the Netherlands Astronomers' Club, which was held at the Leiden Observatory on 15 April, 1944, he presented the results of his calculations predicting that the hyperfine transition of neutral atomic hydrogen, emitting at a wavelength of 21.106 cm, should be observable (see van de Hulst 1945). The minutes of this meeting are reproduced in facsimile and in translation on pages 84-85. Van de Hulst's prediction was based on his realization that the number of hydrogen atoms in the ground state along a line of sight traversing the entire Milky Way would be very large. He also recognized that the density of atoms would be high enough for the hyperfine transition to be stimulated by encounters; indeed, transitions stimulated by encounters between atoms would occur much more frequently than spontaneous ones from isolated atoms.

The high expectations for the importance of radio astronomy led to the establishment, on 23 April, 1949, of the Netherlands Foundation for Radio Emission from Sun and Milky Way (Stichting Radiostraling Zon en Melkweg, SRZM), with Oort as the first chairman. In addition to astronomers, the founders of the SRZM included scientists from the Philips Physics Laboratory in Eindhoven, representatives of the Dutch Post- Telephone- and Telegraph- company (PTT) who were concerned with the effects of solar eruptions on the ionosphere, and meteorologists from the Royal Dutch Meteorology Institute (KNMI) who had similar interests. Several radar dishes of the Würzburg-Riesen class, which had been part of the German radar defenses deployed along the North Sea coast during the War, were transported from their locations in the dunes to the PTT central transmission station near the town of Kootwijk. One of these 7.5-m dishes was made available to the SRZM for use as a radio telescope.

The HI line was first detected by H.I. Ewen and E.M. Purcell in 1951 (on Easter morning) at Harvard University. Work in Kootwijk had proceeded diligently in early 1951, but had been delayed some months by a fire in the receiver cabin. When the receiver had been re-built, C.A. Muller and J.H. Oort succeeded in detecting the spectral line on 11 May, 1951. Within weeks, J.L. Pawsey confirmed Christiansen's and Hindman's measurement of the line in Australia. In an admirable display of scientific cooperation, all three early detections were published side-by-side in the same, 1 September, 1951, issue of *Nature*.

During the next five years, new receivers were developed in the laboratory at Kootwijk. Meanwhile, the 7.5-m telescope was used nearly full-time to make the first maps of neutral atomic hydrogen in the plane of the Galaxy, and to demonstrate fundamental aspects of galactic rotation. The first results of this pioneering work were published in 1954 (see van de Hulst, Muller, & Oort, and Kwee, Muller, & Westerhout, as well as other articles in Vols. 12 and 13 of the *Bulletin of the Astronomical Institutes of the Netherlands*). Before the Kootwijk telescope was effectively superseded by larger facilities, it had also provided important data supporting the first estimates of the thickness of the gaseous disk of the Milky Way, had led to measures of the characteristic optical depth and gas temperature of the interstellar HI, and had contributed to the determination of the direction of the center of the Milky Way and of the fundamental planes of galactic longitude and latitude.

As early as 1945, Oort had argued to the Royal Dutch Academy of Sciences that a radio telescope with a diameter as large as 25 meters should be built to map the Galaxy thoroughly in the 21-cm line. At that time the War had only recently ended, and such an ambitious project could not be realized. But a decade later, after the successful work done with the Kootwijk telescope, Oort, as chairman of the SRZM, did manage to obtain a grant for building a 25-m telescope.

The location of the site for the new observatory was partly motivated by the requirement of radio silence. A nature reserve near the village of Dwingeloo (in the relatively sparsely populated province of Drenthe in the northern Netherlands) was expected to remain undisturbed by radio interference in the foreseeable future. The telescope was built at the edge of this reserve, where the forest meets the heath. An important consideration behind approval of the project by the authorities of the nature reserve was the observatory's intent to minimize the presence of motorized vehicles in the area.

The Dwingeloo telescope received 'first light' in November, 1955, when a rare lunar occultation of the Crab Nebula (Tau A) occurred. Although the telescope was then not yet fully operational, the occultation was successfully observed at a frequency of 400 MHz, with pointing and tracking done manually. The official opening of the telescope took place on 16 April, 1956, when Queen Juliana pressed a button setting in motion what was then the largest telescope in the world. The Dwingeloo telescope has now been in continuous operation for almost 40 years, longer than any other radio telescope.

4. The Dwingeloo Telescope

Accounts of various aspects of the background of the Dwingeloo 25-m telescope and its scientific achievements have been given by Kleibrink (1957), Westerhout (1961), van Woerden, Brouw, & van de Hulst (1980), Spoelstra (1981), Sullivan III (1982), van Herk, Kleibrink, & Bijleveld (1983), and Hartmann (1994), as well as in the annual reports of the Netherlands Foundation for Radio Emission from Sun and Milky Way, now renamed the Netherlands Foundation for Research in Astronomy.

The Dwingeloo telescope has contributed to a wide range of astronomical research. Among its principal contributions made in 21-cm studies of our own Galaxy include the discovery of the 3-kpc arm, studies of the galactic center, work directed at mapping the Galaxy-at-large and determining its rotation and general kinematic structure, discovery of the high-velocity-cloud and intermediate-velocity-cloud classes of objects, as well as studies of individual regions of particular astrophysical interest. Hartmann (1994) has listed the Ph.D. theses which were based on observations made with the 25-m telescope.

The Dwingeloo radio telescope has a reflecting mirror with a diameter of 25 meters and a focal length of 12 meters. The antenna surface is constructed of 372 triangular frames each with sides 1.8-m long. Each frame was covered originally with a 15×15 mm wire mesh, woven from steel wire with a diameter of 1.5 mm. Crossing wires were welded together and the entire surface galvanized. The dish moves up and down in elevation in a horizontally-mounted support structure. The support structure is constructed as a robust steel frame resting on four wheels, each 80 cm in diameter. It can rotate in azimuth over a circular steel track 16 meters in diameter. The weight of the dish alone is 18 tonnes. The total weight of the telescope is 120 tonnes, half of which is carried by a central pivot bearing. When constructed,

Figure 1. *The 25-meter Dwingeloo radio telescope of the Netherlands Foundation for Research in Astronomy. The observations in this Atlas utilized this telescope for a continuous five-year period, compensating for the modest size of the antenna by a substantial investment in time.*

the receiver feed was suspended at the focus of the telescope by a single central pole of 15 cm diameter, guyed with three steel wires to the edge of the dish. The single-channel Dicke-switched receiver weighed 25 kg; it was not cooled, and had a system temperature of about 400 K.

During the 20 years following its official opening, three major modifications were made to the telescope structure. The first modification was required by the new front-ends which had been developed with feeds that better illuminated the mirror and had lower sidelobe responses. They were, however, bigger (blocking more of the beam) and heavier (weighing about 150 kg) than the old front-end and could not be supported by the single pole. In late 1961, the single pole supporting the receiver was replaced by a more stable tripod construction.

The second structural modification involved re-surfacing the antenna in order that research carried out using the telescope could be extended to wavelength regimes shorter than the 21-cm one. In 1969 a new 7.7×7.7 mm mesh made from 0.8-mm diameter stainless-steel wire was applied to the triangular panels, allowing observations at wavelengths as short as 6 cm. To stabilize the parabolic shape of the dish, the frame of each panel was kinked in the middle of each side. Deviations from a perfect parabolic shape were now less than 1 mm.

The third modification followed the development of cryogenically cooled receivers, which again increased the size and weight of the front-ends beyond the capacity of the feed-support structure. In 1973–74, the tripod was replaced by a quadripod. The Dwingeloo telescope was now also suited to test front-ends for the Westerbork Synthesis Radio Telescope. The relevance of these structural modifications to the HI survey presented in this Atlas is discussed below.

In 1974, the Dwingeloo telescope was equipped with a prototype of the WSRT multi-frequency front-end receiver (Casse, Woestenburg, & Visser 1982). It was a cryogenically-cooled system, with a 256-channel digital auto-correlator spectrometer back-end, which also was a prototype of a new WSRT instrument. A wide range of bandwidths could be selected. The system temperature was about 40 K, half that of the previous front-end. The receiver was improved in 1981, when the two-step parametric amplifier was replaced by a helium-cooled FET amplifier with a noise temperature of 13 K. The total system temperature is now about 35 K when observing directions of low HI column density. The receiver is equipped with a single channel which is matched to the linearly polarized dipole antenna.

The 256-channel spectrometer was replaced in 1988 by the 1024-channel prototype of the Dwingeloo Auto-correlation Spectrometer (DAS; see Bos 1989), which had been developed at the NFRA as the common-user spectrometer for the British/Dutch/Canadian James Clerk Maxwell Telescope on Hawaii. Equipped with a receiver of the WSRT class and with the DAS back-end, the Dwingeloo telescope was thus outfitted with state-of-the-art electronics; this was the instrumental setup pertaining at the time of our survey.

5. Observing strategy and parameters

The availability of a 1024-channel back-end guided the choice of the bandwidth. To cover the entire velocity range expected from galactic HI, including all but the most extreme high-velocity clouds, we observed at 5-MHz bandwidth, covering a velocity interval of 1055 $\mathrm{km\,s^{-1}}$ centered at $v_{\mathrm{lsr}}=0$ $\mathrm{km\,s^{-1}}$ (central channel). After reduction, the effective useful velocity range was $-450<v_{\mathrm{lsr}}<+400$ $\mathrm{km\,s^{-1}}$, sampled at an interval of 1.03 $\mathrm{km\,s^{-1}}$ between channels. (Radial velocities in this project were expressed relative to the Local Standard of Rest, defined in terms of the Standard Solar Motion of 20 $\mathrm{km\,s^{-1}}$ toward $(\alpha,\delta)_{1900}=(18^{\mathrm{h}}, 30°)$.)

The half-power beam width (*HPBW*) of the Dwingeloo telescope at 1420 MHz is 36 arcminutes, or 0°.6. A fully sampled survey would require positional spacings of 0°.3. A 5°×5° region would then contain about 300 pointed observations, and the total number of observations for the survey would have exceeded 550,000. We decided instead to sample the sky with a 0°.5 (true-angle) spacing. At 85% of the *HPBW*, or 60% Nyquist sampling, the estimated total number of grid points was reduced by a factor of 2.8.

The choice of the integration time, t_{int}, was influenced by two principal factors. The integration time was to be long enough to improve the signal-to-noise ratio over that of earlier work. The completed survey is characterized by an *rms* noise of about 0.07 K. On the other hand, the integration time was to be short enough to allow mapping the entire accessible sky in a reasonable amount of time. The expected number of observations was of the order 2×10^5. This implied a net total integration time of about 140 days per net minute of integration time per spectrum. An additional overhead time for each spectrum was about 90 seconds. We were thus led to a choice of $t_{\rm int}$=180 seconds, implying an estimated net telescope time of 625 days. As the survey progressed, we realized the importance of repeated observations of single positions, particularly for adequate tuning of the stray-radiation correction procedure. In addition, a variety of unforeseen software, hardware, and weather problems lengthened the expected time commitment. As it turned out, the Dwingeloo telescope was dedicated to this project for a period of five years.

The HI spectral line contributes only a small fraction of the total power received and recorded by the telescope system during an observation. Most of the power originates from instrumental noise, from the far sidelobes of the antenna looking at the ground, and from true cosmic background continuum radiation. The various amplifiers and filters in the receiver system introduce frequency-dependent gain variations, shaping the bandpass of the signal that enters the spectrometer. Removing the bandpass characteristics from the observations requires determining what the bandpass shape would be if *no* spectral-line emission were present. There are three general observing modes designed to accomplish this: beam-switching, load-switching, and frequency-switching.

In the *beam-switching* mode, a signal spectrum ('on source') is compared with a reference spectrum ('off source') close by in position (typically a few beam widths away) and hopefully devoid of emission. For 21-cm spectral-line observations this method generally is not useful, as not a single line of sight has ever been found devoid of HI emission.

In the *load-switching* mode, instead of switching to a patch of emission-free sky, a reference spectrum is obtained from a cold matched resistor in a cryogenic system. A white-noise signal is added to balance the output power with that of the signal spectrum. However, gain variations over the frequency band are likely to be different in the reference spectrum than in the signal spectrum, introducing baseline problems in the combined spectrum.

The *frequency-switching* mode is the most commonly-used technique for galactic 21-cm spectral-line observing. Tuning the receiver away from the spectral window containing the HI line, a reference spectrum may be obtained; as the frequency switch takes place at the first local oscillator, the shape of the bandpass is affected by the consequent amplifiers, filters, and oscillators in very much the same way as the signal spectrum. An important disadvantage of the frequency-switching mode is that this mode doubles the required integration time.

We chose the so-called total-power variation of the frequency-switching mode. The local oscillator was tuned to set the frequency for the central channel to that corresponding to $v_{\rm lsr}$=0 km s^{-1}. A total-power measurement consists of two cycles, S and $(S+N)$, yielding a correlation function and total power for the signal cycle, S, for the duration of $t_{\rm int}$, and a total power for the signal+noise cycle, $S+N$, for the duration of $t_{\rm noise}$. Reference spectra were obtained by regularly observing total-power spectra toward selected grid points at a central frequency that was a full bandwidth (5 MHz) higher than the $v_{\rm lsr}$=0 km s^{-1} setting. The instrumental baseline of a number of survey spectra could then be accounted for by a single reference spectrum.

About 10% of all survey grid points measured were such references. In addition to the observations made on the survey grid, certain standard directions were observed in frequency-switching mode. These standards served as stability calibrators for the absolute intensity-scale conversion, from antenna temperature to brightness temperature, and as consistency calibrators for the stray-radiation cor-

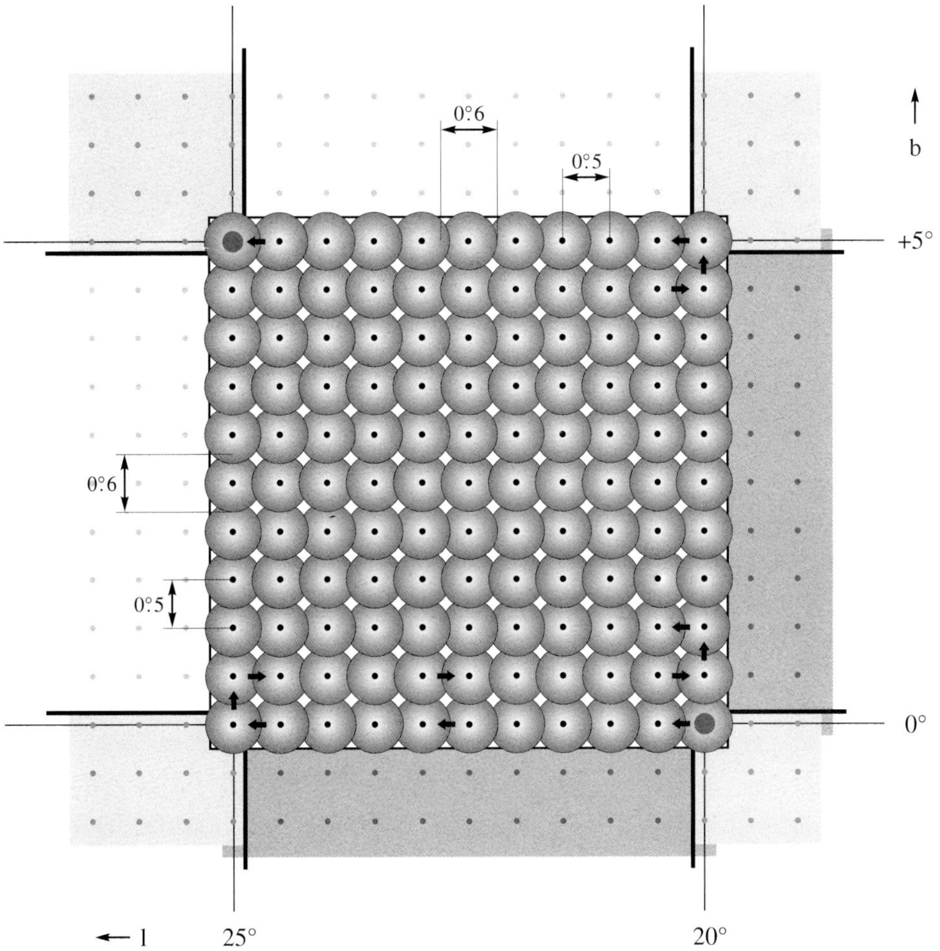

Figure 2. *Distribution of survey grid points in a 5°×5° box. In this particular example, l_{min}=20° and b_{min}=0°. The longitude spacing here is Δl=0°.5 (see Table 1); Δb is equal to 0°.5, as for all boxes. Each grid point is marked by a dot surrounded by a circle indicating the size of the HPBW. The arrows and the overlap of the circles indicate the direction in which the grid was traversed. Reference spectra were observed at the first and last grid points (indicated by the large dots). In this example, eight References and 121 Sources were measured. This box has 40 grid points in common with its four neighboring boxes; these positions were subsequently measured again, yielding a total of 36 doubly and four quadruply observed directions.*

rection procedure described below. They also provided a safeguard against general problems such as pointing errors and gain variations.

Code was written to schedule the observations optimally. The entire sky accessible from Dwingeloo (longitude $6^h23^m48^s$ east, latitude +52°48′48″) was divided into galactic-coordinate 'boxes', each measuring 5°× 5°. If the center of a box was north of declination −32°.5 , it was marked for observing. The strategy was to observe low-declination boxes as early during the survey as possible. In the final stages of the data taking, remaining boxes were largely circumpolar; these were scheduled by hand, as were the galactic polar-cap regions (at $|b|$>85°) and the positions that for some reason needed to be re-observed.

Within each marked box a uniform rectangular grid was specified, as illustrated in Figure 2. The objective was to observe all the grid points in a particular box in a single contiguous run. The latitude separation, Δb, was constant for all boxes, and equal to 0°.5 (85% of the HPBW). The longitude separation, Δl, was con-

Table 1. *Longitude separation of the grid points in an observing 'box', as function of the galactic latitude of the center of the box*

$\|b\|$	Δl
0° ... 35°	0°.5
35° ... 45°	0°.6
45° ... 55°	0°.7
55° ... 60°	0°.8
60° ... 65°	1°.0
65° ... 70°	1°.1
70° ... 75°	1°.4
75° ... 80°	1°.9
80° ... 85°	2°.8
85° ... 90°	5°.0

stant within each individual box, and chosen such that the true angular sky separation was 0°.5 or less. For $|b|>85°$, the longitude interval was 5°.0. The longitude sampling interval for the different latitude strips is given in Table 1. A box at $|b|<35°$ had 121 grid points; boxes at $|b|>35°$, fewer.

Observing a box consisted of the following steps. Pointed (i.e., 'stop-and-stare', or 'point-and-shoot') observations in total-power mode were made at each grid point. First, four reference spectra were measured at the initial grid position (l_{min}, b_{min}). Next, the grid lattice points were observed, increasing galactic longitude before latitude. On moving to the next latitude, the longitude direction was reversed to reduce the slewing time of the telescope. After taking the last spectrum at (l_{max}, b_{max}), four reference spectra were again taken. The scheduling code avoided observing positions lying closer than 25° to the Sun, or closer than 10° to the Moon. Before initiating observations of a new box, selected standard sources were scheduled. In addition, once every 24 hours, an observation of either Virgo A or Taurus A was scheduled.

An observation of a single grid point required about 270 seconds. The noise cycle time and the overhead added 90 seconds to the net integration time of 180 seconds. The overhead time included the telescope slewing time between grid points, and telescope settling time, but was largely due to the initialization procedure of the auto-correlator. Steering the telescope to the first grid point in a box proceeded at about 0°.5 per second. The total overhead time required was thus substantial, namely about 50% of the net integration time. Observing an entire box, involving 121 lattice points plus eight references, required about 10 hours to complete if the system operated optimally.

It was convenient for a variety of practical reasons to schedule the observing for periods of about a week's duration; we called observations made during a single such period a 'batch'. Routine observing began in November, 1989, about a year had been spent debugging the DAS and preparing the telescope and the observing and operating procedures. Between November, 1989, and November, 1992, 126 batches were completed. During these three years, the telescope was fully devoted to the survey; periods of maintenance, system failure, and poor weather conditions increased the gross observing time. In May, August, and September, 1993, three additional batches were observed to complete the data-taking phase of the project.

6. Initial data reduction

The installation of the DAS spectrometer required that entirely new data-flow and reduction procedures be established for this project. The following describes the reduction procedure; discussion of the stray-radiation correction is deferred until section 8. The data were reduced in four steps. First, the correlation function was converted to a power spectrum. Second, the telescope pointing accuracy was verified, controls were made on the stability of the system, correction for atmospheric extinction was applied, and correction was made for the shape of the bandpass. Third, the intensity scale was converted to brightness temperature, the stray-radiation contribution removed, a polynomial baseline subtracted, and interference spikes eliminated. Finally, a sine wave was fit to the baselines and subtracted in order to account for residual ripples in the baselines due to standing-wave reflections within the telescope structure, and spectra with special demands were manually reduced.

The Dwingeloo telescope was controlled by an HP1000/21MXE minicomputer. It had a single CPU, 192 kbytes of memory, and was equipped with a 125 Mbyte hard disk and a 1600-bpi magnetic tape unit. The computer was linked to the back-end and telescope control through a 16-bit multiplexer. The first reduction step was performed on a MicroVAX computer in Dwingeloo. All subsequent data reduction (except the stray-radiation correction) was carried out on a Personal Computer (PC) in Leiden. Much of the reduction was carried out using the DrawSpec package written by H.S. Liszt (1987, and subsequent updates).

During the second step of the reduction, measurements of continuum sources were extracted from each batch and used to check the pointing of the telescope. Once every 24 hours, a continuum source of known flux was observed. The initial intention was to calibrate the noise diode using these data. It soon became apparent, however, that the stability of the noise generator gave accuracies better than the uncertainties in the continuum calibration data. The continuum-source data were used subsequently primarily as pointing checks.

At any given time, either Virgo A, at $(\alpha,\delta)_{1950}=(12^h28^m7^s, +12°9')$, or Taurus A (the Crab Nebula), at $(\alpha,\delta)_{1950}=(5^h31^m31^s, +21°59')$, is observable from Dwingeloo. Once a day, two 'crosses' centered on one of these sources were measured as a check on the pointing accuracy of the telescope. The sequence off–on–off (source) with a separation of 2° (true angle) in right ascension, was followed by the same sequence in declination. This was repeated for an angular separation of 0°.3. A first-order pointing check was obtained by comparing the symmetry of the two sets of four off-source fluxes.

Tests proved that the Dwingeloo instrument was extremely stable over long periods of time. Standard sources were observed several times a day to monitor gain fluctuations. One of the three standard fields recommended by the International Astronomical Union (IAU; see van Woerden 1970), namely S7 at $(l,b)=(132°, -1°)$, was observed with regularity; some 7% of the total telescope time was devoted to measuring this field. Another standard field, S8 at $(l,b)=(207°, -15°)$, was observed for several days on a few occasions, in order to provide data for the absolute temperature calibration gauge, and to determine the atmospheric extinction correction.

There was no indication that the gain of the system varied systematically on a time scale of less than a few weeks. For every batch, the S7 spectra were integrated over the velocity interval [−142, +50] km s^{-1}. The mean values and the standard deviations were determined by iteratively discarding values deviating more than 5 sigma from the mean. Typically, less than 1% of the measurements were excluded on this basis.

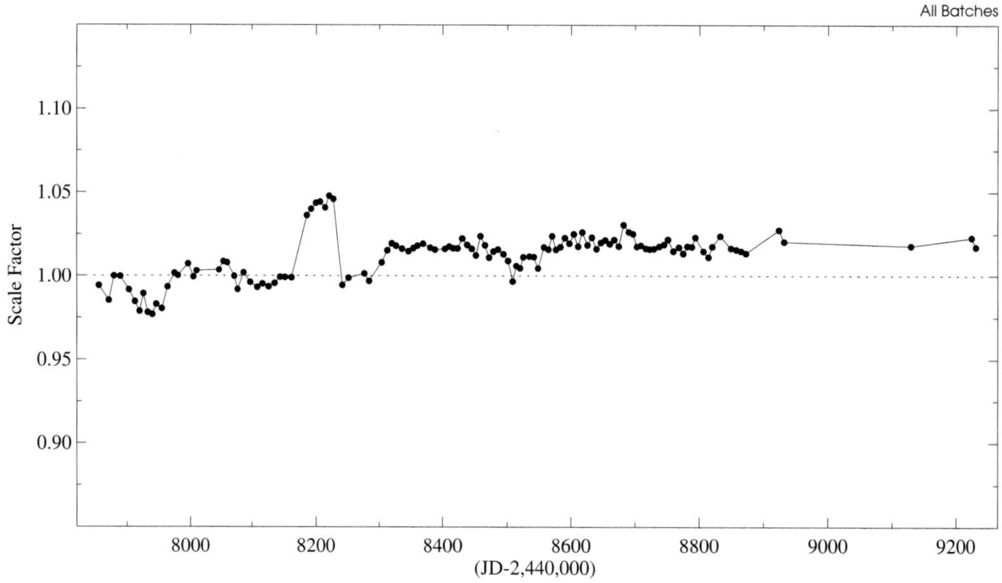

Figure 3. *Stability of the Dwingeloo telescope over the entire observing period. The relative scale factors for all batches are plotted against Julian Date. Unity was gauged using two carefully selected batches, and is seen offset from the mean value. The long-term instrumental stability of the instrument was good.*

The S7 and S8 measurements in batches 14 and 15 were used to gauge the conversion of antenna temperature, T_a, to brightness temperature, T_b. The mean integrated intensity over all S7 measurements in these batches was set equal to unity. The mean integrated S7 intensities for all the other batches were used to calculate relative scale factors to normalize the temperature scale uniformly over the entire observing run. Figure 3 shows these scale factors as a function of the mean Julian Date for each of the 129 batches. (The data plotted in Figure 3 do not contain the subsequent stray-radiation correction; the measures of stability were even smoother after that correction.)

The data were next corrected for the effects of atmospheric extinction. The survey aimed at covering as much of the sky as possible to the declination limit of -30°. This required that occasional pointings be made at elevations as low as the physical limit of the telescope would allow, namely $3^\circ\!.5$. At such extreme elevations, the airmass column, varying as sec(z), increases rapidly, and atmospheric refraction and extinction become non-negligible (see Williams 1973). (Atmospheric *refraction* was accounted for automatically in the telescope-pointing software. In requesting a measurement at galactic coordinates (l, b), a coordinate transformation was performed that calculated the actual horizontal position (Az, El) of the source, and the elevation was corrected for atmospheric refraction, yielding the apparent elevation of the source.)

The effect of atmospheric extinction was measured by observing standard source S8 from rising to setting, continuously for several days. The integrated intensity for the velocity interval $[-75, +90]$ km s^{-1} was plotted against elevation, and a model was fit to the data. If $S(z)$ is the flux measured at zenith distance z, and S_0 the flux outside the atmosphere, then $S(z)=S_0 \cdot d^{-X(z)}$ where d is the attenuation at the zenith for $\cos(z)=1$, and $X(z)$ is some function of zenith distance. We determined $X(z)$

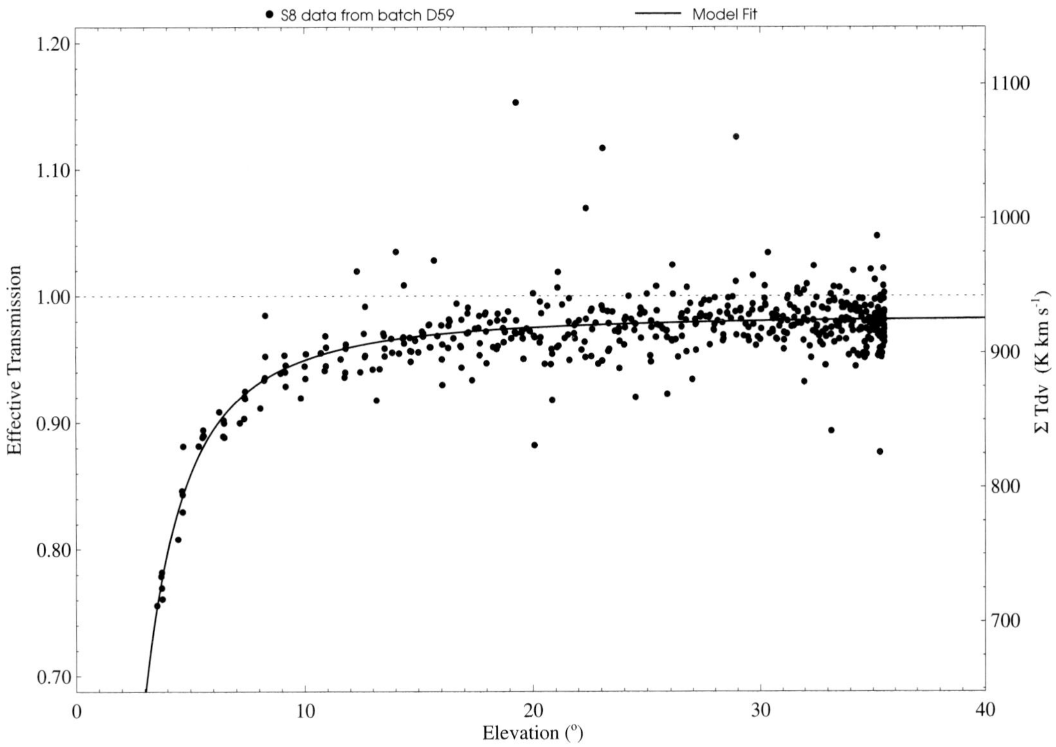

Figure 4. *Atmospheric extinction correction necessitated by the air-mass column through which the 21-cm radiation reaches the telescope. (Atmospheric refraction was automatically accounted for in the algorithm that calculated the pointing of the telescope.) The points represent the intensities integrated over* $v_{lsr}=[-75,+90]$ *km* s^{-1} *plotted against elevation, for all S8 measurements which were collected over the course of a few days. S8, at* $(l,b)=(207°,-15°)$*, was observed from rising to setting, providing data at elevations ranging from 3°.5 (the physical limit of the telescope) to 35°.5 (culmination). The effective transmission (left axis) was gauged against the absolute intensity scale (right axis) by defining perfect transmission as* $1/d\sum Tdv \equiv 1$.

as a third-order function of sec(z) by a least-squares fit to the S8 data. The S8 data and the transmission function are shown in Figure 4.

The shape of the bandpass was the focus of the following reduction step. It was obtained from reference spectra observed near, in both position and time, each source position in the survey. For each position, a single reference spectrum was created by averaging eight spectra (weighted by $1/T_{sys}$) from a file of references, choosing those nearest in time, both before and after the source was observed. The before and after positions proved necessary because the shape of the bandpass (especially at low elevations) is quite sensitive to elevation.

The system temperature, T_{sys}, is the equivalent temperature of the accumulated noise from the receiver, attenuations in the waveguides and coaxial cables, the atmosphere, and radiation that enters through the far sidelobes pointing to the ground. The value of T_{sys} was determined for each observation by comparing the total power in the spectrum with the total power generated by a calibrated noise source.

Ground radiation increases the system temperature, but also affects the shape of the bandpass. The mean reference was Hanning smoothed, reducing the *rms* noise level to approximately $1/5$ that of a typical source spectrum. The source S and reference R were combined to yield the bandpass-corrected spectrum, $S'=T_{sys}\,(S-R)/R$,

Table 2. *Calibration measurements of S7 and S8. The spectra were integrated over the velocity intervals indicated, and the maximum temperatures were determined. The results are compared with data from the Hat Creek telescope (which had a beam width similar to that of the Dwingeloo dish), and with smoothed spectra observed with the Effelsberg 100-m telescope. The results indicate that a uniform brightness temperature scale has been established*

	S7		S8	
velocity range ($\mathrm{km\,s^{-1}}$)	[−56.8, −45.3]		[−5.1, +22.3]	
	$\int T dv$	T_{max}	$\int T dv$	T_{max}
Dwingeloo	1089	100	833	71
Hat Creek	1095	–	869	–
Effelsberg†	1050	96	856	71
T_b scale ratios	$\int T dv$	T_{max}	$\int T dv$	T_{max}
Dwingeloo/Hat Creek	0.995	–	0.959	–
Dwingeloo/Effelsberg†	1.037	1.042	0.973	1.00
	($\mathrm{K\,km\,s^{-1}}$)	(K)	($\mathrm{K\,km\,s^{-1}}$)	(K)

† Smoothed to 36′ resolution

where T_{sys} is here the weighted average system temperature of the references that were used to create R. Intensities near the extreme edges of both the source and reference spectra are close to zero; after the bandpass removal, the intensities in these channels are unreliable. The outermost channels were discarded, leaving effective velocity coverage $-450<v_{lsr}<+400$ km s^{-1}. The low *rms* noise in R ensures that the noise level in S' is about equal to that of the original source spectrum S.

The next reduction step involved converting the observed antenna-temperature intensity scale, T_a, to the astrophysically more important brightness temperature, T_b, which represents the brightness on the sky. T_a is the convolution of the antenna pattern with T_b. This implies that for the conversion of T_a to T_b, the properties of the telescope must be known. The conventional conversion of T_a to T_b was done by using a scaling factor, K, to calculate $T_b=KT_a$; the value of K was determined by the parameters of the antenna. In the simplest approach, K has commonly been taken as $1/\eta_{MB}$, where η_{MB} is the main beam efficiency. Williams (1973) pointed out that K is a function of the brightness distribution illuminating the antenna pattern, rather than a constant; later, Kalberla, Mebold, & Reif (1982) argued that the absolute HI brightness-temperature scale for a radio telescope can only be accurately determined by taking the contribution of stray radiation into account.

The internal calibration of the survey was gauged to the measurements of the standard sources S7 and S8 in batches 14 and 15. Integrating the spectra over the velocity intervals given by Williams (1973), we were able to compare the intensity scale of this survey with that for the Hat Creek survey of Weaver and Williams (1974) and with the intensity calibration used for the 21-cm data observed with the Effelsberg 100-m telescope. The results are shown in Table 2.

During the next stage of the reduction, a polynomial baseline was fit to the emission-free regions of the spectrum, and subtracted. The method of frequency switching introduces a zero-level offset as well as a baseline curvature, because the reflected power, the gain, and the system temperature are all frequency dependent. Removing both the offset and the curvature has generally been done by fitting a low-order polynomial to spectral regions judged, usually by visual inspection, to be free from emission features. Manually fitting a baseline to each spectrum individually was not feasible because of the sheer number of spectra in the survey, and in any case such fitting would introduce an undesirable level of subjectivity.

An automated baseline-subtraction routine was developed. If a spectrum were to contain no signal, and consist only of noise, a baseline should be fit to all channels except the outermost ones, where the intensities are unreliable. Because, at the very least, every line of sight shows substantial emission near $v_{\rm lsr}=0$ km s^{-1}, the central channels should never be included in a baseline fit. The regions in fact used in the baseline procedure were chosen and modified in an iterative procedure, operating on a Hanning-smoothed copy of the spectrum; the details of the procedure and examples of its application are described by Hartmann (1994). The baseline procedure aimed at preserving real, albeit weak and broad, features such as high-velocity clouds in the profile wings; it was also designed to avoid contamination by spurious interference spikes. The baseline regions determined in the final iteration were used to fit a polynomial of order three (and never higher) to the original, unsmoothed spectrum.

7. Radio interference

After the baseline corrections had been applied, efforts were directed to detect, model, and remove radio interference spikes. Radio interference is a source of annoyance to all radio astronomers; although the 1420-MHz band is protected by international regulations, many spurious signals pollute this frequency range. Because the spectral-line receiver transforms the observed frequencies in three steps to the video band, each of the intermediate frequency regimes is also a new potential target for interfering signals. Finally, in the spectrometer, yet more signal transformations are carried out. The electronic components of the front-end and back-end are not themselves free of suspicion; an unstable power supply actually in the control room of the telescope generated such bad interference that all spectra from a couple of days of observing were rendered useless.

The interference environment of the Dwingeloo telescope turned out to be more benign than we had expected when the project began, but nevertheless required vigilance. Many spectra were contaminated by spurious signals, in most cases by very narrow spikes typically a few channels wide and generally rather easy to recognize for what they were. Sometimes, however, interference damaged spectra beyond repair.

The detection and removal of interference from a spectrum were carried out in three stages. First, narrow spikes, only one or two channels broad, were tracked down and removed. It was a rather simple matter to identify such narrow spikes using an automated algorithm. Next, each spectrum was analyzed for the presence of sinc-pattern interference. Finally, the much more troubling contamination by broad-bandwidth interference was sought.

We believe that we have removed most of the interference signals from the final spectra entering the Atlas. The cautionary remark should be made that this involved modification of some spectral values. Elimination of single-channel spikes did not affect the spectrum in any significant way. However, the subtraction of a sinc-interference model was not always harmless. Usually, the *rms* noise of the spectrum remained higher than average, and in some cases the baseline was affected. In the further processing of the data, spectra that originally contained strong interference were rejected, regardless of the success of the spike removal, if multiple observations for that position were available. Indeed, this possibility provided much of the motivation for making many multiple observations.

A strong monochromatic signal ($\Delta\nu$<4.88 kHz, the channel spacing) picked up by the antenna causes a sinc-pattern ($\sin(x)/x$) to be superimposed on the spectrum. The auto-correlation function has only a finite time delay, and therefore yields the convolution of a single cosine component (the Fourier transform of a δ-function), and a block function (the Fourier transform of which is a sinc function). When a sinc spike was detected (but not necessarily successfully removed), the spectrum was flagged. Subsequent operations on the reduced spectra took this flag into account. Spectra showing very strong sinc interference were discarded, and the positions re-observed.

The existence of broad-bandwidth interference was confirmed using a spectrum analyzer connected to the front-end of the telescope. When a sudden increase in the momentary total power was registered, either on the real-time digital display or pen recorder, or heard in the audible noise signal to which the total power was also converted, the spectrum analyzer displayed an increased featureless noise modulation across the entire passband. Bursts of this kind usually lasted less than a second, and occurred at irregular intervals.

A search for the origin of this type of narrow interference was carried out by Leiden graduate students Maartje Sevenster and Frank van den Bosch in 1992. They considered a large data sample, containing the material in 55 batches (about 40% of the total number of spectra in the entire survey spectra). All spectra (both Sources and References) observed between 22 November, 1990, and 4 February, 1992, were searched for the presence of interference spikes; Sevenster and van den Bosch sought patterns underlying these artificial signals which might reveal their origin. At some level 90% of all spectra have identifiable interference. It turned out that experiments on the DAS in the Dwingeloo electronics laboratories were the cause of almost all of the narrow-band interference detected by the 25-m telescope, which is located at a distance of some 100 meters from the NFRA electronics labs.

8. Stray-radiation correction

In an ideal radio telescope, the only radiation recorded is that which is incident on the antenna from the direction in which the telescope is pointing, and that radiation is undisturbed by any structure supporting the antenna feed, or by the feed itself at the focus of the mirror. A parabolic reflector like the Dwingeloo one, with the aperture partially blocked by the feed and its support structure, is certainly not perfect in this respect. Radiation may be received directly into the feed, and also scattered off the feed-support legs; these legs also cast shadows on part of the antenna pattern. Using modern receivers, the sensitivity of the antenna in every direction allows the ubiquitously distributed HI radiation to be received, in a single observation, from every direction on the sky.

8.1 Background

The response of the antenna as a function of direction is described by the antenna pattern. The pattern consists of a main beam and sidelobes. The angular extent of the main beam is usually given as the *HPBW* of the antenna. A measure for the concentration of the power pattern in the main beam is given by the main beam efficiency, η_{MB}, where $\eta_{MB}=\Omega_{MB}/\Omega_{\mathrm{a}}$ and where Ω_{a} represents the solid angle in which all power over $4\pi\,\mathrm{rad}^2$ is received. All directions outside the main beam are referred to as the sidelobes or stray pattern, and the corresponding stray-pattern solid angle

Ω_{SP} is $\Omega_{SP}=\Omega_a-\Omega_{MB}$. The ratio $\eta_{SP}=\Omega_{SP}/\Omega_a$ is called the stray factor. Consequently, $\eta_{MB}+\eta_{SP}=1$.

A perfect radio telescope would be sensitive in the main beam only, and thus have $\eta_{MB}=1.00$. No such telescope exists. Although the stray pattern is very much less sensitive than the main beam, the total contribution from the $4\pi\,\mathrm{rad}^2$ solid angle can be considerable. The problem is greatly reduced, although not eliminated, for horn antennas which have an aperture unblocked by the receiver or its feed, or by the support structure. The 20-ft horn telescope of the AT&T Bell Laboratories in Crawford Hill, approaches this ideal; it has $\eta_{MB}=0.92$ (Stark *et al.* 1992) and most of the remaining sensitivity is in nearby sidelobes.

Parabolic telescopes following new designs, such as that of the 100-m Green Bank Telescope currently under construction by the National Radio Astronomy Observatory, involve off-axis placement of the receiver system and thus a largely unblocked aperture; such telescopes will experience significantly reduced stray-radiation effects. The Dwingeloo telescope is, of course, a blocked-aperture parabola of early design; at the sensitivity levels afforded by the modern electronics with which the telescope was equipped for this project, the effects of stray radiation are severe and warrant correction.

Because the main-beam efficiency of the Dwingeloo telescope is 0.69, 31% of the radiation can theoretically be received in the sidelobes. Because the telescope is ground based, half of the antenna pattern is always pointing to the ground. Although some suspicion exists that the ground may in some cases reflect 21-cm waves, in general the antenna pattern touching the ground only increases the system temperature. The other half of the pattern is directed towards the HI sky, and will receive 21-cm emission from a vast expanse.

We describe in this section how the stray-radiation contribution was calculated for every spectrum observed in the survey, using a modification of the correction algorithm developed by Kalberla (1978). The details of this procedure as applied to the Dwingeloo data are discussed by Hartmann *et al.* (1996); here we suffice by describing the results of these calculations.

Van Woerden (1962) was the first observer to interpret variations in HI spectra as due to stray radiation. After carefully reducing spectra observed in Dwingeloo during different seasons, he concluded that the spectra were contaminated by emission that was received outside the main beam of the antenna. He coined the phrase 'stray profile' for this spectral emission contribution, and estimated that about half of the contaminating emission was contributed by the 'spillover ring', the region of the antenna pattern where radiation could enter the feed directly, i.e. without reflecting off the telescope dish. Although it was not possible in the 1960s to attempt a full correction for the effects of stray radiation (because of the lack of knowledge then about the true antenna pattern, the lack of an HI input sky, and the insufficiency of computers), van Woerden was able to estimate the stray-radiation contribution by numerical convolution of rough estimates of the antenna pattern and of the HI sky at the time of observing. Although his model calculation successfully explained the observed discrepancies in the spectra of different epochs, no correction to the data was actually applied.

Raimond (1964) was the first to use a computer to apply the correction. On a primitive early computer at the University of Leiden, he calculated the stray profiles for more than 500 HI spectra observed in a region surrounding two stellar associations in Monoceros, lying near the galactic equator.

After the existence of stray radiation had been empirically established, it was largely ignored for the next decade. When the sensitivity of HI receivers had been significantly improved in the early 1970s, the importance of stray radiation was recognized anew. Mebold and Hachenberg in Bonn had noticed discrepancies between repeated observations at high galactic latitudes made with the Effelsberg 100-m telescope. Their findings motivated Kalberla (1978) to address the problem of stray radiation in a general way and to develop a formal correction procedure. The procedure for accurately calculating stray profiles involved two requirements, namely an input HI sky, and a detailed model of the antenna pattern which would sample this sky. The availability of the Hat Creek HI surveys (Weaver & Williams 1973; Heiles & Habing 1974), which together gave the first detailed coverage of the northern sky, fulfilled the first of these two requirements.

In order to satisfy the second requirement, Kalberla measured the antenna pattern of the 100-m Effelsberg telescope (HPBW=9′) to a radius of 2° from the main beam axis, and created an empirical model of the far sidelobes. For this model, he inspected the characteristics of the Dwingeloo 25-m antenna pattern, which had been extensively measured by Hartsuijker *et al.* (1972). Prominent features in this stray pattern could be explained in terms of various structural components of the telescope. This enabled Kalberla to predict the behavior of similar structures for the Effelsberg antenna, and to create a model, albeit one with many free parameters. Repeated observations at different hour angles and at various epochs provided the material necessary to determine the relative sensitivities of the various components of the model. The empirically tuned model consisted of a spillover ring, four stray cones (from radiation scattered off the feed support legs), and blockage ('shadowing') of the pattern by the support legs. Additionally, four small components were found that were caused by reflections off the roof of the apex cabin.

The high efficiency in the main beam of the Bell Labs horn antenna motivated a major sky survey ($\delta \geq -40°$) of galactic HI using this telescope (Stark *et al.* 1992). A study of the far-sidelobe contamination of the published survey data was made by Kuntz & Danly (1992). They showed that two known far sidelobes of the telescope (one at $\theta=-18°$ from the optical axis, probably caused by either the drive wheel or the weather cover of the horn; the other at $\theta=70°$, probably caused by diffraction from the edge of the reflector) caused stray radiation that was generally comparable to or less than the baseline uncertainties (~0.05 K). The sensitivity of the Bell Labs survey represented an order of magnitude improvement over what was reached in the Hat Creek data.

Lockman, Jahoda, & McCammon (1986) effectively demonstrated the need for stray-radiation corrections to HI spectra observed with a low-noise receiver, especially in directions of low total HI column densities. They were able to exploit the availability of the Bell Labs data by developing a method of correcting the stray-radiation contamination in spectra observed with the National Radio Astronomy Observatory (NRAO) 140-ft telescope. The higher-angular-resolution 140-ft HI maps were smoothed with the beam pattern of the Bell Labs telescope, and then compared with the data from the Bell Labs survey. Differences between the two maps were interpreted as due to stray radiation in the 140-ft telescope data. By applying this bootstrapping method, the stray-radiation contamination in the original spectra was reduced by an order of magnitude (see also Murphy 1993). The main practical limitation to the bootstrapping method may reside in the 2°.5 effective spatial resolution of the Bell Labs data, as well as in its coarse effective velocity resolution of $\Delta v=10.0$ km s^{-1}. The main advantage of this method is that no knowledge is required about the antenna pattern of the telescope for which the spectra are corrected. A particular disadvantage, however, and one which largely rules out application of this method to observations made with the Dwingeloo telescope, is that only the *mean* stray radiation can be determined; fluctuations due to changes in the position angle of the telescope are not accounted for. This effectively restricts the application of the Lockman *et al.* algorithm to telescopes mounted, unlike the Dwingeloo one, equatorially; for such telescopes, but not for those with an altitude/azimuth mounting, the antenna pattern has a fixed orientation with respect to the sky. Furthermore, the correction procedure will only work properly when a field of at least some 2° in extent is observed; the method is unsuited for the determination of the stray radiation in single, pointed observations.

To overcome these limitations, we corrected the HI data in this Atlas using a modification of the Kalberla approach developed for the Effelsberg 100-m telescope (Kalberla 1978; Kalberla, Mebold, & Reich 1980; see also Hartmann *et al.* 1996). This required an input HI sky, which we constructed from the uncorrected survey data, and the determination of the antenna pattern parameters. The antenna pattern was divided into three parts: main beam (*MB*), near sidelobes (*NSL*), and far sidelobes (*FSL*).

8.2 Input sky

The input HI sky comprised the Leiden/Dwingeloo survey itself, reduced in all aspects except for the stray-radiation correction. (The correction could therefore only be applied after the entire accessible sky had been observed.) The new survey improves upon the Berkeley surveys in all relevant parameters except positional resolution, which is comparable. The advantages over the Bell Labs survey lie in the higher velocity resolution and greater dynamic range; our better spatial resolution is irrelevant, because the input sky was created as $2°\times2°$ cells. Re-binning the Dwingeloo data to the same resolution improves its sensitivity beyond that of the Bell Labs survey. The Bell Labs data are, of course, initially a better estimate of the true sky brightness (at least for $T_b \lesssim 40$ K; see Kuntz & Danly 1992) due to the lack of stray radiation received by the horn reflector. However, it was assumed that the spectral details of the stray radiation would disappear when the individual contributions of all spectra in the stray-pattern solid angle Ω_{SP} were convolved with the antenna pattern.

After carrying out the initial reduction of the Dwingeloo spectra described above, further preparation for the stray-radiation correction involved creating a homogeneous all-sky data cube from the reduced spectra, binned into cells of size $2°\times2°$ or smaller, with the size of the cells principally determined by the computer power required to implement the algorithm. The spectra binned into these cells were averaged with unit weights and clipped to 512 channels covering a velocity range of $|v_{\rm lsr}| \leq 264$ km s^{-1}. The input sky consisted of 8071 spectra, each representing a cell of specified position and solid angle.

8.3 Antenna pattern of the Dwingeloo Telescope

Hartsuijker *et al.* (1972) published a nearly complete map of the Dwingeloo antenna pattern as it was in 1969. By joining the 25-m telescope with a nearby 7.5-m Würzburg antenna, they created an interferometer and used it to measure the antenna response at 1415 MHz. For some 19,000 points in the pattern, representing about 60% of a complete sphere, the response to strong radio sources was measured. (Mapping the antenna pattern to that completeness was a major effort, requiring some 1500 hours of observing time.)

Although the feed-support structure of the telescope was changed from three to four legs shortly after the Hartsuijker *et al.* experiment, their maps were still of value for determining the current antenna pattern characteristics. We used, in addition, other available data on the calibration of the Dwingeloo receiver and feed. These included hot/cold calibrations at 1611 MHz, made using a scaled feed. We

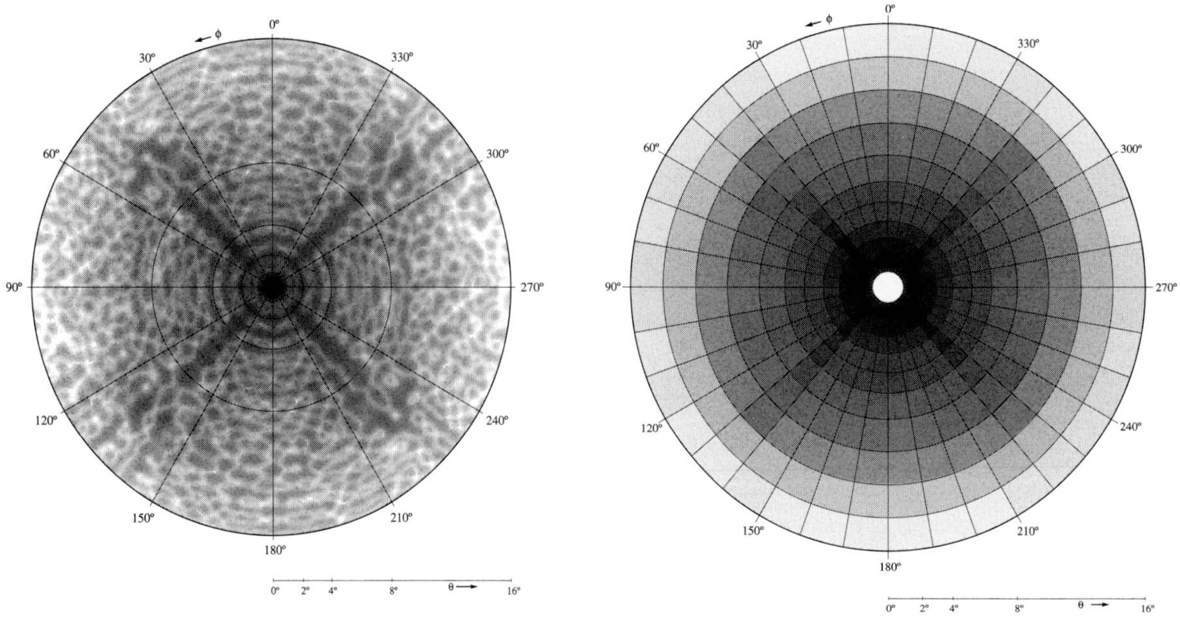

Figure 5. ***(left)*** *Mean WSRT* NSL *pattern, generated from holographic measurements of 12 telescopes (van Someren Greve 1991) and used as an approximation for the* NSL *pattern* ($0\overset{\circ}{.}9\leq\theta\leq16^\circ$) *of the Dwingeloo 25-m dish. The image is logarithmically scaled between –70 dB and –20 dB, relative to the main beam sensitivity. Concentric diffraction lobes are visible at* $\theta=2^\circ$, 3°, *and* 4°. *Radiation scattered off the four feed-support legs causes regions of higher sensitivity, centered on* $\phi=45^\circ$, *at* 90° *intervals.* ***(right)*** *Model of the* NSL *pattern* ($0\overset{\circ}{.}9\leq\theta\leq16^\circ$) *of the Dwingeloo telescope, derived from the observed mean WSRT pattern. The 468 cells represent the elements of the kernel function for which the computer implementation was carried out. The image is logarithmically scaled. The concentric diffraction lobes (at main-beam distances* $\theta=2^\circ$, 3°, *and* 4°*), and the enhanced stray regions caused by the feed-support legs (at antenna azimuth* $\phi=45^\circ$ *modulo* 90°*), can be identified.*

also measured, ourselves, the total-power response of strong continuum sources (Cas A, Cyg A, Vir A, Tau A, and the Sun) away from the main beam in order to verify some characteristics of the antenna pattern.

When Hartsuijker *et al.* measured the antenna pattern of the Dwingeloo telescope within a radius of 6° from the beam axis ($\theta=0^\circ$), the main beam was found to be slightly elliptical, with *HPBW*s of $0\overset{\circ}{.}57\times0\overset{\circ}{.}62$. No sharp 'null' separated the main beam from the first diffraction sidelobe. To determine the present total *NSL* response (defined to extend to $\theta=16^\circ$) of the Dwingeloo telescope, we used holographic beam measurements of 3C84 made with the *WSRT* at 6-cm wavelength by van Someren Greve (1991). The amplitudes of the measurements for 12 telescopes were averaged, auto-correlated, and Fourier transformed. After scaling with respect to the wavelength, these data were used as the basis for a model of the Dwingeloo *NSL* pattern.

Shown on the left in Figure 5 is a gray-scale image of the mean *WSRT* pattern, logarithmically scaled between 70 dB and 20 dB below the main-beam sensitivity. There are two principal justifications for the generalization of the mean *WSRT* pattern to that of the Dwingeloo telescope. First, the geometry of the telescope in Dwingeloo is quite similar to that of those in Westerbork (except for the mountings). Second, as mentioned above, the feed-support structure of the Dwingeloo telescope was modified to accommodate a new generation of front-ends built for the *WSRT*. This changed the aperture efficiency and the main-beam efficiency from the values determined earlier. The Dwingeloo feed is currently of similar design to that of the

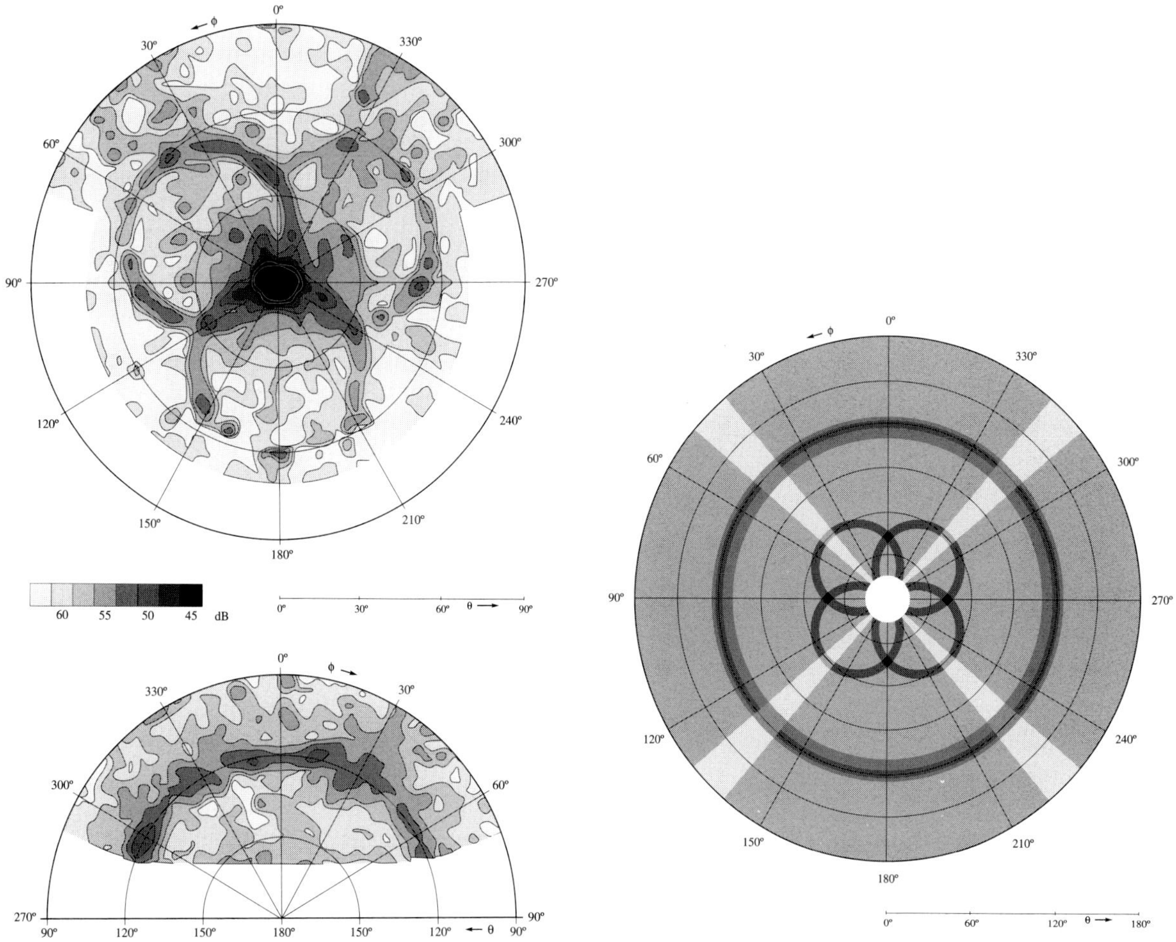

Figure 6. ***(upper left)*** *Forward half (0°≤θ≤90°) of the antenna pattern of the Dwingeloo telescope, corresponding to the situation in 1968 (Hartsuijker et al. 1972). Radiation scattered off the three feed-support legs causes the stray cones of enhanced sensitivity, centered on θ=30°. Regions of lower sensitivity are found where the support legs block (shadow) the feed at ϕ=0°, 120°, and 240°.* ***(lower left)*** *Rear half (90°≤θ≤180°) of the antenna pattern of the Dwingeloo telescope corresponding to the situation in 1968. The most prominent feature is the spillover ring, centered at θ=120°, which is about 5 dB more sensitive than the surrounding average; this structure accounts for 4.5% of the total sensitivity of the antenna pattern.* ***(right)*** *FSL model of the antenna pattern (16°≤θ≤180°) for the Dwingeloo telescope, corresponding to the situation during our project. The model was derived from the geometry of the telescope. The antenna pattern published by Hartsuijker et al. was extrapolated to determine the global characteristics of the main features: the spillover ring, the four stray cones, and the shadowing of the support legs.*

WSRT feeds. We calibrated the sensitivity of the model by observing the total-power response of Cas A in the NSL pattern of the Dwingeloo telescope.

The antenna pattern *beyond* $\theta=16°$ was not determined by direct observations. The option to connect the 25-m telescope to a reference antenna, creating a two-element interferometer no longer existed. (The Würzburg antenna was dismantled during the course of our work, and donated by the NFRA to the Deutsches Museum von Meisterwerken der Naturwissenschaften und Technik in Munich.) Without interferometry, the Dwingeloo antenna pattern could not be determined below a level of about −30 dB, whereas sensitivity to −60 dB is needed for an accurate stray-radiation determination. This requirement can be argued as follows. Assume a homogenous HI sky distribution, and a homogeneous FSL pattern, which has a constant sensitivity of –60 dB compared with that of the main beam. At any moment, half of the FSL pattern is directed towards the sky, and therefore $\Omega_{\rm sky}=2\pi\,{\rm rad}^2$ at –60 dB will receive radiation. This integrates to ~3% of the power received in the main beam.

We derived a model of the FSL pattern based on an extrapolation of the measurements made by Hartsuijker *et al.* to the current geometry of the Dwingeloo telescope. They had found two major features that contributed to the FSL antenna pattern in 1972, namely a spillover ring and three stray cones corresponding to radiation scattered off the legs of the feed-support tripod.

The forward half ($\theta \leq 90°$) of the Dwingeloo antenna pattern shown in Figure 6 (upper left) was completely mapped for ($\theta \leq 70°$) and over 40% of the ϕ-range, out to $\theta=90°$. The stray cones are prominent, and clearly indicate the three-legged feed support structure for the situation in 1968. The centers of the cones are at $\theta=30°$, for $\phi=60°$, 180°, and 300°, and their sensitivity is about 6 dB higher than the surrounding average. The three support legs (each about one wavelength in diameter) made an angle of 30° with the reflector axis, and were directed towards $\theta=150°$ at $\phi=0°$, 120°, and 240°. For $50° \leq \theta \leq 90°$ in these ϕ-directions, the pattern shows regions of low sensitivity caused by shadowing of the feed by the support legs. The difference from the average sensitivity of the areas in-between the leg shadows is about 4 dB.

Figure 6 (lower left) shows the rear half of the antenna pattern. The spillover ring is dominant. It is centered around $\theta=120°$, and has a mean sensitivity about 5 dB above the surrounding regions. The edge of the reflector was at $\theta=125°$.

Unfortunately, the Hartsuijker *et al.* data could not be used directly as a FSL model, because the telescope dish had been re-surfaced in 1969, and the feed-support structure changed from three to four legs in 1974. The re-surfacing of the dish was not expected to cause significant changes in the FSL pattern; the relevant characteristics are determined by the feed-support structure (stray cones), and the position of the rim (spillover). We derived a model of the stray cones and the spillover ring from the present telescope geometry, using the earlier quantitative data as a guideline for the extent and sensitivity of these features. The model was refined by including shadowing from the support legs. In its final form, the model consists of 23 elements, each described by its extent (in antenna coordinates) and sensitivity. The stray cones were accounted for separately, and added to the total FSL contribution.

Figure 6 (right) shows the FSL model over the entire range $16° \leq \theta \leq 180°$. A general background level is shadowed by the four feed-support legs. The sensitive areas are the spillover ring and the stray cones. In regions where the stray cones overlap, their contributions are added. The FSL model was calibrated by observing continuum radiation from the Sun entering in the stray cones and in the spillover ring.

8.4 Implementing the correction

For every individual HI spectrum entering this Atlas, the NSL and FSL profiles were calculated separately. For each correction, the input sky was separately consulted so that the situation at the date and time appropriate to each individual observation could be accounted for.

The salient points of the numerical calculation of the stray-radiation correction from the sidelobe models can be illustrated for the case of the far-sidelobe correction. The spectra of the input sky were converted to antenna coordinates, and convolved with the FSL features with which they coincided. The relevant observational parameters, namely the galactic coordinates and the date and time of observing, were read from the survey spectrum; the FSL contribution was calculated by processing all

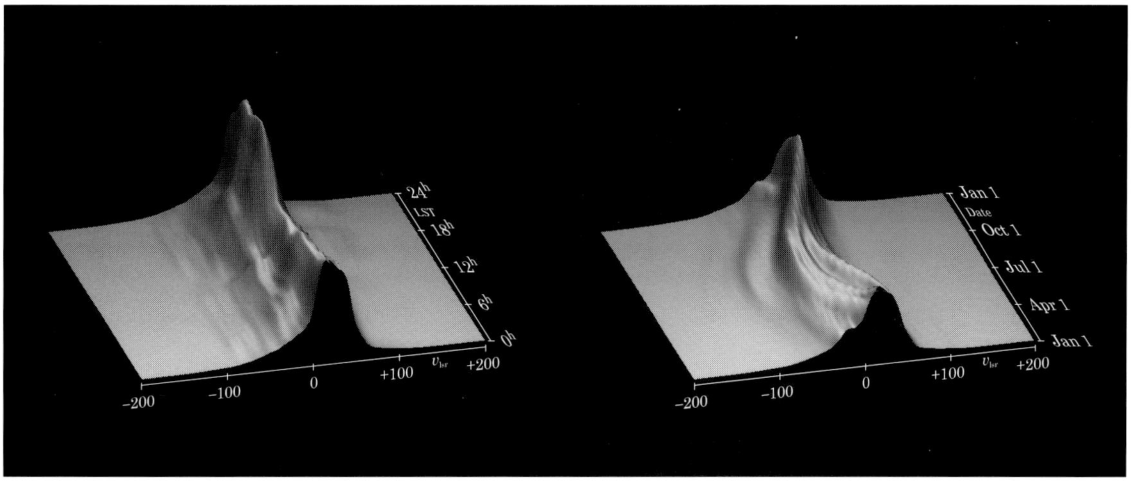

Figure 7. ***(left)*** *Diurnal variations in the FSL stray-radiation contamination expected at (l,b)=(160°,+50°) over a 24-hour (sidereal time) period. FSL profiles were calculated for 1 January, 1994, in steps of one hour in LST. The spectra are displayed as a surface plot. The peak temperature in the FSL spectra ranges from 0.41 K (LST=13^h) to 0.78 K (LST=17^h).* ***(right)*** *Seasonal variations in the FSL stray-radiation contamination expected at (l,b)=(160°,+50°) over a year. The FSL profiles were calculated at LST=6^{h}04^m (culmination) in steps of about 15 days. The spectra are displayed as a surface plot. Peak temperatures range from 0.28 K (15 April) to 0.60 K (15 October). Note the curvature in velocity of the peaks over the course of a year; this is due to the changes in the LSR velocity correction because of the Earth's revolution around the Sun.*

the spectra from the input sky in a loop that carried out the following operations. The position of the input-sky spectrum was converted to (α,δ), and then to (Az, El) for the time of observation. The position was checked for obstruction by the horizon (which is not necessarily at El=0°: the algorithm contained a model to account for obstructions, for example by woods). If the input-sky cell was 'up', it was converted to antenna coordinates (ϕ, θ). The list of 23 *FSL* features was searched for positional coincidence. There was always a contribution from the overall background, but in some cases three features overlap (for example at (ϕ, θ)=(45°, 60°), where the background, a stray cone, and a leg shadow are coincident; see the right-hand panel of Figure 6). The contributions of all cross sections (the convolution between input-sky cells and *FSL* features, weighted with the *FSL* pattern power and the solid angles) were added. The result was corrected for the *lsr*-velocity shift, and accumulated in the *FSL* correction profile.

The Kalberla algorithm calculates the stray-radiation contribution on the basis of a model of the sky and of the antenna pattern, using only the position and time of observing of the input spectrum. This enabled us to generate stray profiles for positions and times that were never actually observed. Such simulations allowed us to study the behavior of the antenna under controlled circumstances. Dummy spectra were assigned positional and time parameters, which were used as input to the correction program. Figure 7 illustrates the use of the correction procedure in 'simulation mode'. The left-hand panel shows the variability of the *FSL* component over a 24-hour period of sidereal time (*LST*). For (l,b)=(160°, +50°), a *FSL* stray profile was calculated at every full hour in *LST* for the date 1 January, 1994. The results are displayed as a surface plot. The right-hand panel shows the seasonal variability of the *FSL* component. Again, *FSL* corrections were calculated for (l,b)=(160°, +50°), but this time the *LST* was kept constant at 6^{h}04^m (culmination of the source), while the date was varied in roughly 15-day steps.

We note that the amplitude of the FSL correction shown in Figure 7 is a full order of magnitude more than the *rms* noise in the HI spectra in the survey. We note also that the large and varying widths of the FSL profiles provide an additional mandate for the correction. Even if a particular direction would have had no HI whatsoever within the main beam, the uncorrected observations would have displayed HI intensities with the spectral characteristics indicated in Figure 7.

8.5 Examples of the stray-radiation correction

We show here examples of the stray-radiation correction. More attention will be given to the FSL profiles than to the NSL ones, because the far-sidelobe situation is more variable with time. We discuss inconsistencies between corrected spectra and propose an additional FSL component, originating from reflections of the HI sky on the ground.

As the Dwingeloo telescope has an (Az, El) mounting, the orientation of the antenna pattern with respect to the sky is constantly changing. The FSL contribution is mainly determined by the position of the galactic plane (where the HI intensities are everywhere high) with respect to the most sensitive FSL features, namely the spillover ring and the stray cones. These cross sections dominate the total integrated stray intensity received, while the spectral distribution depends on the difference in *lsr* velocity between the directions of the principal cross sections and the direction of the main beam. For an equatorially mounted telescope, the FSL time variability depends only on the elevation at the time of observing. The elevation determines if cross sections point to the ground, that is, whether or not the bright sky cells are below the horizon and thus whether or not the sensitive FSL features (still 'pointed' at those sky cells) are projected on the ground.

Repeated observations made with identical main-beam pointings, but at various hour angles, will involve the sensitive FSL features in the antenna pattern always pointing in the same direction. Only the position with respect to the horizon determines if the spectrum will be contaminated by these, otherwise constant, cross sections.

In order to judge the quality of the FSL antenna-pattern model, and eventually to fine-tune its parameters, four positions at intermediate and high latitudes were observed repeatedly during the course of the survey. The spectra toward these directions are characterized by a low-intensity, narrow profile, rendering them particularly susceptible to FSL contamination. Many observations of these positions were made, representing the entire range of FSL contaminations for different hour angles and seasons. Comparing the corrected observations with each other provides a direct indication of the accuracy of the correction procedure. If the corrections were perfect, then all corrected spectra should be identical, within the uncertainty of the spectral noise.

There are two ways of comparing results. One way involves calculating a mean over all corrected spectra, and examining the residuals. Another way involves comparing corrected spectra that are most dissimilar. In either case, any discrepancies must be explained and, if possible, accounted for by an improved FSL pattern.

Example I: (90°, +40°). Two observations of the test position $(l,b)=(90°, +40°)$ are shown in Figure 8. The uncorrected spectra are plotted in the upper panel of each pair. The calculated NSL and FSL contributions are shaded; the total amount of stray radiation (NSL+FSL) is also shown. The lower panel of each pair shows the corrected spectra and the total stray profiles. The NSL profiles are almost identical for the two observations; the differences are dominated by the FSL contribution. Note, however, the presence of NSL stray radiation near $v_{\rm lsr}$=−115 km s^{-1}. It is received in the NSL pattern from the ~1 K feature that is present in the spectrum at the same velocity. Before stray-radiation correction, the June, 1990, spectrum contained a spectral feature near $v_{\rm lsr}$=−65 km s^{-1} and the March, 1992, spectrum had a 0.5 K feature near $v_{\rm lsr}$=+70 km s^{-1}. Both features are shown to be due to stray radiation, and have disappeared in the corrected spectra.

It is appropriate for this example to discuss in some detail the manner in which the stray radiation originates in the far sidelobes. Figure 9 supports this discussion. Each panel in this figure displays stray spheres, representing the projection of the full 2π rad^2 of the HI sky that was above the horizon at the moment of observing. The center of the stray sphere corresponds to the zenith (El=90°), and the edge to the horizon (El=0°). The position of the dot (the main beam) reflects the (Az, El) coordinates of the observation, as given at the bottom. The white region around the main beam is the NSL pattern (radius=16°), which was calculated separately. In the shaded area, the FSL pattern receives radiation from the sky. (The stray spheres are projected such that the antenna-pattern features remain essentially constant in shape; this accounts for the relative distortion of the stray spheres.)

The top-left panel shows the stray spheres of the convolution of the HI sky brightness with the FSL antenna pattern, integrated over [−250, +250] km s^{-1} (the full range over which the stray-radiation contamination was calculated). Brighter pixels correspond to higher cross sections. Clearly visible are the main components of the FSL pattern (spillover ring, stray cones, and leg shadows), especially where they intersect the galactic plane. The total integrated intensities are indicated by Σ (in units of K km s^{-1}), which is proportional to $N_{\rm HI}$ over the relevant velocity range.

The six panels of stray-sphere pairs on the right of Figure 9 show how the FSL radiation is distributed over different velocity intervals. The scaling is relative to the total integrals; integrated intensities over the indicated velocity intervals are given.

The individual frames enable detailed analysis of the origin of the spectral components caused by the FSL contamination. The component near $v_{\rm lsr}$=−60 km s^{-1} (Figure 8a), is seen in both the [−200,−60] km s^{-1} and the [−60,−20] km s^{-1} stray spheres, and is mainly caused by the galactic plane gas emitting into the stray cones. The $\Delta v_{\rm lsr}$ sphere shows that the velocity shift is negative in that area, but not as large as −60 km s^{-1}. That implies that the component originates from (negative) intermediate-velocity gas that is shifted to more negative values.

The FSL component near $v_{\rm lsr}$=+70 km s^{-1} seen in Figure 8b is clearly present in the [+60, +200] km s^{-1} frame. It has the highest FSL contribution (Σ=11.2 K km s^{-1}) of all frames shown. The bright regions are identified with the spillover ring; the cross section is highest where this ring overlays the galactic plane. In that region, the $\Delta v_{\rm lsr}$ is about +70 km s^{-1}, which implies that the (high intensity) galactic plane emission near zero velocity is directly shifted to the velocity of the FSL emission feature.

Both observations shown in Figure 8 were made at roughly the same azimuth and elevation, but in different seasons. The orientation of the galactic plane with respect to the FSL pattern was quite similar, but the *lsr* velocity correction had changed due to the different location of the Earth in its orbit around the Sun.

Example II: (160°, +50°). Another illustration of the details of the stray radiation entering the FSL antenna pattern is shown by the observations in Figure 10, with the discussion supported by Figure 11. In this example, the pointing of the telescope was quite different for the two observations of (l,b)=(160°, +50°); this difference is directly reflected in the stray spheres. The high elevation (73°) of the November, 1990, observation placed the main beam near the zenith, and projected the four stray cones entirely onto the sky. The spillover ring, however, was completely on the

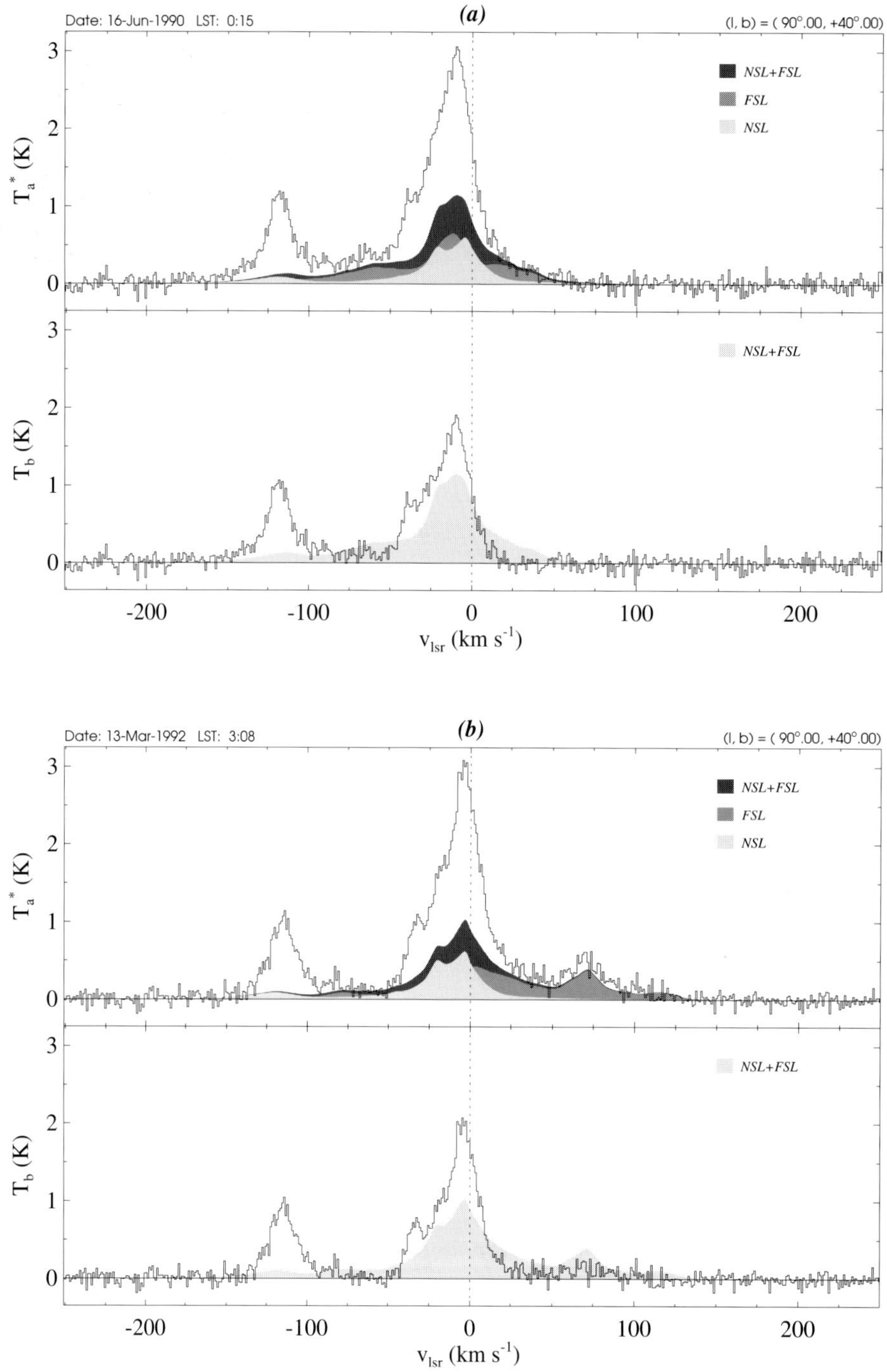

Figure 8. *Stray radiation in two spectra observed towards the direction $(l,b)=(90°,40°)$.* ***(a)*** *Spectrum observed in June, 1990. The uncorrected spectrum and the stray profiles are shown in the upper panel. The* NSL *contribution contained a component that was caused by the HVC near $v_{lsr}=-115$ km s^{-1}. After correction (lower panel) there is hardly any emission left between the HVC and the low-velocity gas. The wing near $v_{lsr}=-60$ km s^{-1} was due to the* FSL *stray radiation.* ***(b)*** *Spectrum observed in March, 1992. A prominent feature of 0.5 K peak temperature is seen in the uncorrected spectrum (upper panel) near $v_{lsr}=+70$ km s^{-1}. The calculated* FSL *profile proved this to be completely attributable to stray radiation. After correction (lower panel) the feature has disappeared. There seems to be no emission between $v_{lsr}=-50$ km s^{-1} and the HVC near $v_{lsr}=-115$ km s^{-1}. The details of the origin of the* FSL *stray-radiation contamination can be seen from Figure 9.*

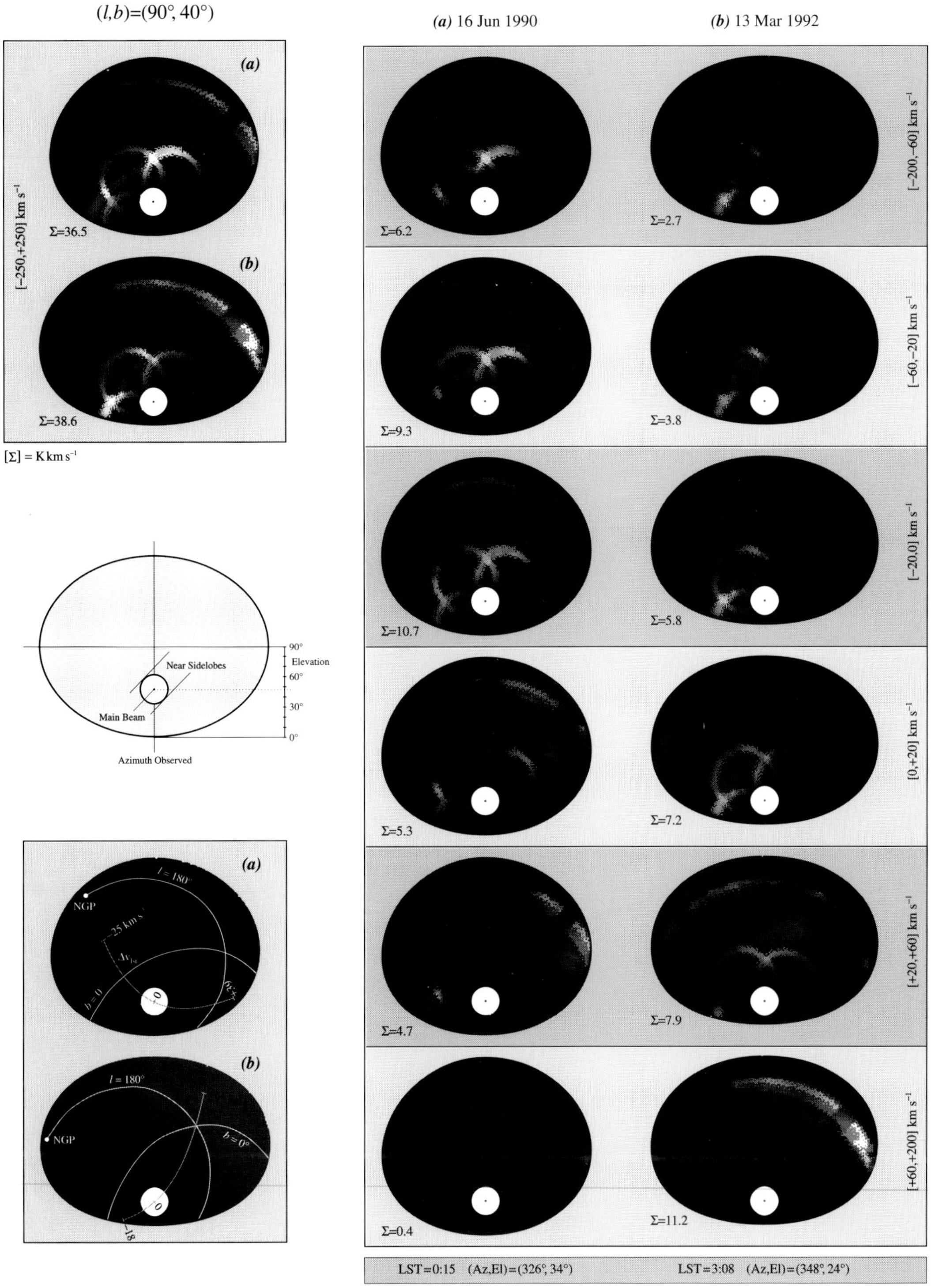

Figure 9. *Stray-radiation spheres for two observations of (l, b)=(90°, +40°).* ***(upper left)*** *Convolution of the FSL pattern with the HI sky, integrated over the entire velocity range of the correction.* ***(center left)*** *Orientation of the spheres. The center corresponds to the zenith; the circumference represents the horizon. Each sphere displays the full $2\pi rad^2$ of the visible sky at the time of observing. The position of the main beam (dot) represents the (Az, El) of the observation.* ***(lower left)*** *Lsr-velocity correction with respect to the direction of the main beam. The galactic plane and north pole are indicated, as is the line l=180°.* ***(right)*** *Six pairs of stray spheres, each for a distinct velocity interval. The scaling is relative to the total FSL stray radiation. The integrated intensities are indicated in units of K km s^{-1}. The sensitive features in the FSL pattern and the bright regions on the sky are clearly visible. The interpretation of the stray spheres in relation to the spectra shown in Figure 8 is given in the text.*

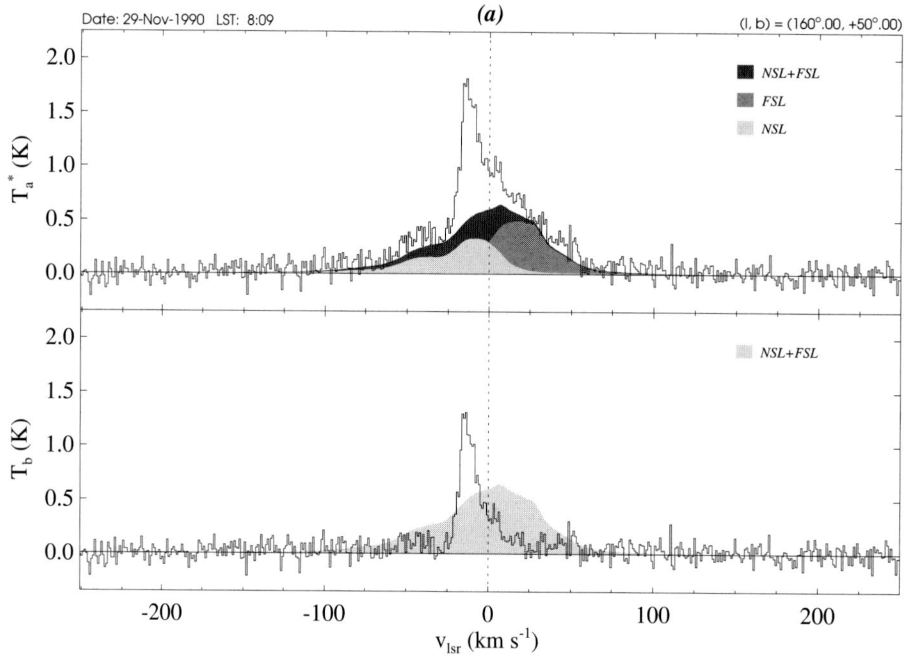

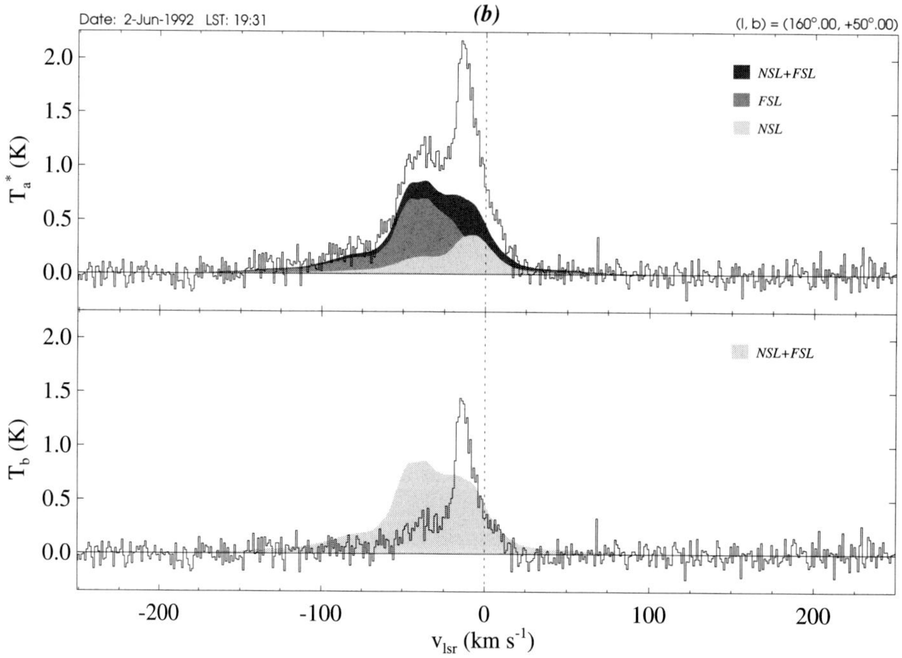

Figure 10. *Two dissimilar spectra resulting from separate observations towards the direction (l,b)=(160°, 50°). The* FSL *profile dominated the stray radiation at high $|b|$. The stray radiation can exceed 50% of the total emission in directions of low column density. This is seen in the lower panel in each pair, where the stray profiles (shaded) contain more emission than the main beam; the far sidelobes dominate the shape and intensity of the uncorrected spectrum.* ***(a)*** *Observation made in November, 1990, at high elevation (73°). The* FSL *profile was less prominent than in (b), because the spillover ring was on the ground rather than towards the sky.* FSL *emission with a peak temperature of 0.5 K is found near +115 km s^{-1}.* ***(b)*** *Spectrum in the same direction observed in June, 1992, at an elevation of 21°. Here, the spillover ring was directed towards the sky. A negative-velocity shoulder is present in the uncorrected spectrum, with a peak temperature of about 0.7 K near v_{lsr}=−40 km s^{-1} and extending to v_{lsr}=−150 km s^{-1}. Figure 11 clarifies the difference in* FSL *contamination between the two spectra.*

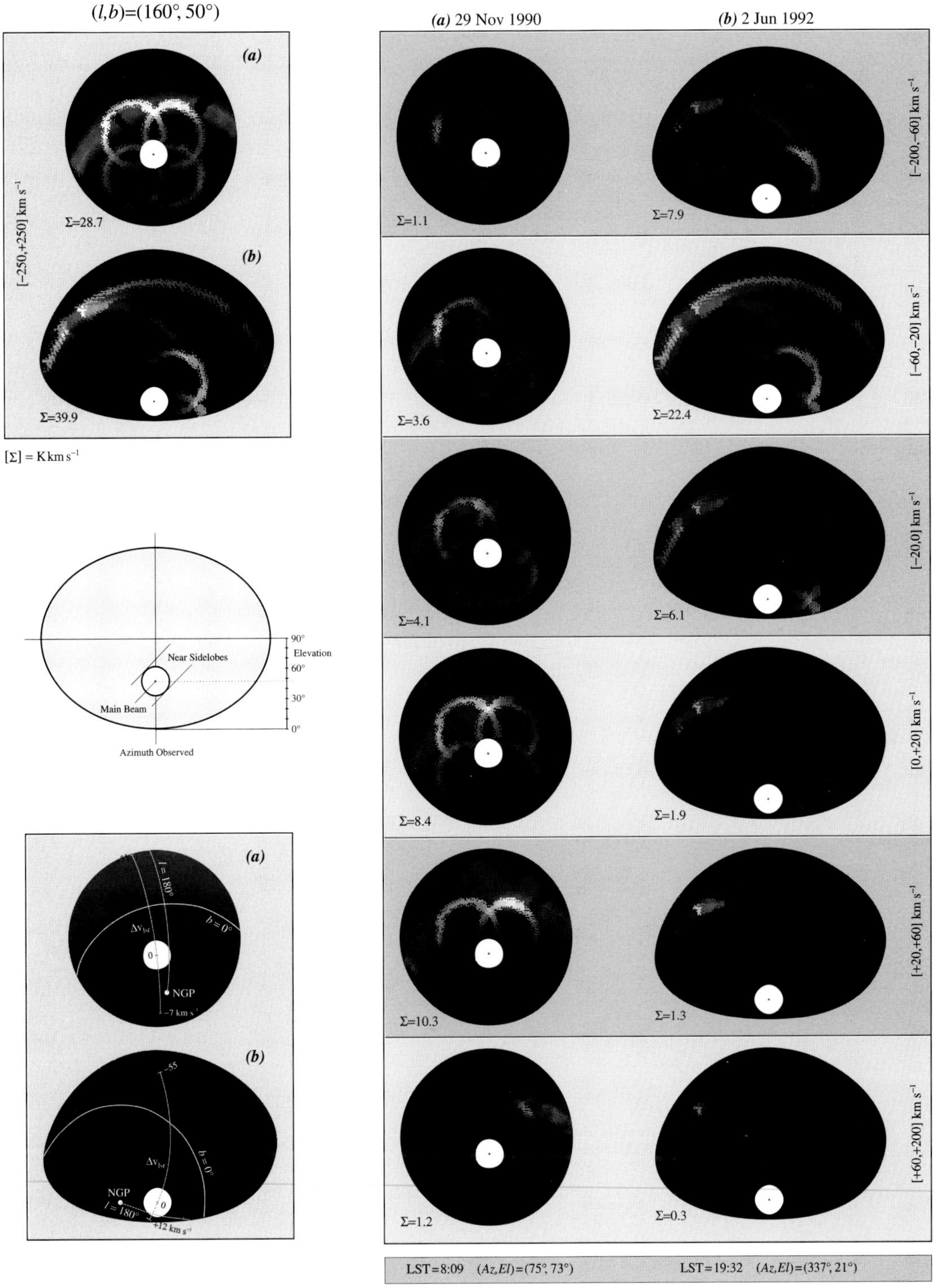

Figure 11. *Stray-radiation spheres for two observations of* $(l,b)=(160°,+50°)$*.* ***(upper left)*** *Convolution of the FSL pattern with the HI sky, integrated over the entire velocity range of the correction.* ***(center left)*** *Orientation of the spheres. The center corresponds to the zenith; the circumference represents the horizon. Each sphere displays the full* $2\pi rad^2$ *of the visible sky at the time of observing. The position of the main beam (dot) represents the (Az, El) of the observation.* ***(lower left)*** *Lsr-velocity correction with respect to the direction of the main beam. The galactic plane and north pole are indicated, as is the line* $l=180°$*.* ***(right)*** *Six pairs of stray spheres, each for a distinct velocity interval. The scaling is relative to the total FSL stray radiation. The integrated intensities are indicated in units of* $K\ km\,s^{-1}$*. The sensitive features in the FSL pattern and the bright regions on the sky are clearly visible. The main text explains the interpretation of the stray spheres in relation to the spectra shown in Figure 10.*

ground. For the June, 1992, observation, half of the spillover ring, and only two of the stray cones, were directed towards the sky. Also, the orientation of the galactic plane was distinctly different in the two cases.

Figure 10a shows a prominent *FSL* component of about 0.5 K peak intensity, near v_{lsr}=+25 km s^{-1}. The stray spheres at [0, +20] km s^{-1} and at [+20, +60] km s^{-1} reveal that the galactic plane was received in the stray cones, at a moderately positive velocity. The total contribution from these velocity intervals is 65% of the total *FSL* stray radiation.

Quite the opposite situation is displayed in Figure 10b. Very prominent *FSL* stray radiation is seen near v_{lsr}=−40 km s^{-1}, reaching a peak intensity of about 0.7 K. From the stray spheres in Figure 11, it is clear that 56% of the total stray radiation was received in the interval [−60,−20] km s^{-1}. The major contribution came from the spillover ring, especially where it overlayed the galactic plane. The *lsr* velocity correction agrees with the assumption that the zero-velocity HI gas from the galactic plane was shifted by this amount, and was received in the spillover ring.

These examples demonstrate the following characteristics of the *FSL* stray-radiation contribution:

1. The total integrated intensity received in the *FSL* antenna pattern is highly variable. In Figure 10b, it is 40% higher than in Figure 10a.
2. The velocity distribution of the *FSL* radiation depends predominantly on how the $v_{lsr}\approx 0$ km s^{-1} emission from near the galactic plane is shifted with respect to the main beam.
3. The elevation of the observation determines whether sensitive parts of the *FSL* pattern are 'up' or 'down'. In Figure 11a, the spillover ring was fully on the ground, and thus did not receive sky emission.

The complicated interplay between *FSL* sensitivity, sky brightness, and velocity shift, makes it quite difficult to trace the exact origin of stray components. For example, comparison of the lower panels in Figure 10a and 10b shows that there was additional emission near v_{lsr}=−40 km s^{-1} in the June, 1990, observation. This implies that either the Figure 10a spectrum was over-corrected, because it shows less emission, or that the Figure 10b spectrum was not corrected enough. In either case, the difference indicates the worst-case limitation to our procedure. The prime precaution in the *FSL* correction algorithm was to guard against over-corrections. A too-high estimate of the *FSL* contribution would create an interval of negative intensity in the spectrum upon subtraction. Such intervals are unacceptable, but easily detected.

Some remarks on the *NSL* contribution are also appropriate. The *NSL* pattern was defined as covering the region $0\overset{\circ}{.}9 \leq \theta \leq 16°$; it represents only 0.5% of the entire antenna pattern. Nevertheless, *half* of the total stray-pattern *sensitivity* is contained in this solid angle. The *NSL* contribution is almost constant with time, mainly because the pattern is highly symmetrical, and will therefore not yield a different convolution when the input sky is rotated with respect to the main beam. The *NSL* contribution largely mimics the shape of the spectrum in question. Thus, although correct recognition of the *NSL* contribution is crucial to accurate determination of total column depths, for example, or of peak brightness temperatures, the near sidelobes do not introduce spurious additional spectral components in the manner of the far sidelobes. The *NSL* contamination generally broadens spectral features.

The combined *MB*+*NSL* region of the antenna pattern can be regarded as the point-spread function of the telescope. Therefore, the effect of correcting a spectrum for the *NSL* stray-radiation contamination is, in principle, a deconvolution to a pencil-beam (main beam) response. The net result of the *NSL* corrections is sharper images.

The examples shown in Figure 8 and Figure 10 indicate that the *NSL* contribution is more-or-less constant with time, and that for intermediate galactic latitudes, some 50% of the total stray radiation is contained in the *NSL* profile. The situation is quite different near the galactic plane. There, the intensities are high, and so the *NSL* pattern receives more power than at high latitudes. The *NSL* profile was typically responsible for about 15% of the total emission received from directions near the galactic equator. In these directions of high intensity the *FSL* contribution was relatively insignificant.

8.6 Possible reflections off the local terrain

A direct indication of the instantaneous stray radiation can be obtained by pointing the telescope toward the Moon, thereby blocking the main beam. Because of the ubiquity of galactic HI, there is a lunar occultation at all times. Several observers have demonstrated the large residual 21-cm signal remaining when the main beam is fully blocked by the Moon, and have interpreted this signal as fully due to stray radiation entering the sidelobes.

Although this explanation no doubt correctly accounts for *most* of the residual signal, it does not account for all of it. Comparing calculated stray profiles with lunar occultation observations, Kalberla *et al.* (1980) found discrepancies that could be explained by emission entering the *main* beam after being *reflected off* the lunar surface. The residuals were consistent with a lunar albedo of 0.07 and a *HPBW* of about 60° (considering the Moon as a transmitting antenna). Direct evidence for lunar 21-cm reflections had been mentioned by Giovanelli & Haynes (see Kalberla *et al.* 1980), who used the Arecibo telescope to observe the Moon at 21-cm. They reported the presence of an HI emission component that was not observed when the Moon was absent. The velocity of the component was found to correspond to the relative *lsr* velocity shift between opposite-pointing directions of the telescope. Therefore, it seemed plausible that HI emission at $v_{lsr}=0$ kms^{-1}, in the direction opposite the telescope main-beam pointing, was being observed reflected off the Moon.

If HI emission reflected off the Moon, with an albedo at 21-cm wavelength of 0.07, could be clearly identified as such, it seemed reasonable to consider whether additional sources of reflected radiation might be relevant. A body of water seen close to glancing incidence is an effective flat mirror to radio waves. This was demonstrated by one of the earliest radio interferometers, which used the signal reflected by the sea as a second path between a source and the receiver (Bolton & Stanley 1948).

The Dwingeloo telescope is located on the northern edge of an extensive, very flat, largely tree-less heath. That landscape is Dutch in its essence: it affords distant horizons, and is frequently wet. Although little is known about the HI reflectivity of soil and vegetation, extensive studies (employed in the field of remote sensing) exist on the subject of back-scattering of radar from a variety of natural objects. Some radar bands are quite close to 1420 MHz, and these data provide a few useful leads. Radar at decimeter wavelengths is back-scattered from grass, trees, and scrubs. It therefore seems likely that it can be reflected in the forward direction as well. The radar signal is strongly dependent on the physical nature of the scattering objects. Moisture is particularly important in determining the efficiency of radar reflectivity; dry grass, wet grass, snow-covered grass, and ice-covered grass all reflect radar waves differently.

The heath stretches unobstructed for about five kilometers south of the telescope, and contains many small pools of water, some open and some largely clogged with aquatic vegetation. The weather at the time of observing might well be an important influence on the instantaneous 21-cm reflectivity. The season may also be of influence, because it determines, on a longer time scale, factors such as the height of the vegetation, the degree of leafiness, and the moisture content of the soil. Since none of this information was available for this project, we experimented with models based on some crude assumptions, and reached some tentative conclusions.

We explored the Dwingeloo heath and sketched a map, paying particular attention to the location of various pools and ponds. A model for ground reflection was based on these sketches. Because of our lack of knowledge on the mean reflectivity of the ground, our initial modeling attempts were made with a 100% reflectivity at glancing incidence, in order to study the exaggerated effects of reflection on the total FSL contribution. Although generally over-correcting the spectra, the model did account for some of the features that were considered left-over stray radiation. The final, though still crude, model we tested contained a parameterization of the terrain to the south of the telescope, with reflectivity at elevation 0° varying smoothly between 1.0 at azimuths centered near 210°, and decreasing to 0.0 at azimuths less than 110° and greater than 260°. We call the reflection profiles calculated with this model the reflection sidelobes, or RSL for short. The assumed reflectivities are high (cf. 0.07 for the Moon), but the results of this empirical model are nevertheless quite supportive of high reflectivities at the lowest elevations.

In Figure 12, the results of the ground-reflection calculations are compared with the results from the regular stray-radiation correction (NSL+FSL) for two ('worst-case') spectra taken from a test suite of observations which we made toward the Lockman *et al.* (1986) hole. The figure shows the effect of the RSL correction on the two spectra, observed toward the same direction. The observed spectra, uncorrected for stray radiation, are shown in Figure 12a. After subtraction of NSL+FSL, as in our standard reduction, the spectra, shown in Figure 12b, are in much better agreement. The differences between these two spectra do, nevertheless, represent an exceptionally poor case, and motivated us to seek the cause of the discrepancy. The two spectra shown in Figure 12c were corrected for the reflected radiation as well as for the standard NSL+FSL contribution; including the RSL contribution clearly reduces the residual discrepancy. The individual contributions of the RSL component are shown in Figure 12b. Evidently, reflection occurs in spectrum [A] at the velocity for which the FSL profile reaches a maximum, and where the residual component was seen in Figure 12b. It is still present, since the RSL component appears to have removed only a narrow portion from the center of the feature. The RSL correction in spectrum [B] was less successful, although still suggestive of the validity of the approach; the intensity and width of the RSL profile match the left-over component, although the modeled central velocities are not in good agreement with the observed structure. We note that the RSL correction applied did not produce the negative residuals that would by themselves indicate an erroneous correction.

The mean HI total column density observed towards the test suite of observations toward the Lockman *et al.* hole was 15.3 (in units of $10^{19}\mathrm{cm}^{-2}$) before any stray radiation correction, 6.2 after the NSL+FSL correction, and 5.1 after the NSL+FSL+RSL correction. The NSL, FSL, and RSL corrections account for, respectively, 28%, 62%, and 10% of the total correction. Although we did not incorporate a correction for radiation reflected from the ground into our standard reduction, we did make use of experiments with the crude reflection model, applied to the spectra in the test suite, in estimating the accuracy of the stray-radiation correction applied to the entire Leiden/Dwingeloo HI survey.

For the 24 spectra in the test suite of observations towards the Lockman *et al.* hole we calculated the mean spectra and the residuals for three different cases, namely for uncorrected spectra (corresponding to T_a^*), for spectra corrected for NSL

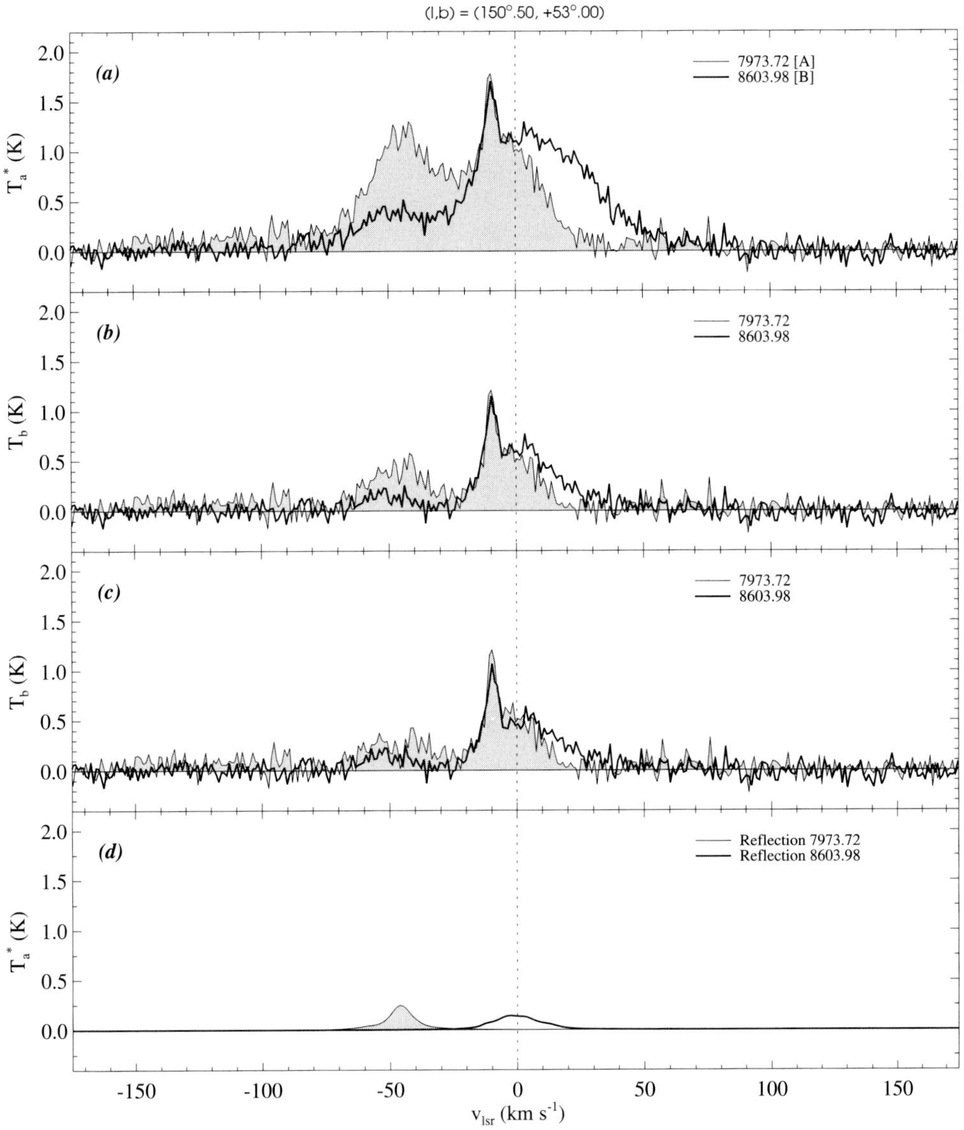

Figure 12. *Two of the most dissimilar spectra (indicated by [A] and [B]) taken from a test suite of observations towards a single position in the Lockman et al. hole.* ***(a)*** *Uncorrected spectra, showing large differences.* ***(b)*** *Spectra after correction for* NSL *and* FSL. ***(c)*** *Spectra after correction for* NSL, FSL, *and* RSL *contamination showing better agreement.* ***(d)*** *Individual* RSL *components calculated for [A] and [B].*

and *FSL* stray radiation (T_b), and for spectra which were additionally corrected for *RSL* contamination (T_b). From the residual spectra we calculated the channel-by-channel *rms* deviations from the mean value as an indication of the velocity-dependent accuracy of the intensities in the individual spectra. The upper-left panel of Figure 13 shows the mean of the uncorrected spectra in the test suite; the corresponding panel on the upper right of Figure 13 shows the *rms* values for all the channels. The peak uncertainties are about 0.35 K near v_{lsr}=−45 km s^{-1}, while the mean *rms* uncertainty over the velocity interval $|v_{lsr}|\leq 100$ km s^{-1} amounts to 0.16 K. After applying the *NSL+FSL* correction, the results are greatly improved, as shown in Figure 13b. The peak and mean *rms* uncertainties are now 0.16 K and 0.10 K,

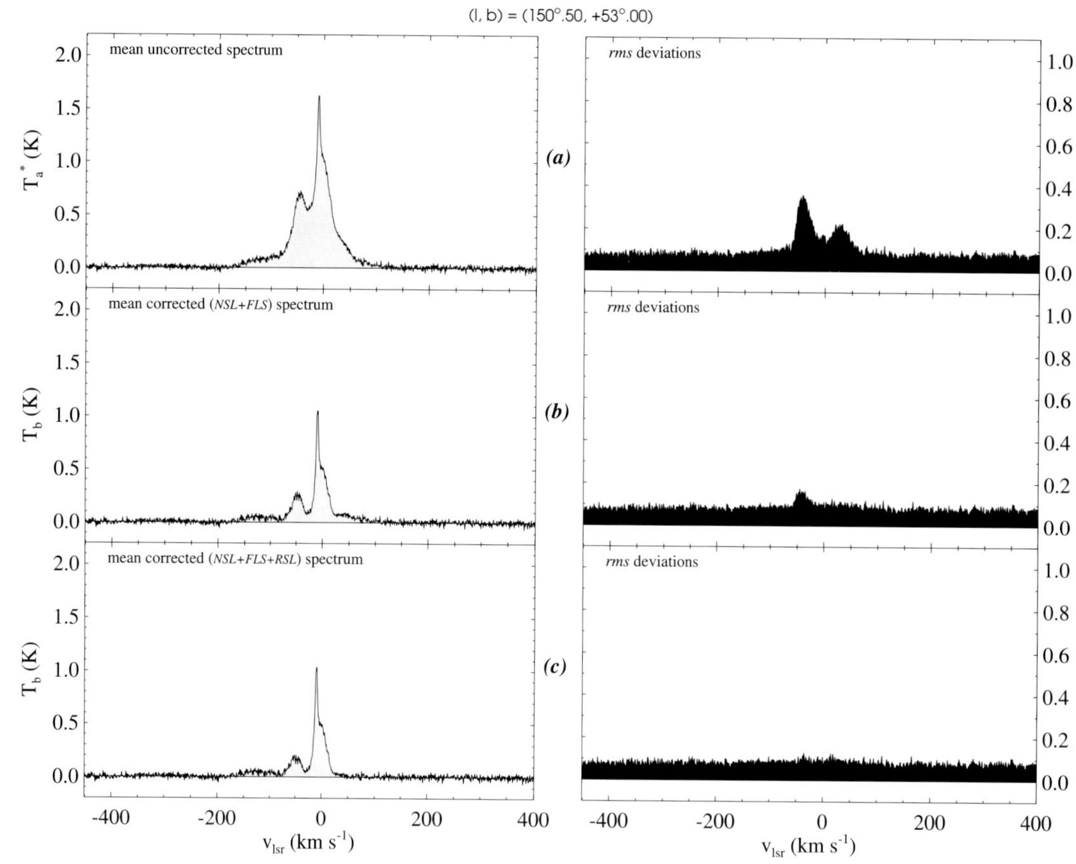

Figure 13. *Mean spectra of 24 observations towards the Lockman et al. hole (left) and channel-by-channel rms deviations of the individual spectra from the mean (right). The mean level outside the velocity range for which stray radiation was calculated is indicative of the spectral noise due to the receiver system and the reduction procedure.* ***(a)*** *Uncorrected data. Without stray-radiation correction the uncertainties for $|v_{lsr}|<100$ km s^{-1} in the spectral intensities of spectra of low column density exceed the rms spectral noise due to the receiver system and the reduction procedure by an order of magnitude.* ***(b)*** *Data corrected for NSL+FSL stray radiation. The correction greatly reduces the residuals in the spectra for $|v_{lsr}|<100$ km s^{-1}. The mean discrepancies over this velocity interval are of the order of the rms spectral noise due to the receiver system and the reduction procedure. Because these data are towards a single direction, they do not represent a general estimate of the accuracy of the correction procedure.* ***(c)*** *Data corrected for NSL+FSL+RSL stray radiation. With the application of the RSL correction, uncertainties for $|v_{lsr}|<100$ km s^{-1} disappear in the general rms noise due to the receiver system and the reduction procedure.*

respectively. The brightness temperature at the velocity of the largest residual is about 0.25 K, compared to 0.7 K in the mean T_a^* spectrum.

Application of the additional correction for ground reflection lowered the mean *rms* uncertainties to the order of the spectral noise expected from the receiver alone and from other aspects of the reduction procedure, for example baselining. Figure 13c shows the mean spectrum on the left and the practically flat *rms* spectrum on the right. We do note, however, that the spectra in this test suite were used to optimize the *RSL* correction algorithm and therefore may not be representative of the general accuracy that can be obtained from applying the reflection correction.

We reduced all spectra entering the survey for the combined *NSL+FSL* contamination; although the reflection correction was calculated for various spectra, it was not applied to the survey as a whole because of uncertainty regarding the parameters which influence this process.

9. Accuracy of the survey spectra

Our estimate of the accuracy of the spectra presented in this Atlas is based on the assumption that repeated observations towards the same line of sight must yield identical spectra. (It was, after all, such an expectation that first led to the practical recognition of the contaminating properties of stray radiation.) Discrepancies between individual spectra and their mean are a measure of accuracy for the entire reduction procedure. We note that such a measure indicates only the relative accuracy; inaccuracies on an absolute scale cannot be determined in this way.

In order to explain the residual errors as dominated by left-over stray-radiation contamination, we must exclude the possibility that they are spurious instrumental effects other than emission received in the sidelobes. We made the a priori assumption that HI emission from the sky (significantly above the *rms* noise level) has positive intensities only, except towards lines of sight where there is absorption against strong continuum background sources. We therefore demanded that the correction procedure never yield negative-intensity intervals in the corrected spectrum. This constraint may have produced spectra in which the true amount of stray radiation was underestimated.

The scheduling of observations in boxes which overlap at their boundaries was a deliberate choice to ensure repeated observations for many lines of sight. Because the priority of observing the boxes depended on the declination of each central position, the observations in the directions where multiples would appear were spread out over the entire observing period. We consider the accuracy derived from comparing the residuals from multiple spectra spaced randomly in time as representative for the entire dataset.

We compared multiply observed spectra in two strips, narrowly confined in latitude but wrapping over the full range of longitude. One of the strips was at high latitude, the other near the galactic equator. We discuss the results for the high-latitude strip below; the results for the galactic-equator strip were consistent with these results.

Repeated observations towards identical lines of sight in the latitude strip $+65° \leq b \leq +70°$ were used to estimate the accuracy of the FSL stray-radiation correction. At high latitudes, the major stray contribution originates from the far-sidelobe region of the antenna pattern. From the test suite of observations toward the Lockman *et al.* hole, we had inferred that in directions of low column density approximately equal amounts of radiation are received by the main beam and by the far sidelobes. It is therefore evident that uncertainties in the FSL profile determination will be mainly responsible for the inaccuracies in the corrected spectra.

A total of 456 pairs of repeatedly observed spectra were averaged, and the *rms* deviations calculated. The mean of all averaged spectra is shown in the left-hand panel of Figure 14. This spectrum has little astrophysical relevance (but it does give an indication of the generally low intensities in this latitude strip). Shown on the right in Figure 14 are the channel-by-channel *rms* deviations, σ, averaged over all multiply observed spectra. Channels at velocities outside the range over which the stray-radiation correction was applied ($-250 \leq v_{\rm lsr} \leq +250$ km s^{-1}) are, of course, unaffected and reflect the general accuracy of the reduction procedure. The average uncertainties over the velocity interval $|v_{\rm lsr}| \leq 100$ km s^{-1}, σ_{100}, are about 0.11 K, with a maximum deviation of 0.24 K at $v_{\rm lsr}=0$ km s^{-1}. We compared the value of σ_{100} with the mean *rms* deviation for $|v_{\rm lsr}| \geq 100$ km s^{-1}, which we call the general reduction

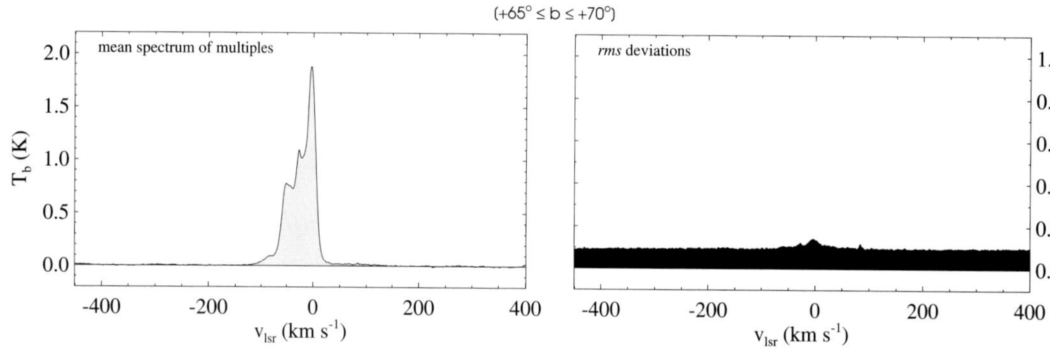

Figure 14. (left) *Mean of the averaged multiply-observed spectra in the latitude strip $65° \leq b \leq 70°$.* ***(right)*** *Mean of the channel-by-channel rms deviations of the individual (multiply-observed) spectra from their averaged profiles. The deviations for $|v_{lsr}| < 100$ km s^{-1} are mainly due to uncertainties in the FSL stray-radiation corrections. No RSL correction was attempted for these spectra. The accuracy of the stray-radiation correction procedure was estimated from such data. The average uncertainties for $|v_{lsr}| < 100$ km s^{-1} are of the order of the rms spectral noise due to the receiver system and the reduction process. (The peak near +85 km s^{-1} is due to interference.)*

accuracy, $\sigma_{\rm red}$. For the high-latitude strip we determined $\sigma_{\rm red}$=0.096 K. We estimated the uncertainties due to the stray-radiation correction, $\sigma_{\rm stray}$, by disentangling $\sigma_{\rm red}$ and $\sigma_{\rm stray}$ from σ_{100}, yielding $\sigma_{\rm stray}$=0.052 K. Using the mean *rms* spectral noise value of 0.07 K determined over the entire survey, the accuracy of the stray-radiation correction procedure as applied to low-intensity spectra was calculated as $\sigma_{\rm stray}$=0.084 K.

The results derived from the test suite of observations toward the Lockman *et al.* hole, from the high-latitude strip multiples, and from an analogous experiment with repeated observations near the galactic equator, lead us to state the general accuracy of the HI data in this Atlas as follows: uncertainties in the stray-radiation corrected spectra due to the *FSL* correction are less than 0.1 K, and the uncertainties due to the *NSL* correction are of the order of 2%. The general accuracy of the brightness temperature calibration of about 1% is responsible for the larger part of the *NSL* errors.

The accuracy of the stray-radiation correction is limited by three factors:

1. *Antenna pattern.* We used the averaged *NSL* response of the *WSRT* telescope antennas as an approximation of the Dwingeloo *NSL* antenna pattern. Determination of the *NSL* pattern of the present Dwingeloo antenna by direct measurement of the *NSL* response would no doubt improve the *NSL* stray-radiation correction, although limitations concerning the accuracy and resolution of the input sky are probably of greater influence. We believe that the feedback procedure employed in tuning the parameters of the *FSL* model produced an accurate final estimate of the *FSL* response of the Dwingeloo antenna. An improved determination of the antenna pattern from direct measurements might yield more accurate *FSL* corrections, but only if the computer implementation of the correction algorithm is calculated to a higher spatial resolution.
2. *Input sky.* The input sky was created from the uncorrected survey data, binned into 2°×2° cells. The size of the cells was determined by the spatial resolution employed in the computer implementation of the correction algorithm. The general error of about 1% in the calibration of the brightness temperature scale

propagates directly into the calculated NSL profiles, and increases the uncertainties in the corrected spectra by approximately 2%. Using an input sky created from binning the *stray-radiation-corrected* survey might yield an improvement in the determination of the stray profiles. For the NSL correction, however, the gain variations will remain the major source of uncertainty.

3. *Computer implementation.* The computational effort involved in the present implementation of the correction algorithm was considerable. Even though the calculations of both the NSL and the FSL stray profiles require merely one CPU second on a DecStation 3000/500 computer for each spectrum observed, increasing the spatial resolution of the correction would increase the computational burden considerably.

We suggest that the most practical improvement over our application of the stray-radiation correction would be reached by using the *corrected* Leiden/Dwingeloo HI survey to create a new input sky of cell size 1°×1°. Observations should also be performed to obtain more detailed knowledge of the characteristics at 21-cm wavelength of reflection from the ground, under various conditions; this is evidently an important but poorly understood phenomenon.

10. Preparation of the complete HI data cube

Choices were necessary regarding the format of the final data cube, which has been stored on the CD-ROM of the Leiden/Dwingeloo HI survey, as well as regarding its display in the Atlas of representative maps. We are aware that physical structures, not uncommonly with elongated, filamentary, or otherwise complicated forms, and furthermore not uncommonly showing strong kinematic gradients, snake through any HI data cube constructed of orthogonal planes of l, b, and $v_{\rm lsr}$, and thus any display based on such orthogonal planes will have its shortcomings.

Flat-plane cuts through such a data cube will not reveal the complete structural nature of the HI features. Nevertheless, flat-plane cuts seem to be the most useful when displaying images of the entire sky and were thus the ones chosen. We describe here the construction of the final data cube.

The results of the final reduction step were separate files of survey spectra, each file corresponding to a separate observing batch. These were split into twenty files of equal Δl (keeping positive- and negative-latitude intervals separate) and then sorted according to the FITS (l, b) ordering scheme (Wells, Greisen, & Harten 1981), whereby l decreases before b increases. Multiple spectra entering the sorted files were then averaged. A mean spectrum was calculated using $1/\sigma^2$ as weights, where σ is the *rms* spectral noise. The final mean spectrum was calculated from the 'good' spectra, using $1/\sigma^2$ to weight the spectral intensities, the system temperatures, and the elevations of the contributing spectra. Integration times were added. This mono-gridding procedure resulted in latitude strips of sorted spectra, at equal longitude separations, appropriately averaged.

The survey was observed at true-angle spatial separations of $\Delta b = 0\overset{\circ}{.}5$ and $\Delta l \leq 0\overset{\circ}{.}5$. To create a homogeneous data cube, the spectra were re-gridded onto a common lattice, with $\Delta l \times \Delta b = 0\overset{\circ}{.}5 \times 0\overset{\circ}{.}5$. Although all spectra at $|b| \leq 35°$ had been observed on such a grid, spectra outside this latitude range had to be linearly interpolated in the l direction. The *rms* noise for the re-gridded spectrum was calculated as the square root of the squares of the weighted individual noise contributions.

The three-dimensional sky cube was created from the re-gridded, mono-gridded spectra. Multiple spectra had to be averaged where the latitude strips were coincident. Some of these spectra were already the means of previous multiples, and many

Table 3. *Summary of the parameters characterizing the final data cube of the Leiden/Dwingeloo HI survey*

# spectra	206,457
disk space occupied	~450 Mbyte
sky cube filling	79%
Δl, [l-range]	0°.5, [0°≤l≤360°]
Δb, [b-range]	0°.5, [−90°≤b≤+90°]
$\Delta v_{\rm lsr}$, [$v_{\rm lsr}$-range]	1.03 km s^{-1}, [−450≤$v_{\rm lsr}$≤+400] km s^{-1}
# channels (effective)	835
sky coverage	all $\delta \geq -30°$
$\langle \sigma_{\rm rms} \rangle$	0.07 K
$\langle T_{\rm sys} \rangle$	40.7 K
$\langle t_{\rm int} \rangle$	244 seconds
$\langle N_{\rm HI} \rangle$	7.82×10^{20} cm^{-2}
$\langle N_{\rm HI} \rangle_{b>+20°}$	2.86×10^{20} cm^{-2}
$\langle N_{\rm HI} \rangle_{b<-20°}$	3.74×10^{20} cm^{-2}

of them had been re-gridded. The overlapping spectra were averaged. The data cube was then created by sorting all spectra, with the averages replacing the single entries, into one large, final file.

The final HI data cube contained 206,457 spectra placed on a regular $(\Delta l \times \Delta b \times \Delta v) = (0°.5 \times 0°.5 \times 1.03 \text{ km s}^{-1})$ grid and spanning the three-dimensional interval $(0° \leq l \leq 360°) \times (-90° \leq b \leq +90°) \times (-450 \leq v_{\rm lsr} \leq +400 \text{ km s}^{-1})$ at $\delta \geq -30°$. The parameters of the data cube are summarized in Table 3. (We note that the mean values tabulated are biased by the re-gridding of the data onto a common lattice, whereby the observations at $|b| > 35°$ are over-represented.)

11. Contamination by external galaxies

The Leiden/Dwingeloo HI survey is intended as a resource for studying the interstellar gas associated with our own galactic system. Although the Dwingeloo 25-m telescope is too small, and our integration times were too short, to expect the survey to be suitable for investigating HI emission from external systems in any detail, we have nevertheless detected many such nearby galaxies, all of them previously known. In some cases their emission signatures mimic what might be expected from a high-velocity cloud, in other cases their emission appears similar to that due to interference of the most troublesome sort, and in yet other cases the external systems are difficult to distinguish from conventional Milky-Way gas. In all cases, the external systems should be identified as such, if only as contaminating nuisances. In the final reduction steps the survey spectra were all inspected as grayscale (*position, velocity*) images. External galaxies caused generally weak, velocity-extended, disruptions of the baselines. In many cases the external galaxies were found at velocities outside the normal range expected for HI associated with the Milky Way. At negative velocities, many HVCs were suspected galaxies at first, when they were seen in individual spectra, but were found as the data became more complete to belong to known extended HVC complexes. Many, but not all, of the HVCs which can be identified in this survey correspond to features listed by Wakker

Table 4. *Properties of external galaxies identified in the survey. The parameters listed in the central group of columns refer to those derived from the Leiden/Dwingeloo survey; the parameters listed in the right-hand group of columns give the intrinsic properties, determined from directed studies tabulated in the RC3*

			Survey parameters						RC3 parameters			
Name1	Name2	PGC	l	b	v_{lsr} (km s^{-1})	*FWHM* (km s^{-1})	T_{max} (K)	$\Sigma T dv$ (K km s^{-1})	l	b	v_{sys} (km s^{-1})	*FWHM* (km s^{-1})
NGC 7793	ESO 349–12	73049	5°.7	−77°.0	240	150	0.28	44	4°.5	−77°.2	230	174
Sag DIG	ESO 594–4	63287	21°.0	−16°.5	−70	35	0.15	5.4	21°.1	−16°.3	−77	32
NGC 6822	DDO 209	63616	25°.0	−18°.5	−40	70	3.5	256	25°.3	−18°.4	−56	63
NGC 6822	DDO 209	63616	25°.0	−18°.0	−65	50	1.4	76	25°.3	−18°.4	−56	63
DDO 210	MCG-2-53-3	65367	34°.0	−31°.5	−120	30	0.11	3.4	34°.0	−31°.4	−137	21
DDO 221	MCG–3–1–15	143	75°.6	−73°.5	−120	65	0.76	52	75°.9	−73°.6	−120	57
DDO 190	UGC 9240	51472	82°.0	64°.5	160	40	0.04	1.9	82°.0	64°.5	153	45
Pegasus	DDO 216	71538	94°.8	−43°.5	−180	20	0.20	4.7	94°.8	−43°.5	−182	23
NGC 6946	UGC 11597	65001	95°.5	11°.5	~85	~170	~0.50	~33	95°.7	11°.7	52	169
NGC 253	ESO 474–29	2789	100°.0	−88°.0	~240	~360	~0.45	>28	97°.4	−88°.0	251	410
NGC 6503	UGC 11012	60921	100°.5	30°.5	~100	~100	0.20	9.6	100°.6	30°.6	44	173
NGC 5585	UGC 9179	51210	101°.0	56°.5	~260	~50	~0.11	~5.5	101°.0	56°.5	305	146
NGC 5474	UGC 9013	50216	101°.0	60°.0	290	40	0.36	14	100°.8	60°.2	277	40
M 101	NGC 5457	50063	102°.0	60°.0	240	~170	1.1	133	102°.0	59°.8	241	143
M 101	NGC 5457	50063	102°.4	59°.5	~240	~165	~1.8	~118	102°.0	59°.8	241	143
NGC 7640	UGC 12554	71220	105°.0	−19°.0	290	70	0.09	7.0	105°.2	−18°.9	369	234
IC 4182	UGC 8188	45314	108°.3	79°.0	330	40	0.24	9.6	107°.7	79°.1	321	40
DDO 168	UGC 8320	46039	110°.6	70°.5	200	50	0.19	11	110°.8	70°.7	195	60
NGC 5204	UGC 8490	47368	113°.6	58°.0	200	110	0.18	21	113°.5	58°.0	204	110
NGC 247	ESO 540–22	2758	114°.8	−83°.5	145	200	0.61	83	113°.9	−83°.6	159	210
IC 10	UGC 192	1305	119°.0	−3°.5	−335	55	0.92	56	119°.0	−3°.3	−344	63
M 31	NGC 224	2557	121°.5	−21°.5	…	…	…	…	121°.2	−21°.6	−300	510
M 31	NGC 224	2557	121°.5	−21°.0	…	…	…	…	121°.2	−21°.6	−300	510
M 31	NGC 224	2557	122°.0	−21°.0	…	…	…	…	121°.2	−21°.6	−300	510
NGC 4236	UGC 7306	39346	127°.4	47°.5	75	50	0.67	36	127°.4	47°.4	0	162
DDO 147	UGC 7949	43129	128°.5	80°.5	335	25	0.12	3.3	128°.4	80°.6	333	30
IC 1613	DDO 8	3844	130°.0	−60°.5	−235	25	2.7	64	129°.8	−60°.6	−230	25
M 33	NGC 598	5818	133°.5	−32°.0	…	…	…	…	133°.6	−31°.3	−179	184
M 33	NGC 598	5818	133°.5	−31°.5	…	…	…	…	133°.6	−31°.3	−179	184
M 33	NGC 598	5818	133°.5	−31°.0	…	…	…	…	133°.6	−31°.3	−179	184
Maffei II	UGCA 39	10217	136°.5	−0°.5	~0	…	…	…	136°.5	−0°.3	−1	305
NGC 4449	UGC 7592	40973	137°.2	72°.5	230	105	0.53	59	136°.9	72°.4	201	136
DDO 125	UGC 7577	40904	137°.2	73°.0	…	…	…	…	137°.8	72°.9	196	28
IC 1727	UGC 1249	6574	138°.0	−34°.0	330	105	0.13	15	138°.0	−33°.9	338	121
IC 342	UGC 2847	13826	138°.0	11°.0	~70	~105	~2.7	~158	138°.2	10°.6	34	151
NGC 1560	UGC 3060	15488	138°.5	16°.0	−35	115	0.55	66	138°.4	16°.0	−36	125
UGC 7356	MCG 8–22–105	39615	138°.6	69°.0	290	90	0.18	17	138°.5	69°.1	272	86
NGC 4068	UGC 7047	38148	139°.0	63°.0	225	35	0.10	3.5	138°.9	63°.0	211	51
UGCA 86		14241	139°.5	10°.5	75	115	0.53	66	139°.8	10°.6	67	99
IC 2574	DDO 81	30819	139°.8	43°.5	85	87	0.39	36	140°.2	43°.6	47	115
Holmberg I	DDO 63	27605	140°.4	38°.5	150	35	0.22	8.6	140°.7	38°.7	136	26
NGC 784	UGC 1501	7671	141°.0	−31°.5	185	90	0.10	10	140°.9	−31°.6	198	96
M 82	NGC 3034	28655	141°.6	40°.5	170	125	0.53	71	141°.4	40°.6	203	146
NGC 3077	UGC 5398	29146	142°.0	42°.0	~10	~70	~0.57	~42	141°.9	41°.7	14	65
M 81	NGC 3031	28630	142°.2	41°.0	~−50	~375	~0.68	~180	142°.1	40°.9	−34	422
NGC 2976	UGC 5221	28120	143°.5	41°.0	~−5	~110	~0.41	~48	143°.9	40°.9	3	97
UGCA 92	MCG 3–41–50	15439	144°.5	10°.5	−105	30	1.7	53	144°.7	10°.5	−99	77
Holmberg II	DDO 50	23324	144°.5	32°.5	160	55	0.67	38	144°.3	32°.7	158	66
UGC 4483		24213	145°.0	34°.5	~120	~165	~0.12	~21	145°.0	34°.4	156	49
NGC 2366	DDO 42	21102	146°.5	28°.5	105	95	0.55	55	146°.4	28°.5	100	96
UGCA 105		16957	148°.5	13°.5	~105	~120	~0.30	~38	148°.5	13°.7	111	118
NGC 2403	UGC 3918	21396	150°.5	29°.0	~135	~250	~0.60	~118	150°.6	29°.2	131	231
NGC 2403	UGC 3918	21396	150°.5	29°.5	~135	~250	~0.60	~118	150°.6	29°.2	131	231
NGC 4244	UGC 7322	39422	154°.5	77°.5	255	195	0.12	7.1	154°.6	77°.2	243	204
NGC 4214	UGC 7278	39225	161°.5	78°.0	290	55	0.45	25	160°.2	78°.1	291	62
NGC 4395	UGC 7524	40596	162°.4	81°.5	320	110	0.47	56	162°.1	81°.5	320	109
DDO 133	UGC 7698	41636	165°.0	84°.0	335	60	0.12	7.2	164°.3	84°.0	333	53
DDO 99	UGC 6817	37050	166°.6	73°.0	235	50	0.13	7.0	166°.2	72°.8	245	37
DDO 43	UGC 3860	21073	178°.0	24°.0	355	20	0.10	2.2	177°.8	23°.9	354	38
DDO 47	UGC 3974	21600	203°.0	18°.5	250	100	0.22	23	203°.1	18°.5	270	56
Sextans B	DDO 70	28913	233°.4	44°.0	295	35	0.25	9.6	233°.2	43°.8	301	41
Sextans A	DDO 75	29653	246°.0	40°.0	315	45	0.57	28	246°.2	39°.9	324	63
NGC 3109	DDO 236	29128	262°.0	23°.0	400	>90	~1.8	>124	262°.1	23°.1	404	116

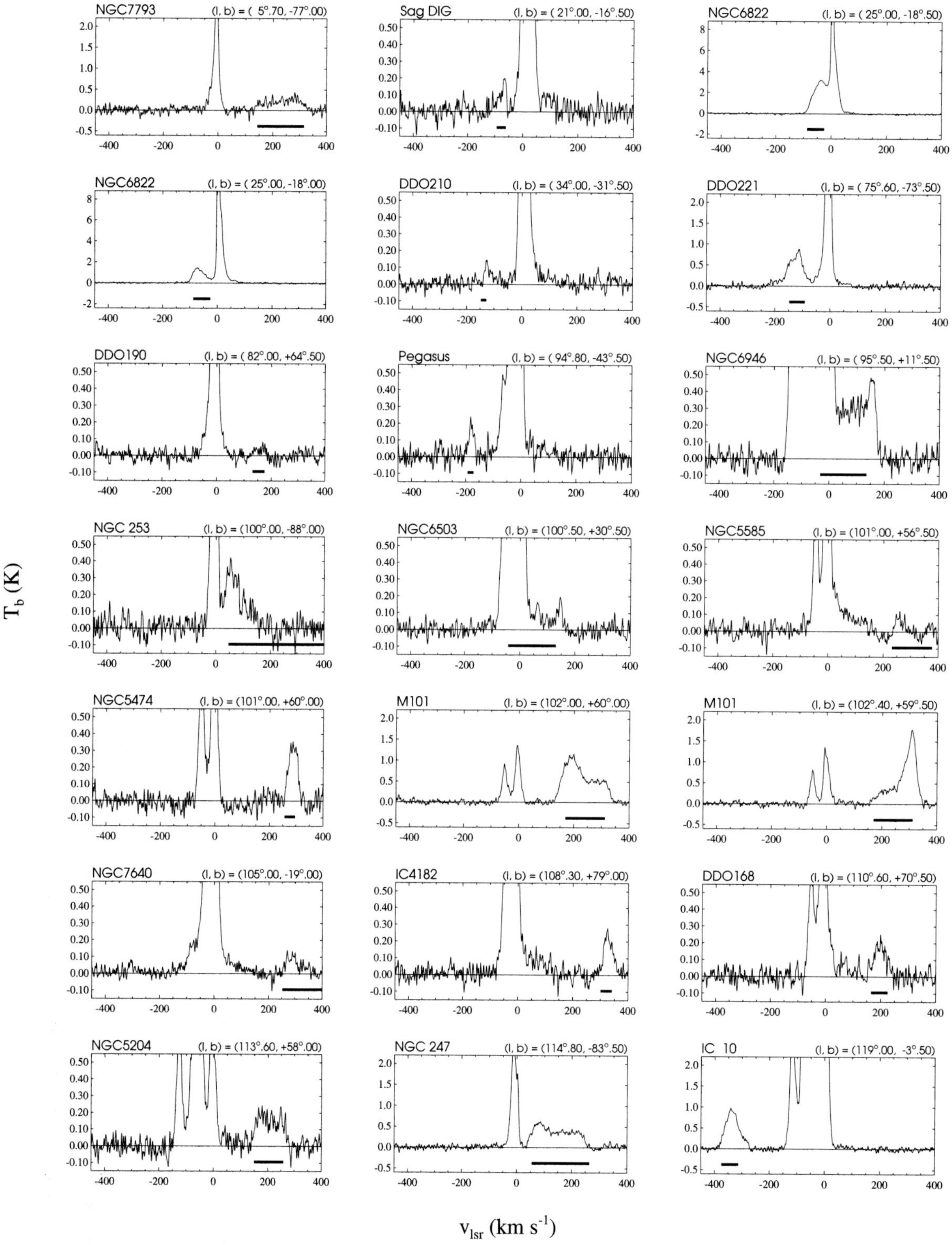

Figure 15. *Spectra from the Leiden/Dwingeloo survey showing contamination by external galaxies. The entire velocity range is shown; three different intensity scalings were used in order to display the galaxies effectively. The properties of the galaxies, both as listed in the RC3 and as derived from these spectra, are listed in Table 4. The bar underlining each galaxy's contribution represents the FWHM, centered at the appropriate systemic velocity; the parameters FWHM and v_{sys} are the intrinsic ones tabulated in RC3. These spectra continue on the following two pages.*

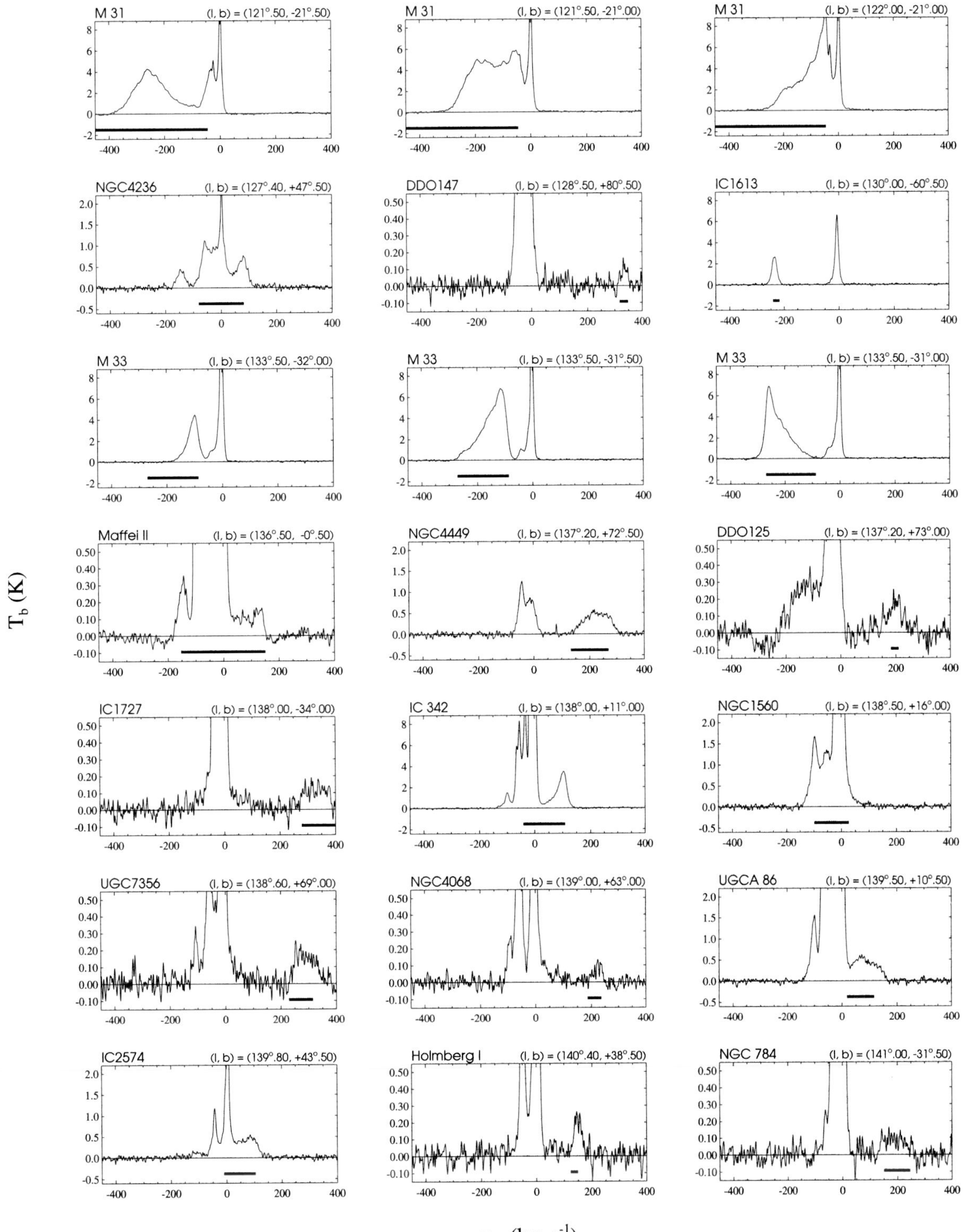

Figure 15. *– Continued*

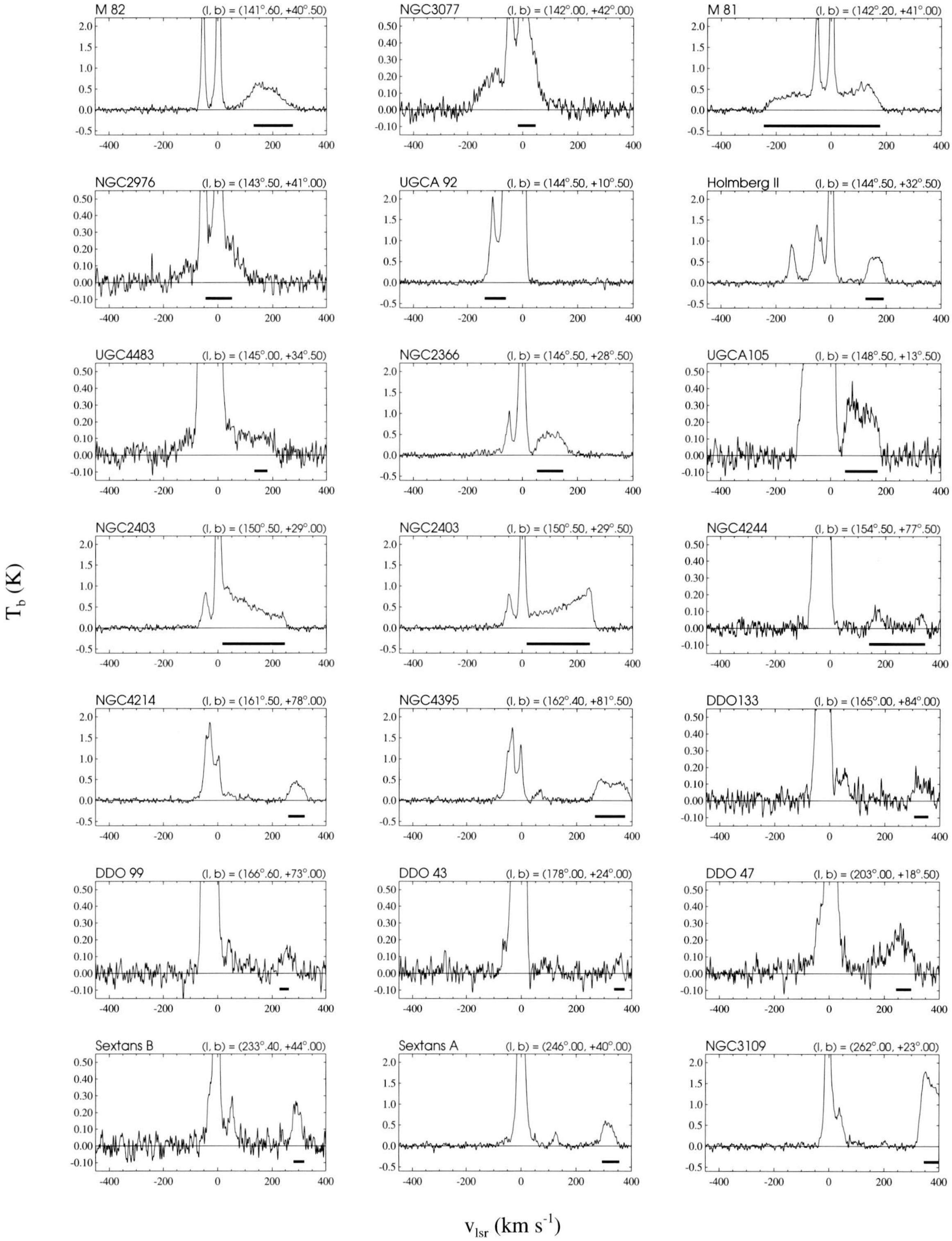

Figure 15. *– Continued*

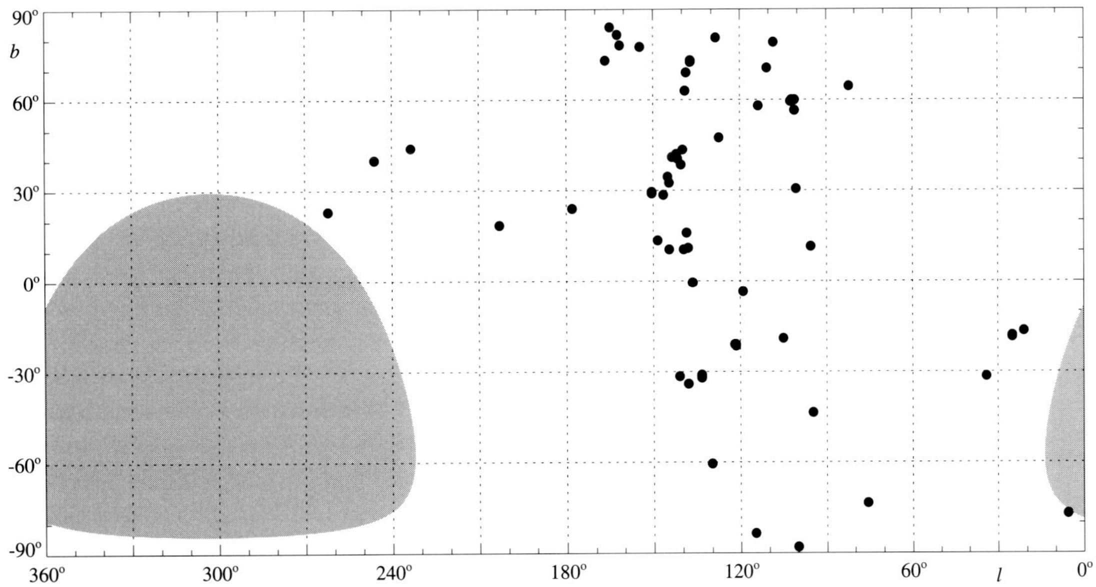

Figure 16. *Distribution on the sky of external galaxies found in the Leiden/Dwingeloo survey. The general properties of these galaxies are listed in Table 4; the spectra are shown in Figure 15. The galaxies are concentrated towards the supergalactic plane. Attention is called to these galaxies here principally in order that their contribution to the data may not be confused with emission from HI associated with the Galaxy, such as, for example, with a high-velocity cloud.*

(1990). Isolated positive-velocity emission features proved to be mostly external galaxies. External galaxies with systemic velocities near $v_{\rm lsr}$=0 km s^{-1} were commonly revealed by their broad emission profiles, placing, as it were, the emission from the Milky Way on a pedestal.

The external galaxies identified in our survey are listed in Table 4. The identifications are given, along with their characteristics, as they contaminate the survey spectra. We note that the parameters determined from our survey serve to identify the external systems as potential contamination of data which otherwise represents only gas associated with our own Galaxy. The *intrinsic* HI properties of these nearby external galaxies have been determined from detailed work done on larger telescopes; these intrinsic properties as given in the *Third Edition of the Reference Catalog of Bright Galaxies* (RC3, see de Vaucouleurs, de Vaucouleurs, & Corwin Jr. 1991), are also tabulated.

The individual survey spectra by which the galaxies were identified are shown in Figure 15. The entire effective velocity range ($-450 \le v_{\rm lsr} \le +400$ km s^{-1}) is plotted, and the *FWHM* velocity extent of each galaxy (as listed in the RC3) is indicated by a horizontal bar placed beneath the galaxy profile and centered at the appropriate systemic velocity (also from the RC3). As a compromise between objective and individual scaling, three different intensity ranges were used for the various galaxy spectra.

No previously unknown galaxy was found in the survey. Upon the completion of the observations for our galactic HI survey the Dwingeloo telescope embarked upon a large-scale search for external galaxies obscured in the Zone of Avoidance, as part of a collaborative effort involving astronomers in Cambridge, Groningen, and Leiden. Kraan-Korteweg *et al.* (1994) have recently reported the detection of 'Dwingeloo 1', a nearby barred-spiral galaxy behind the Milky Way, towards

$(l,b)=(138.^\circ5, -0.^\circ1)$, centered at a systemic *lsr* velocity of +112 $\mathrm{km\,s^{-1}}$. With the benefit of hindsight, Dwingeloo 1 can be identified on the appropriate spectra in our survey. Inspection of the individual spectra plotted in Figure 15 will make it clear that external systems constituted a practical challenge when encoding the automated baseline-removal routine. Most of the external systems identified were first flagged during the reduction system as having a poorly determined baseline; the spectra toward Dwingeloo 1 were so flagged, but we did not identify the cause of the problem.

The sky distribution of the external galaxies contaminating this survey is plotted in Figure 16; it shows the galaxies concentrated in the direction of the supergalactic plane.

12. Atlas of moment maps of the HI sky

A problem which commonly plagues observers of large datasets concerns the manner in which to display and publish the data. The pictorial presentation of HI surveys has evolved, as the amount of information involved has increased and has been influenced by the available means of presentation. We have chosen to show a representative selection of images which are sufficiently complete to allow those interested to browse through the data. Specific use of the data will require quite specific displays, which interested users will provide themselves; for such purposes, we also have prepared a complete version of the data on a CD-ROM.

The deployment of astrophysically relevant structures in a three-dimensional data cube of spectra is notoriously difficult to display, and, partly as a consequence of this, also very difficult to determine. Even sophisticated data-analysis software, although capable of displaying 3-D data cubes, falls short when trying to visualize a 450-Mbyte dataset. (We did find that a program such as IDL works well for small sub-cubes, cut from the main data cube.) The most common way to present an HI data cube involves creating orthogonal slices through it. Cutting the cube perpendicular to the velocity axis yields (l, b) slices (or moment-map sky images), each showing the distribution on the sky of emission originating from a particular narrow-velocity interval. A collection of moment-map sky-image slices comprises the Atlas presented here.

We made a variety of sky images, with velocity intervals ranging from 1 $\mathrm{km\,s^{-1}}$ (slightly less than one channel width) to the full effective velocity coverage of the survey (850 $\mathrm{km\,s^{-1}}$). The velocity intervals for the sky maps shown in this Atlas were chosen to be representative of various aspects of the HI morphology. Three different intervals (2, 10, and 50 $\mathrm{km\,s^{-1}}$) were used to characterize the entire velocity extent of the spectra. A few additional, very broad, slices were added, to cover, for example, velocities where the HI emission is largely contributed by weak, broad, high-velocity clouds.

Choice of the most physically relevant velocity interval will depend, case by case, on the particular aspects of the interstellar medium under consideration. We have found in our survey numerous small features with widths (FWHM) as narrow as 3 $\mathrm{km\,s^{-1}}$. Other HI features, such as those associated with dust cirri, commonly show sharp kinematic gradients (for which $\Delta v_{\rm lsr}=2$ $\mathrm{km\,s^{-1}}$ would even be rather coarse) but a total velocity extent of some 25 $\mathrm{km\,s^{-1}}$. Examples of very broad structure are given by some of the HVC complexes, which can range over more that 100 $\mathrm{km\,s^{-1}}$.

The sky images provide intensities integrated over the appropriate velocity range, given in units of ($\mathrm{K\,km\,s^{-1}}$). Assuming optical thinness, which evidently is valid in regions of most (but certainly not all) slices, these values may be converted to column densities ($N_{\rm HI}$ in units of 10^{19} $\mathrm{cm^{-2}}$) on multiplication by 0.18224 $(\mathrm{K\,km\,s^{-1}})^{-1}\mathrm{cm^{-2}}$.

Rectangular sky images give a satisfactory view of the hydrogen distribution at low and intermediate galactic latitudes. For high latitudes, and especially near the galactic polar caps, the rectangular images are, of course, severely distorted. The principal advantage of extending the rectangular images to $|b|=90°$ is that all data can be displayed in a single map. Better views of the high-latitude regions would be given by projections onto polar diagrams, centered on $b=-90°$ and $b=+90°$, and extending to $b=0°$. Although the data in the polar caps would be well represented in such plots, now the regions near the galactic equator would be heavily distorted.

We tried to overcome the limitations of rectangular and polar-cap projections by representing the sky images also as projections onto a sphere, displayed from different external points of view.

The viewing angle of this orthographic projection was determined by specifying the point $(l,b)_{\perp}$ for which the (external) viewing line is perpendicular to the sphere. Viewing lines defined by two longitudes ($l=60°$ and $l=180°$) and three latitudes ($b=-45°$, $b=0°$, and $b=+45°$) sufficed to cover the entire observed sky distribution. The $b=0°$ spherical projections allow comparison with the rectangular projections, whereas the tilted spheres ($|b|=45°$) bring out the details of the polar cap regions.

Decisions were necessary regarding scaling and coloring of the images. Possibilities include a naive, but objective scaling (e.g. self-scaling of individual images), a scaling specifically chosen to display optimally a particular feature on a particular image, and a general scaling which would afford some continuity between images. Depending on which region is of prime interest, the scaling may be varied to emphasize some areas at the expense of others.

Approximately 200 different shades of gray can be distinguished in a high-quality gray-scale image. The dynamic range of intensities in a sky image depends on the velocity interval, but an order-of-magnitude estimate can be obtained by considering the (single) channel map at $v_{lsr}=0$ km s^{-1}. The intensities in that image range from less than 1 K, in the Lockman *et al.* hole, to more than 100 K near the galactic plane over the full 360° longitude range. A *linear* 200-step gray scale covering this range would thus be quantized in 0.5 K increments; such an increment is about seven times the typical *rms* spectral noise, and thus is clearly insufficient to reveal details in both the high- and low-intensity structures. The situation for linear scaling becomes worse for broader velocity intervals.

We therefore applied a *logarithmic* scaling to all sky images, and mapped that onto a linear 256-step gray-scale palette. For the velocity interval $-70 \leq v_{lsr} \leq +50$ km s^{-1} we created images integrated over 2 km s^{-1}. The image values, $\Sigma(l,b)$, were clipped between 0.2 K km s^{-1} (*ClipLow*) and 200 K km s^{-1} (*ClipHigh*), and log($\Sigma(l,b)$) was mapped in 256 steps onto a gray-scale palette, with log(*ClipLow*) corresponding to black and log(*ClipHigh*) to white. Selected images were also prepared to be printed in color. For the color images we used the same scaling, but mapped the logarithmic intensities onto a 'fire' color palette. The superiority of this color palette over other (perhaps more florid) ones, arises from the intuitive association between 'hot' and high column density. The palette was created over 256 steps in the RGB color model, using the following sequence: (1) increase R from 0% to 70%; (2) increase R from 70% to 100%, and G from 0% to 50%; (3) increase G from 50% to 100%, and B from 0% to 100%. The region for which no material is present ($\delta<-30°$) was given a solid gray (90% black) filling, to contrast with observed areas of low integrated emission.

The sky image integrated over the velocity interval $-450 \leq v_{lsr} \leq -150$ km s^{-1} covers the most negative-velocity HVCs. The velocity interval $-150 \leq v_{lsr} \leq -70$ km s^{-1} is represented by the sky images in 10 km s^{-1} wide slices. These intervals represent the less-extreme negative velocities exhibited by the HVCs; the transition to the IVC regime near -70 km s^{-1} is rather arbitrary. Sky images integrated over 2 km s^{-1} intervals are shown for the range $-70 \leq v_{\rm lsr} \leq +50$ km s^{-1}. Although these slices are quite narrow, the pictures do show considerable localized changes from one to the next. These velocity intervals incorporate many anomalous-velocity structures at higher $|b|$, as well as many aspects of the conventional galactic disk at lower $|b|$. Sky images integrated over 10 km s^{-1} intervals are shown for the range $+50 \leq v_{\rm lsr} \leq +100$ km s^{-1}. The contrast between the distributions shown in these plates and those shown in the complementary negative velocities is striking. The extreme positive velocities are shown in sky images covering $+100 \leq v_{\rm lsr} \leq +200$ km s^{-1} and $+200 \leq v_{\rm lsr} \leq +400$ km s^{-1}. Most of the emission in this latter image emanates from the weak HVCs clustered at positive latitudes near the border of our survey at $\delta = -30°$, and from external galaxies.

Perusal of the images presented in the Atlas or extracted from the data cube will reveal several characteristics which deviate from the appearance one might readily attribute to the structure of interstellar HI. The deviant characteristics are not necessarily blemishes in the data. Consider the occasional occurrence of isolated dark pixels. These 'fly specks' are due to HI absorption against background sources of continuum radiation. Absorption against the extragalactic radio source Cyg A is evident, for example, near $(l,b) = (76°, +5.^\circ5)$ in the image integrated between $-90 \leq v_{\rm lsr} \leq -80$ km s^{-1}. Absorption against the galactic supernova remnant Cas A is evident near $(111.^\circ5, -2°)$ in the image integrated between $-52 \leq v_{\rm lsr} \leq -50$ km s^{-1}, and in numerous others. The dark pixel near $(184.^\circ5, -5.^\circ5)$ seen in the image integrated between $+10 \leq v_{\rm lsr} \leq +12$ km s^{-1} is due to absorption against Tau A (the Crab Nebula); partial absorption against the Orion Nebula is seen near $(209°, -19.^\circ5)$ in the image integrated between $0 \leq v_{\rm lsr} \leq +2$ km s^{-1}. Absorption against Sgr A occurs over a wide velocity range, but is obscured in many of our atlas maps by the grid lines. Thus, these various 'fly specks' are not imperfections in the data.

Some of the images which pushed the display down to the noise limit of the data do, however, show residual imperfections. The appearance of the image integrated between $+200 \leq v_{\rm lsr} \leq +400$ km s^{-1} reflects the two aspects of the observing scheduling which leave such traces. The observing strategy scheduled the observations in boxes measuring $5° \times 5°$. An estimate of the shape of the bandpass for all spectra contained in a box was obtained from reference spectra which were observed at a full-bandwidth higher frequency. Variations in the accuracy of bandpass estimates for neighboring boxes as well as variations in the residual sidelobe contamination for neighboring boxes observed at different hour angles are the cause of the low-intensity residual emission which can be seen in many of the sky images. We note, however, that because of the logarithmic scaling down to the spectral *rms* noise level such low-intensity residuals are enhanced.

13. CD–ROM

The complete HI survey data are contained on the Compact Disk accompanying the Atlas. This CD-ROM was mastered according to the ISO 9660 norm and may be accessed from a variety of computing platforms. The disk is organized into three directories: \DATA, \IMAGES, and \ANIMATE.

The directory \IMAGES contains GIF-format color images of all the sky maps shown in the Atlas. Animated image sequences are written in AutoDesk Animator's FLC format in the directory \ANIMATE. The file REC.FLC is an animation of 140 rectangular sky images between -230 km s^{-1} and $+50$ km s^{-1}, in 2 km s^{-1} intervals. Animations of spherically-projected sky images for apparent lines of sight towards $(l,b)_\perp = (140°, +45°)$ and towards $(l,b)_\perp = (140°, -45°)$ are supplied by SPH1.FLC and by SPH2.FLC, respectively, also between -230 km s^{-1} and $+50$ km s^{-1}, in 2 km s^{-1} intervals. When played using an AutoDesk Animator player, the animations show the sky

Table 5. *Sample FITS header from the file* L1000.FIT. *This file contains a (b,v) map at* $l=100^\circ\!.0$

```
SIMPLE  =                    T
BITPIX  =                   16
NAXIS   =                    3
NAXIS1  =                  849
NAXIS2  =                  361
NAXIS3  =                    1
BUNIT   = 'K       '
DATAMAX =     86.651023982821
DATAMIN =     -0.424910583518
BZERO   =     43.113056699651
BSCALE  =     0.0013287138670
BLANK   =               -32768
OBJECT  = 'Leiden/Dwingeloo HI Survey; (b,v) at l=100.0'
TELESCOP= 'Dwingeloo 25-m'
OBSERVER= 'Dap Hartmann'
DATE-OBS= '02/09/93'
DATE    = '21/01/96'
CTYPE1  = 'VELO-LSR'
CRVAL1  =      0.000000000000
CRPIX1  =                  446
CDELT1  =     1030.5513305347
CTYPE2  = 'GLAT-CAR'
CRVAL2  =     -90.00000000000
CRPIX2  =                    1
CDELT2  =     0.5000000000000
CTYPE3  = 'GLON-CAR'
CRVAL3  =     100.00000000000
CRPIX3  =                    1
CDELT3  =      0.000000000000
COMMENT = +-----------------------------------------------+
COMMENT = | The Leiden/Dwingeloo Survey of HI in the Galaxy |
COMMENT = |           Dap Hartmann & W.B.Burton           |
COMMENT = |               Leiden Observatory              |
COMMENT = |        Cambridge University Press 1997        |
COMMENT = +-----------------------------------------------+
END
```

images from negative to positive velocities. The files with the added extension VV (vice versa) play the animation back and forth.

The directory \DATA contains the actual survey data. The data cube was subdivided into orthogonal (b,v) images for all galactic longitudes ($0^\circ \le l \le 360^\circ$; $\Delta l=0^\circ\!.5$; $\delta \ge -30^\circ$). Each image covers the entire range of galactic latitudes ($-90^\circ \le b \le +90^\circ$; $\Delta b=0^\circ\!.5$; $\delta \ge -30^\circ$) and the effective useful velocity range ($-459 \le v_{\rm lsr} \le +415$ km s^{-1}; $\Delta v=1.03$ km s^{-1}). Images are stored in FITS format (see Wells *et al.* 1981) as files L0000.FIT ($l=0^\circ\!.0$), L0005.FIT ($l=0^\circ\!.5$), etc. Each file occupies 616,320 bytes and contains a single 2880-byte header block. A sample FITS header is shown in Table 5. Additionally, there is an *(l, b)* FITS image named TOTAL_HI.FIT, which contains the integrated intensity map over the velocity range $-450 \le v_{\rm lsr} \le +400$ km s^{-1}.

14. Epilogue

The Leiden/Dwingeloo HI survey project began in late 1987 with exploratory observations towards the Sun and Cas A, aimed at establishing aspects of the antenna pattern. The prototype of the DAS spectrometer was initially installed on the Dwingeloo 25-meter telescope in March, 1988, at which time the telescope was turned over on a full-time basis to this project, and the first test spectral-line observations were made. Testing and debugging the auto-correlator were unexpectedly protracted. Routine survey observations were carried out from November, 1989, through November, 1992. In May, August, and September, 1993, the final data were acquired. The gross total telescope time devoted to the survey proper was about 1100 days (three years and one month). The total elapsed telescope time, including DAS debugging, was considerably longer. The total number of spectra measured was 277,996. The total net integration time was 579 days, and the total overhead amounted to 290 days. Thus, the observing efficiency of the Dwingeloo telescope was 79% over a three year period.

Acknowledgements

This project was supported by the Netherlands Foundation for Research in Astronomy (NFRA) with financial backing from the Netherlands Organization for Scientific Research (NWO). We thank Harvey Liszt of the US National Radio Astronomy Observatory (NRAO) for supporting our use of his *DrawSpec* program; Albert Bos of the NFRA for supporting our use of the DAS prototype; Henny Lem and Daniel Moorrees, also of the NFRA, for assistance throughout the observing period; and Peter Kalberla and Ulrich Mebold of the University of Bonn for their collaboration on the stray-radiatuon correction procedure, which is being described in detail elsewhere.

References

Bolton, J.G., & Stanley, G.J. 1948, *Aust. J. Sci. Res. A*, **1**, 58

Bos, A. 1989, *The 1024 Channel Spectrometer for the Dwingeloo Telescope*, Internal Technical Report, **188**, NFRA

Burstein, D., & Heiles, C. 1978, *ApJ*, **225**, 40

Burstein, D., & Heiles, C. 1982, *AJ*, **87**, 1165

Burton, W.B. 1985, *A&AS*, **62**, 365

Burton, W.B. 1988, in *Galactic and Extragalactic Radio Astronomy*, (G.L. Verschuur & K.I. Kellerman, eds.), Springer-Verlag, New York, p. 295

Burton, W.B. 1992, in *The Galactic Interstellar Medium*, (D. Pfenniger & P. Bartholdi, eds.), Springer-Verlag, Heidelberg, p. 1

Casse, J.L., Woestenburg, E.E.M., & Visser, J.J. 1982, *IEEE Trans. Microwave Theory Tech.*, **30**, 201

Cleary, M.N., Heiles, C., & Haslam, C.G.T. 1979, *A&AS*, **36**, 95

Colomb, F.R., Pöppel, W.G.L., & Heiles, C. 1980, *A&AS*, **40**, 47

de Vaucouleurs, G., de Vaucouleurs, A., & Corwin Jr., H.G. 1991, *Third Edition of the Reference Catalog of Bright Galaxies (RC3)*, Selected Astronomical Catalogs Vol. 1, Astronomical Data Center, NASA/GSFC, Greenbelt (CD-ROM)

Dickey, J.M., & Lockman, F.J. 1990, *ARA&A*, **28**, 215

Hartmann, Dap 1994, The Leiden/Dwingeloo Survey of Galactic Neutral Hydrogen, PhD thesis, University of Leiden

Hartmann, Dap, Kalberla, P.M.W., Burton, W.B., & Mebold, U. 1996, *A&AS*, in press.

Hartsuijker, A.P., Baars, J.W.M., Drenth, S., & Gelato-Volders, L. 1972, *IEEE Trans. Antennas Propagat.*, **20**, 166

Heiles, C., & Habing, H.J. 1974, *A&AS*, **14**, 1
Hulsbosch, A.N.M., & Wakker, B.P. 1988, *A&AS*, **75**, 191
Kalberla, P.M.W. 1978, *D*ie Korrektur der Linienprofile der 21-cm Emission bezüglich der Streustrahlung des Antennendiagrams, PhD Thesis (in German; English translation by R.J. Cohen), University of Bonn
Kalberla, P.M.W., Mebold, U., & Reich, W. 1980, *A&A*, **82**, 275
Kalberla, P.M.W., Mebold, U., & Reif, K. 1982, *A&A*, **106**, 190
Kerr, F.J., Bowers, P.F., Jackson, P.D., & Kerr, M. 1986, *A&AS*, **66**, 373
Kleibrink, H. 1957, *Focus*, **43**, 156 (in Dutch)
Kraan-Korteweg, R.C., Loan, A.J., Burton, W.B., Lahay, O., Ferguson, H.C., Henning, P.A., & Lynden-Bell, D. 1994, *Nature*, **372**, 77
Kulkarni, S.R. & Heiles, C. 1988, in *Galactic and Extragalactic Radio Astronomy*, (G.L. Verschuur & K.I. Kellerman, eds.), Springer-Verlag, New York, p. 95
Kuntz, K.D., & Danly, L. 1992, *PASP*, **104** (No.682), 1256
Kwee, K.K., Muller, C.A., & Westerhout, G. 1954, *BAN*, **12**, 211
Liszt, H.S. 1987, *DrawSpec Users Manual* (Manual to the DrawSpec single-dish spectral-line reduction package; privately circulated)
Lockman, F.J., Jahoda, K., & McCammon, D. 1986, *ApJ*, **302**, 432
Murphy, E.M. 1993, The Beam Pattern of the 140-Foot Telescope at 1390 MHz, MA Thesis, University of Virginia
Raimond, E. 1964, Hydrogen Connected with Two Stellar Associations in Monoceros, PhD thesis, University of Leiden
Spoelstra, T.A.Th. (ed.) 1981, *Een Zilveren Spiegel. 25 Jaar Radiosterrenwacht Dwingeloo, 17 april 1956 – 17 april 1981* (in Dutch)
Stark, A.A., Gammie, C.F., Wilson, R.W., Bally, J., Linke, R.A., Heiles, C., & Hurwitz, M. 1992, *ApJS*, **79**, 77
Sullivan III, W.T. (ed.) 1982, *Classics in Radio Astronomy*, Reidel, Dordrecht
Tolbert, C.R. 1971, *A&AS*, **3**, 249
van de Hulst, H.C. 1945, *Nederlandsch Tijdschrift voor Natuurkunde*, **11**, 210 (in Dutch; reprinted in English translation in Sullivan III (1982), paper 34)
van de Hulst, H.C., Muller, C.A., & Oort, J.H. 1954, *BAN*, **12**, 117
van Herk, G. Kleibrink, H., & Bijleveld, W. (eds.) 1983, *De Leidse Sterrewacht. Vier eeuwen wacht bij dag en bij nacht*. Waanders/De Kler, Zwolle (in Dutch, with a summary and figure captions in English)
van Someren Greve, H. 1991, in Proc. Workshop *'Holography Testing of Large Radio Telescopes'*, Nauka, Leningrad
van Woerden, H. 1962, De Neutrale Waterstof in Orion, PhD thesis (in Dutch), University of Groningen
van Woerden, H. 1970, in *Trans. IAU*, Vol. **XIVB**, (C. de Jager & A. Jappel, eds.), p. 217, §8
van Woerden, H., Brouw, W.N., & van de Hulst, H.C. (eds.) 1980, *Oort and the Universe. A sketch of Oort's research and person*. Reidel, Dordrecht
Wakker, B.P. 1990, Interstellar Neutral Hydrogen at High Velocities, PhD thesis, University of Groningen
Weaver, H.F., & Williams, D.R.W. 1973, *A&AS*, **8**, 1
Weaver, H.F., & Williams, D.R.W. 1974, *A&AS*, **17**, 1
Wells, D.C., Greisen, E.W., & Harten, R.H. 1981, *A&AS*, **44**, 363
Wesselius, P.R & Fejes, I. 1973, *A&A*, **24**, 15
Westerhout, G. 1961, *Hemel en Dampkring* **59**, 86 (in Dutch)
Williams, D.R.W. 1973, *A&AS*, **8**, 505

Index to the Atlas

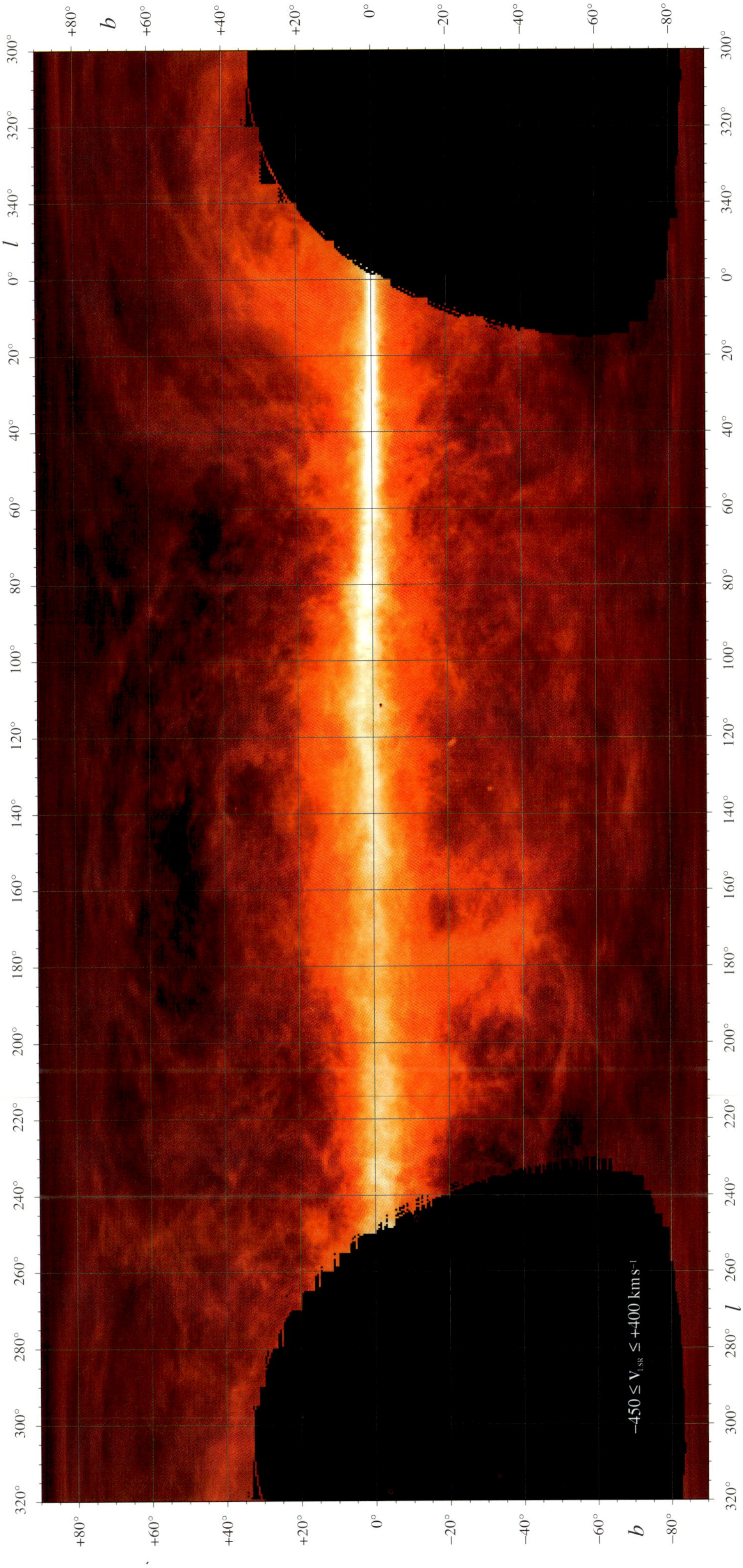
l
b
−450 ≤ V_LSR ≤ +400 kms-1

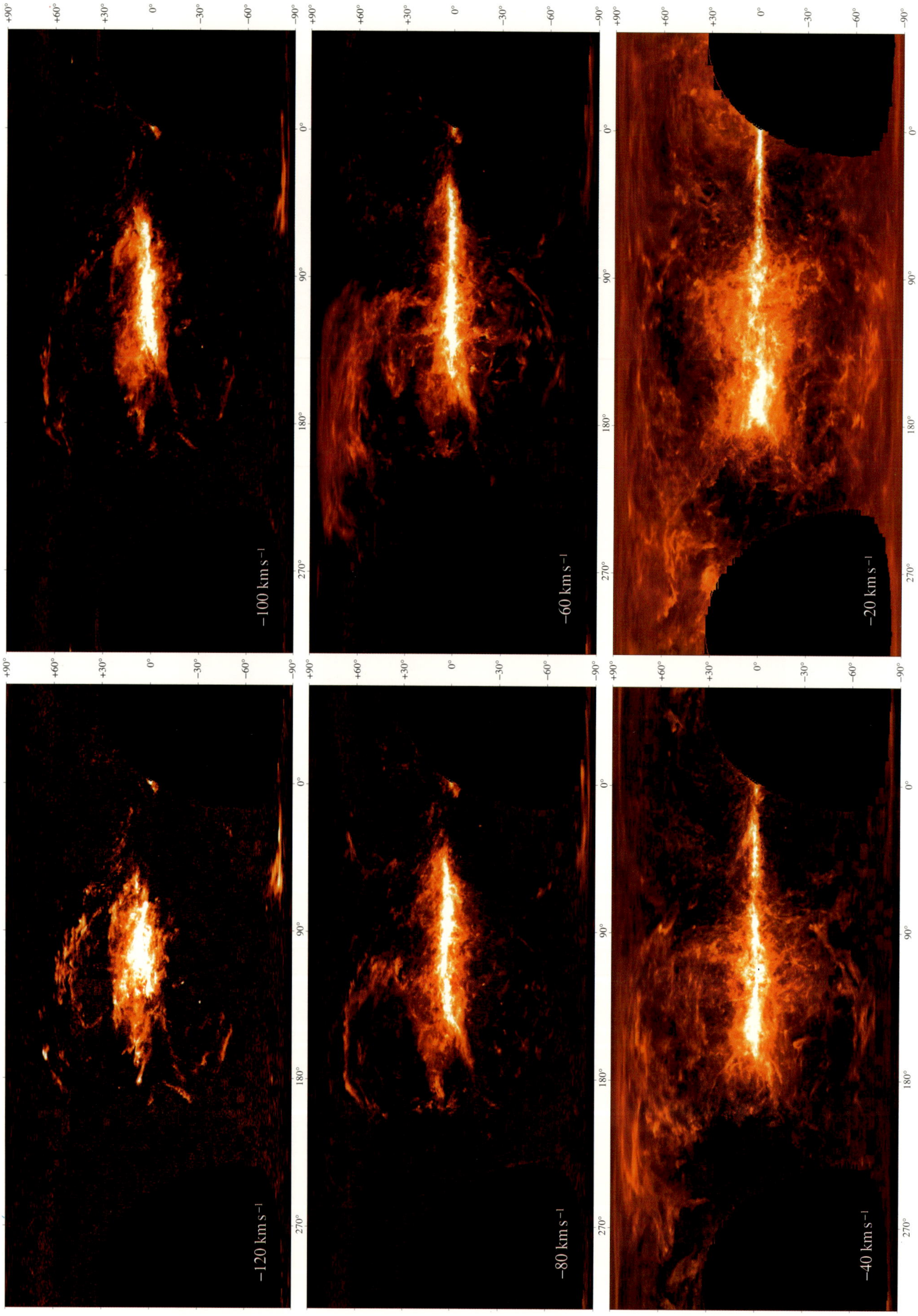

−100 km s−1
−60 km s−1
−20 km s−1
−120 km s−1
−80 km s−1
−40 km s−1

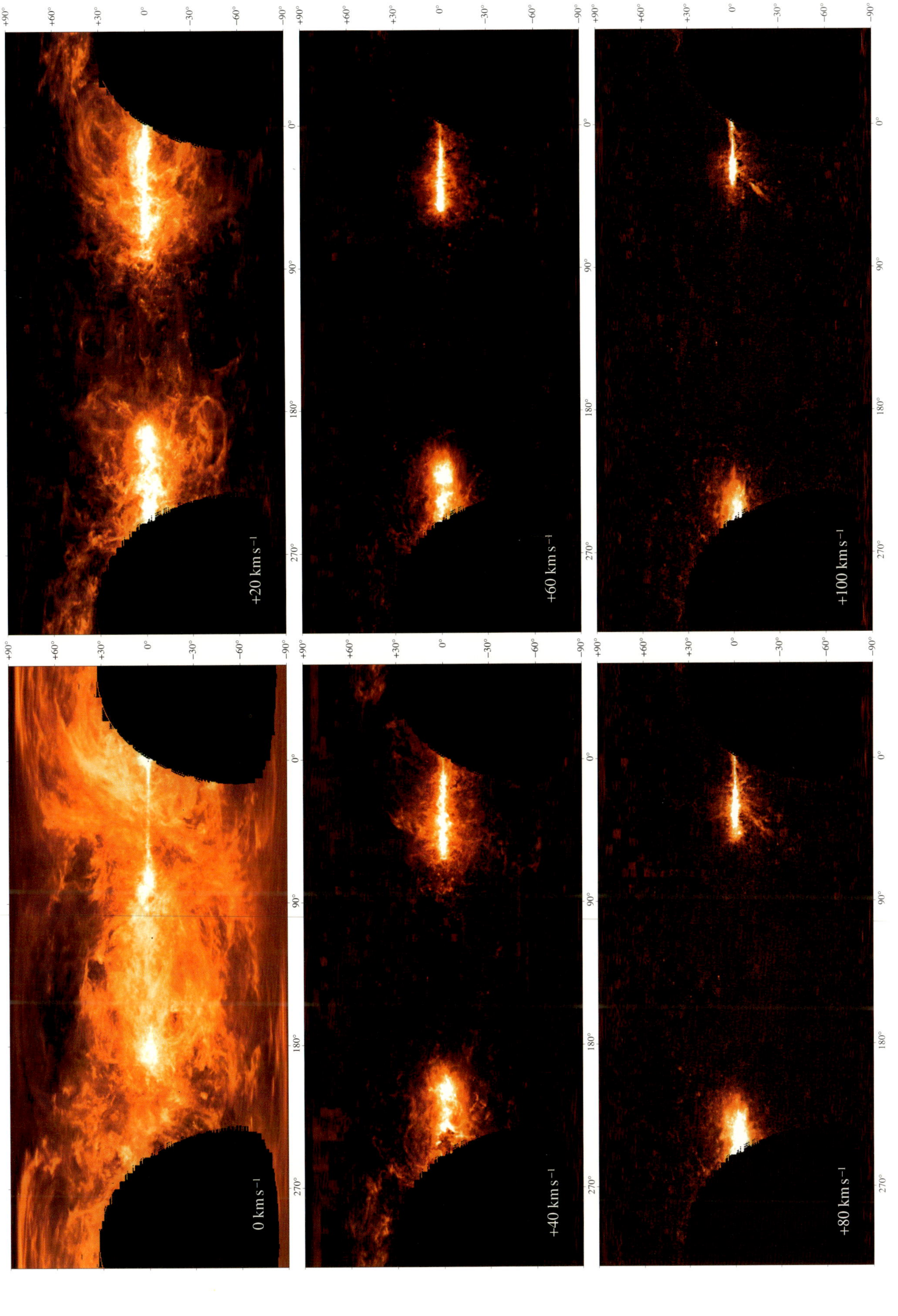
0 km s−1
+20 km s−1
+40 km s−1
+60 km s−1
+80 km s−1
+100 km s−1

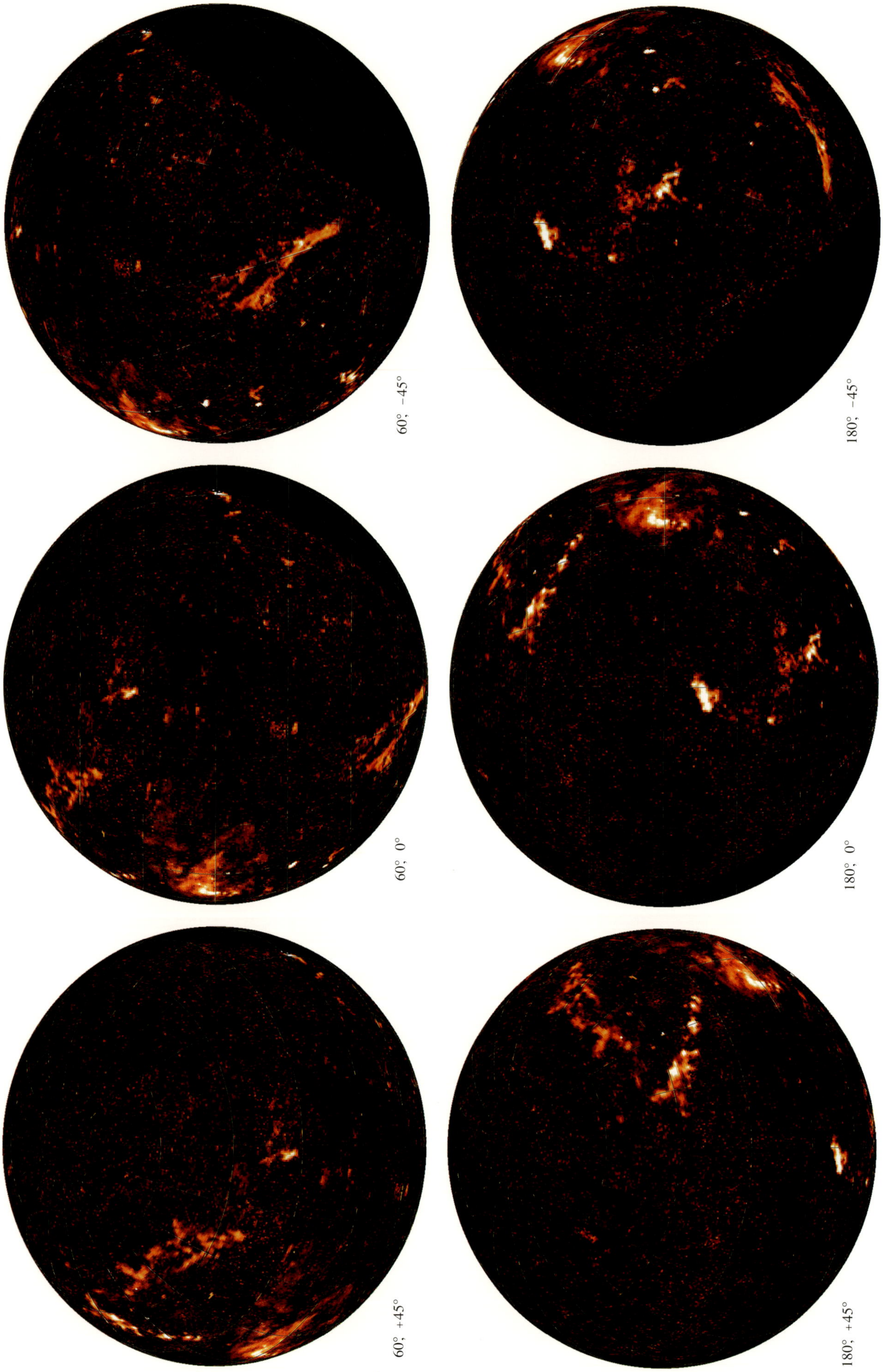

$-450 \leq V_{LSR} \leq -150$ km s^{-1}

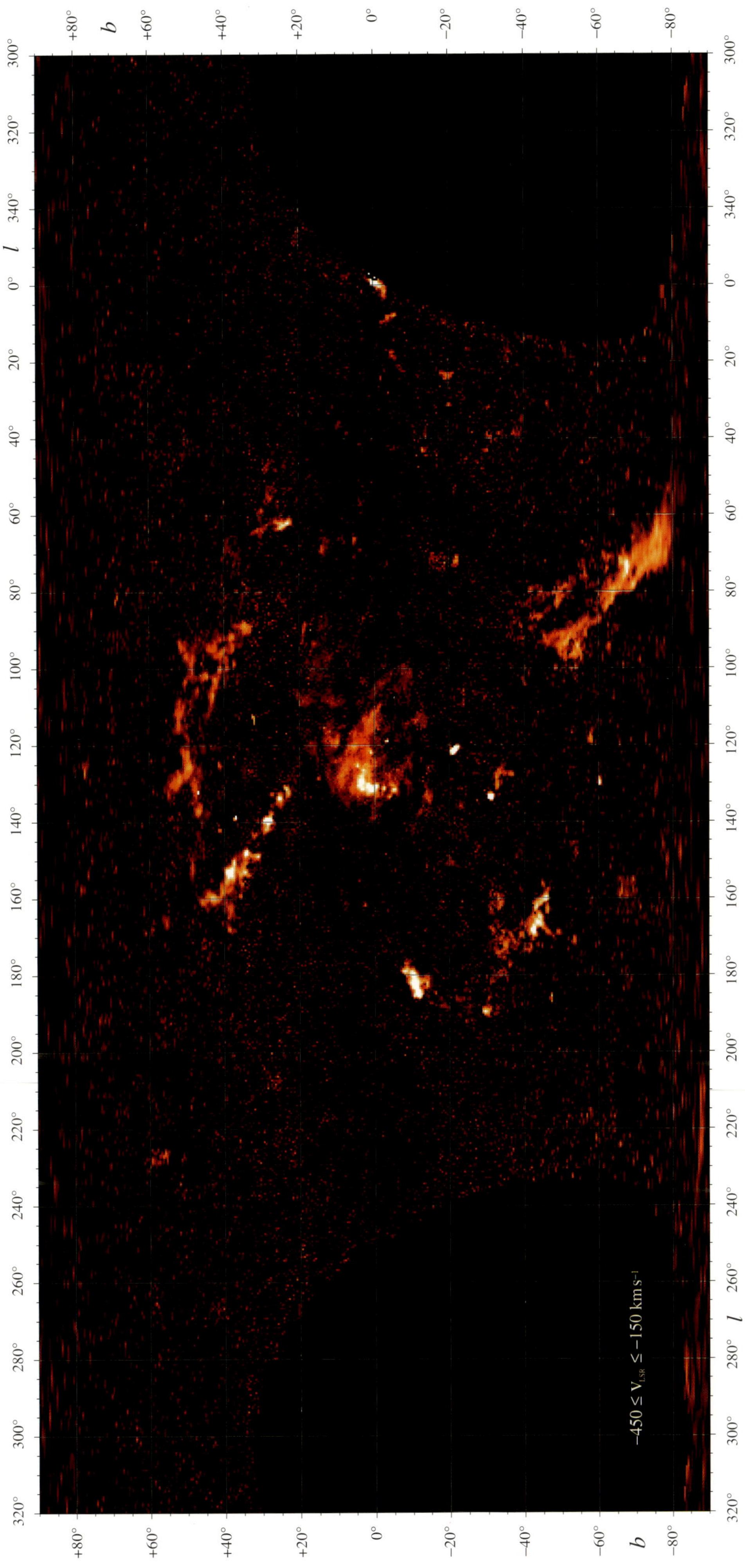
b
l
−450 ≤ V_LSR ≤ −150 km s−1
l
b

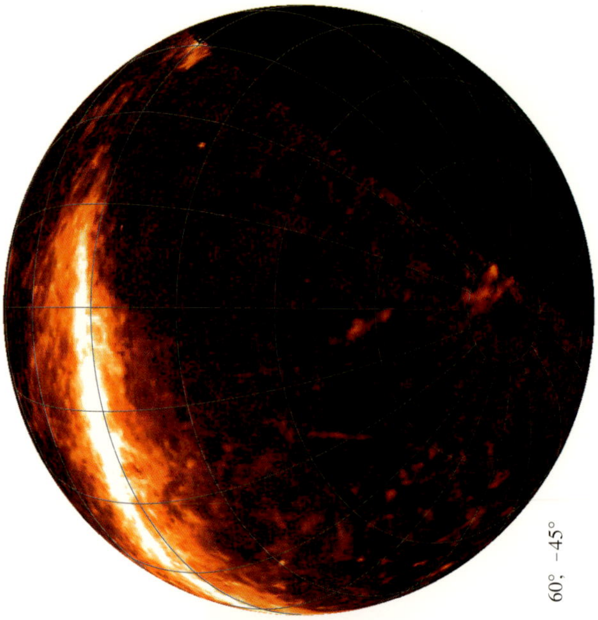

60°, −45°

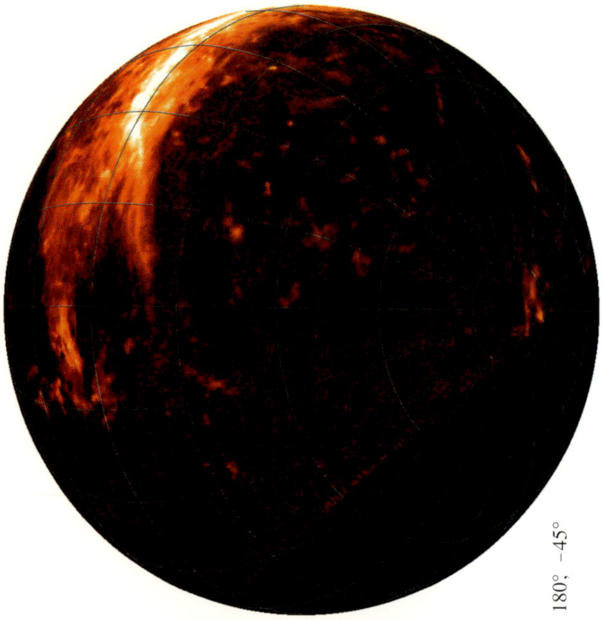

180°, −45°

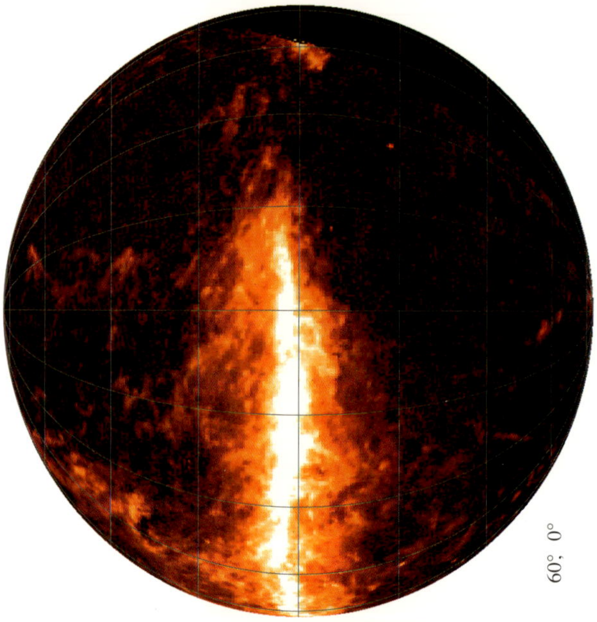

60°, 0°

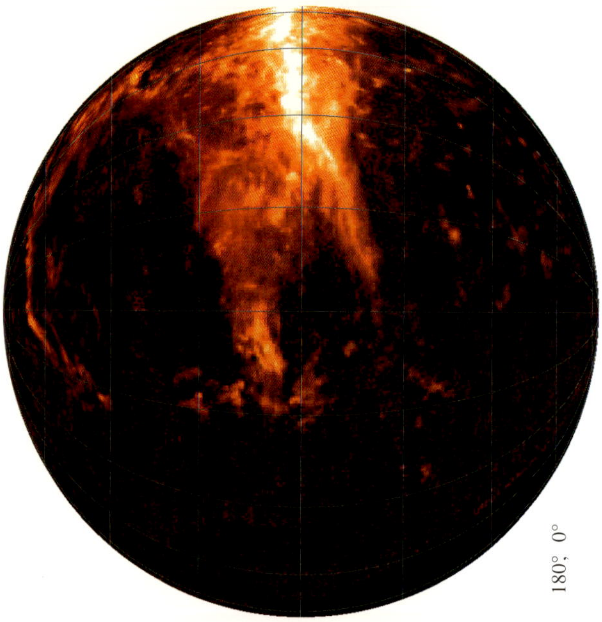

180°, 0°

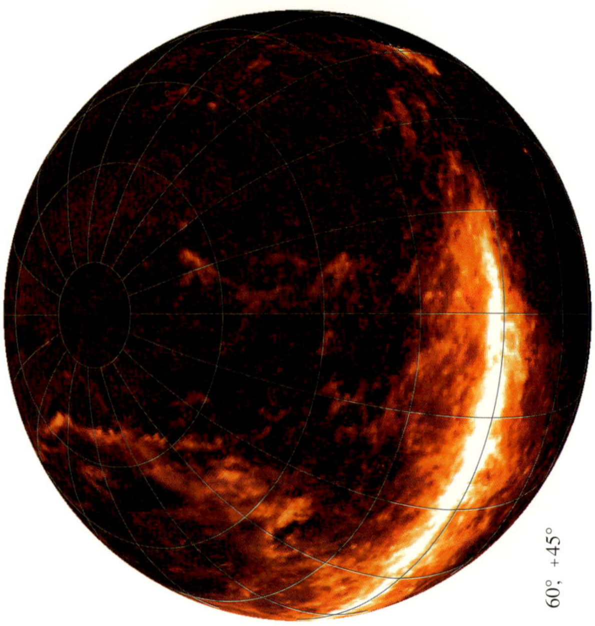

60°, +45°

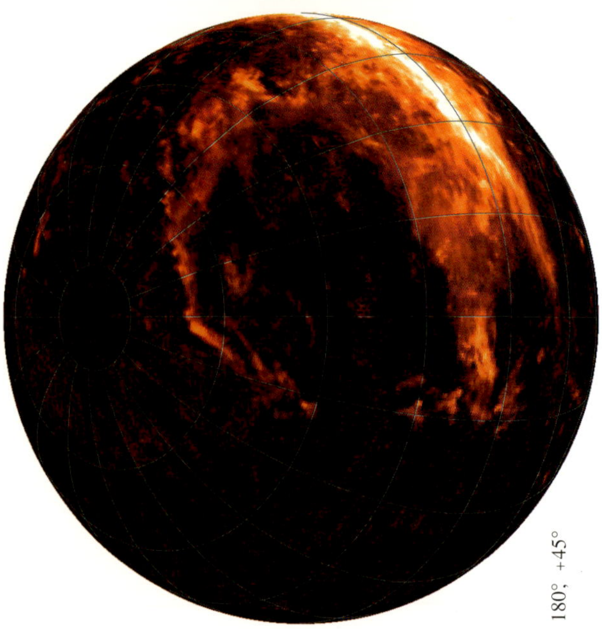

180°, +45°

$-80 \leq V_{LSR} \leq -78$ km s^{-1}

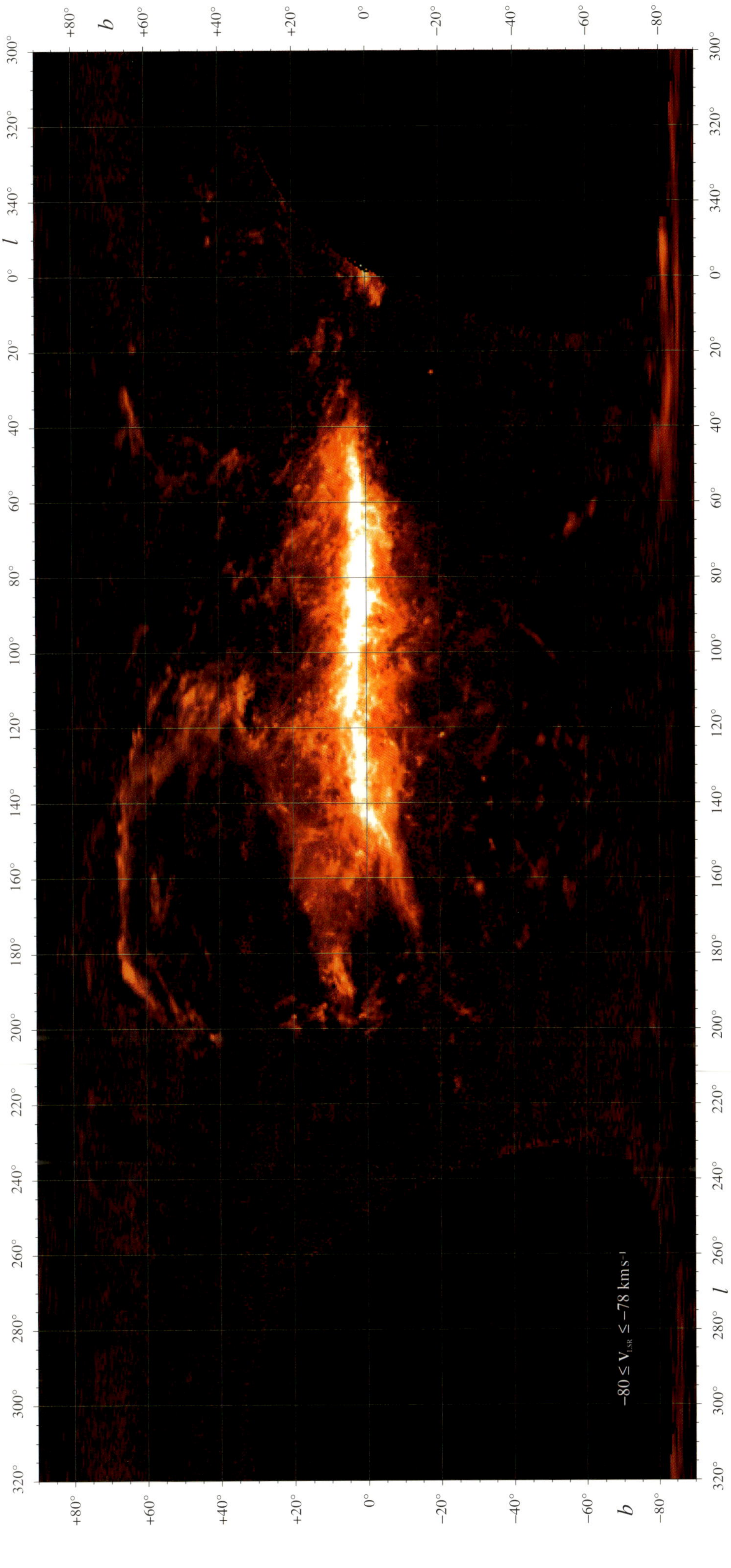
−80 ≤ V$_{LSR}$ ≤ −78 km s^{-1}
l
b
300° 320° 340° 0° 20° 40° 60° 80° 100° 120° 140° 160° 180° 200° 220° 240° 260° 280° 300° 320°
+80° +60° +40° +20° 0° −20° −40° −60° −80°

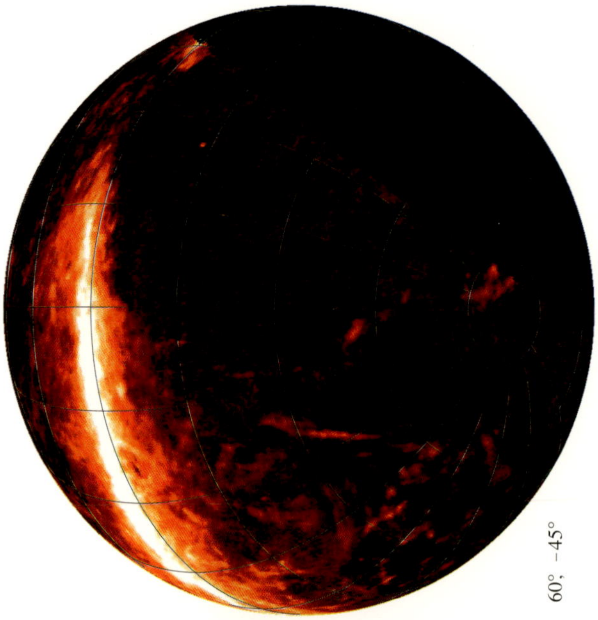

60°; −45°

180°; −45°

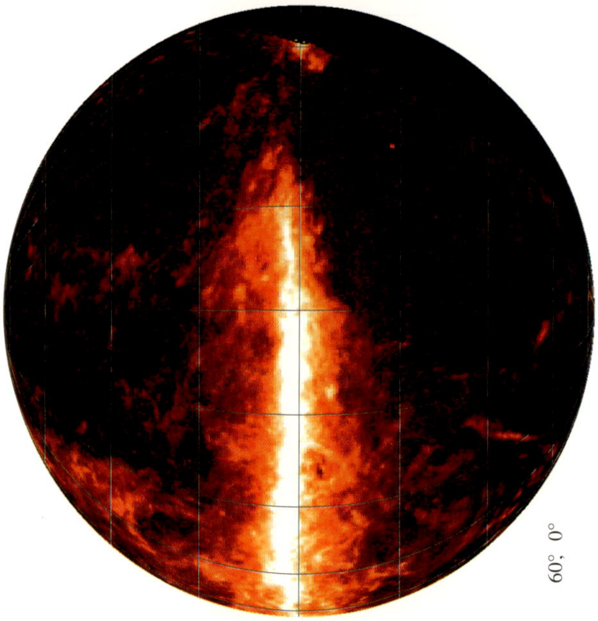

60°; 0°

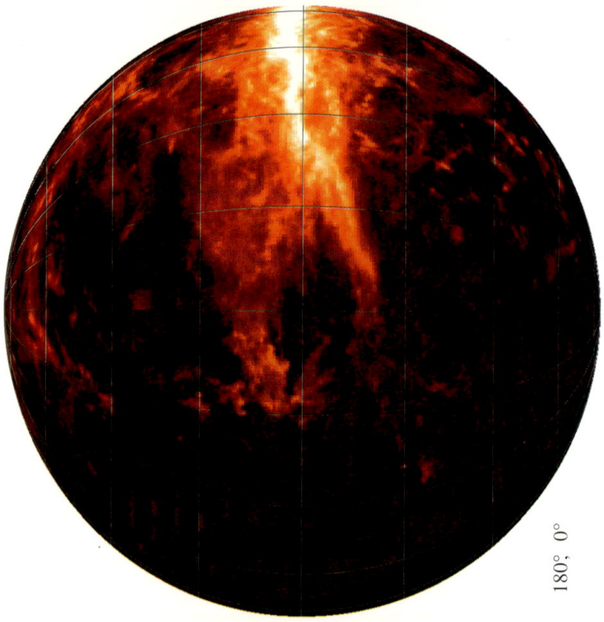

180°; 0°

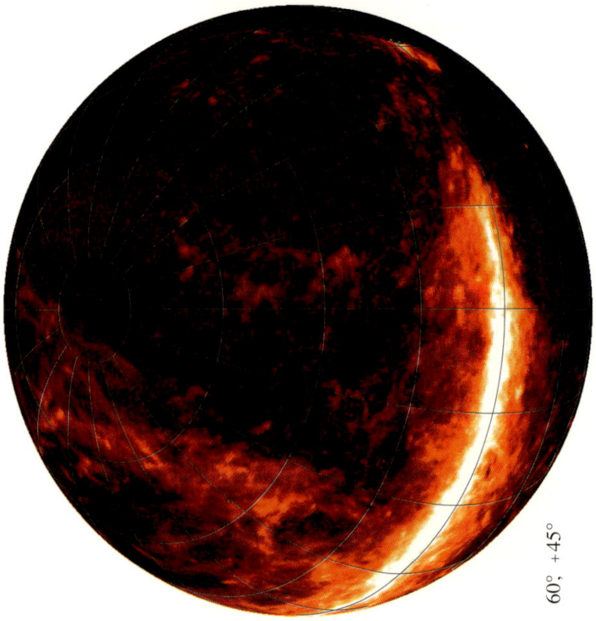

60°; +45°

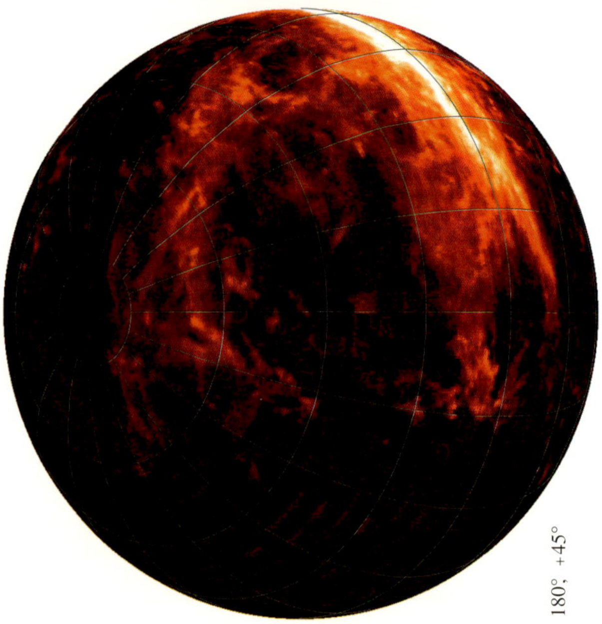

180°; +45°

$-70 \leq V_{LSR} \leq -68$ km s^{-1}

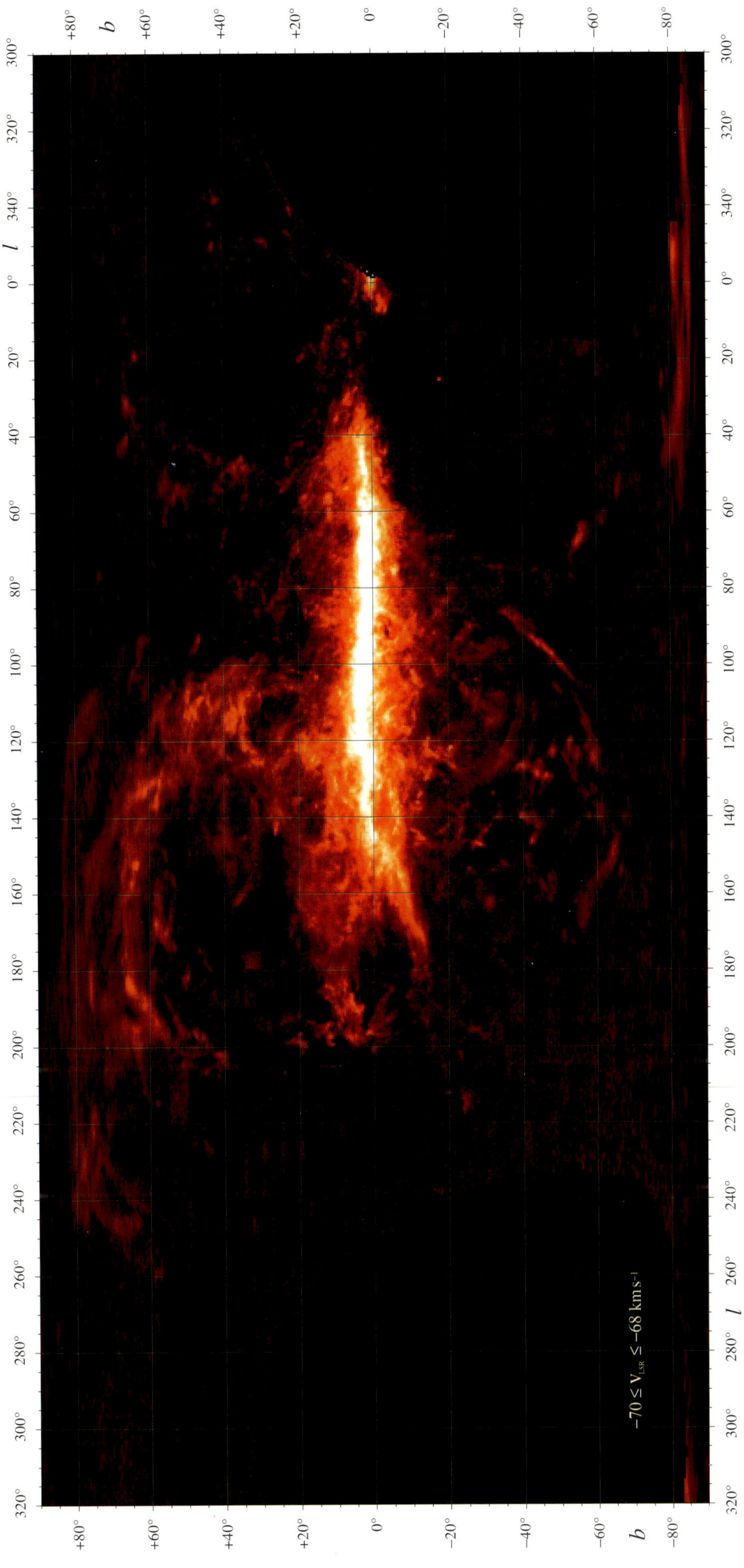
$-70 \le V_{LSR} \le -68$ km s^{-1}

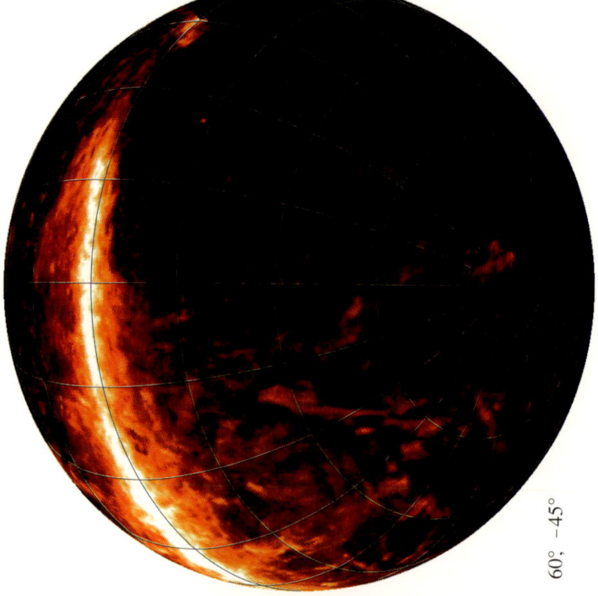

60°; −45°

180°; −45°

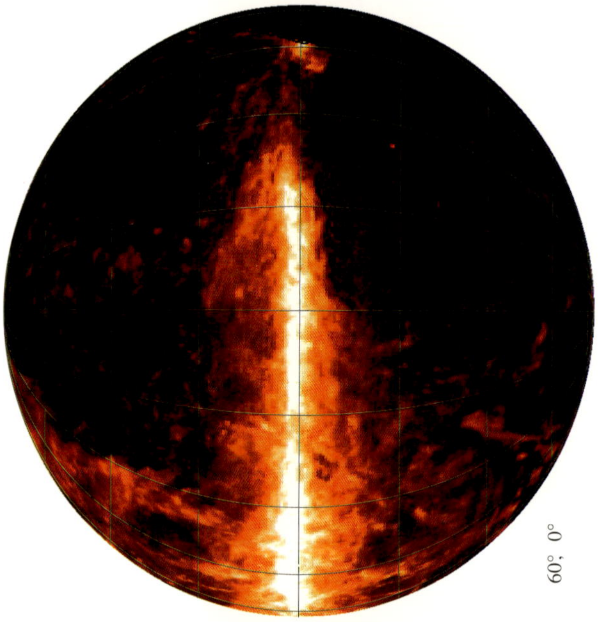

60°; 0°

180°; 0°

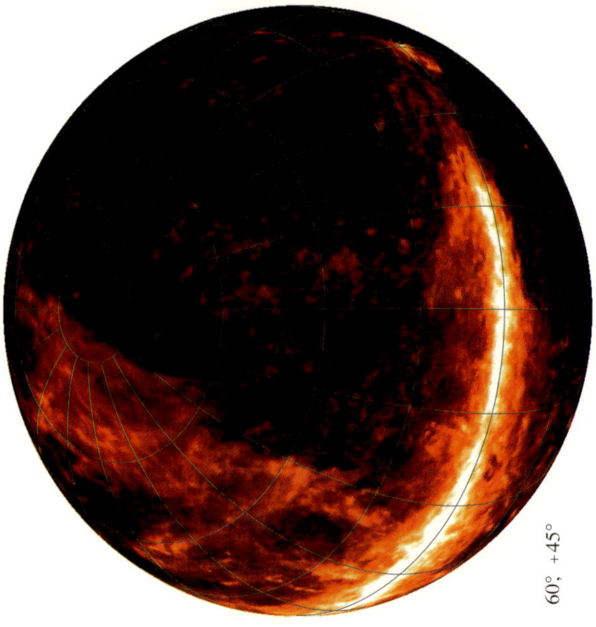

60°; +45°

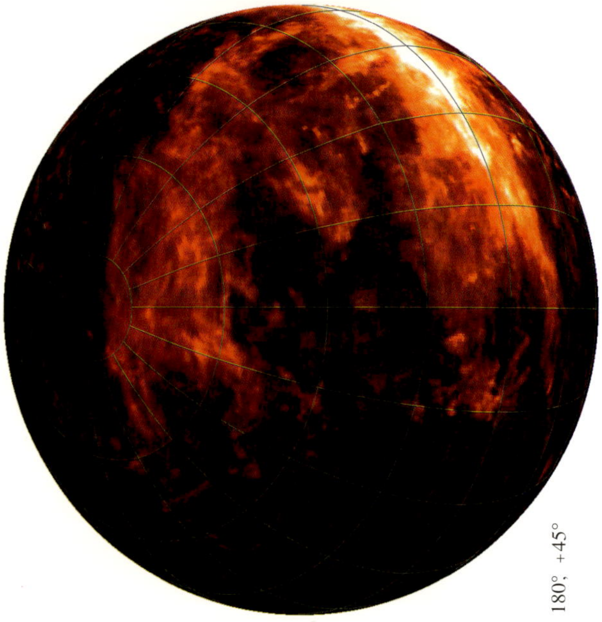

180°; +45°

$-60 \le V_{LSR} \le -58$ km s^{-1}

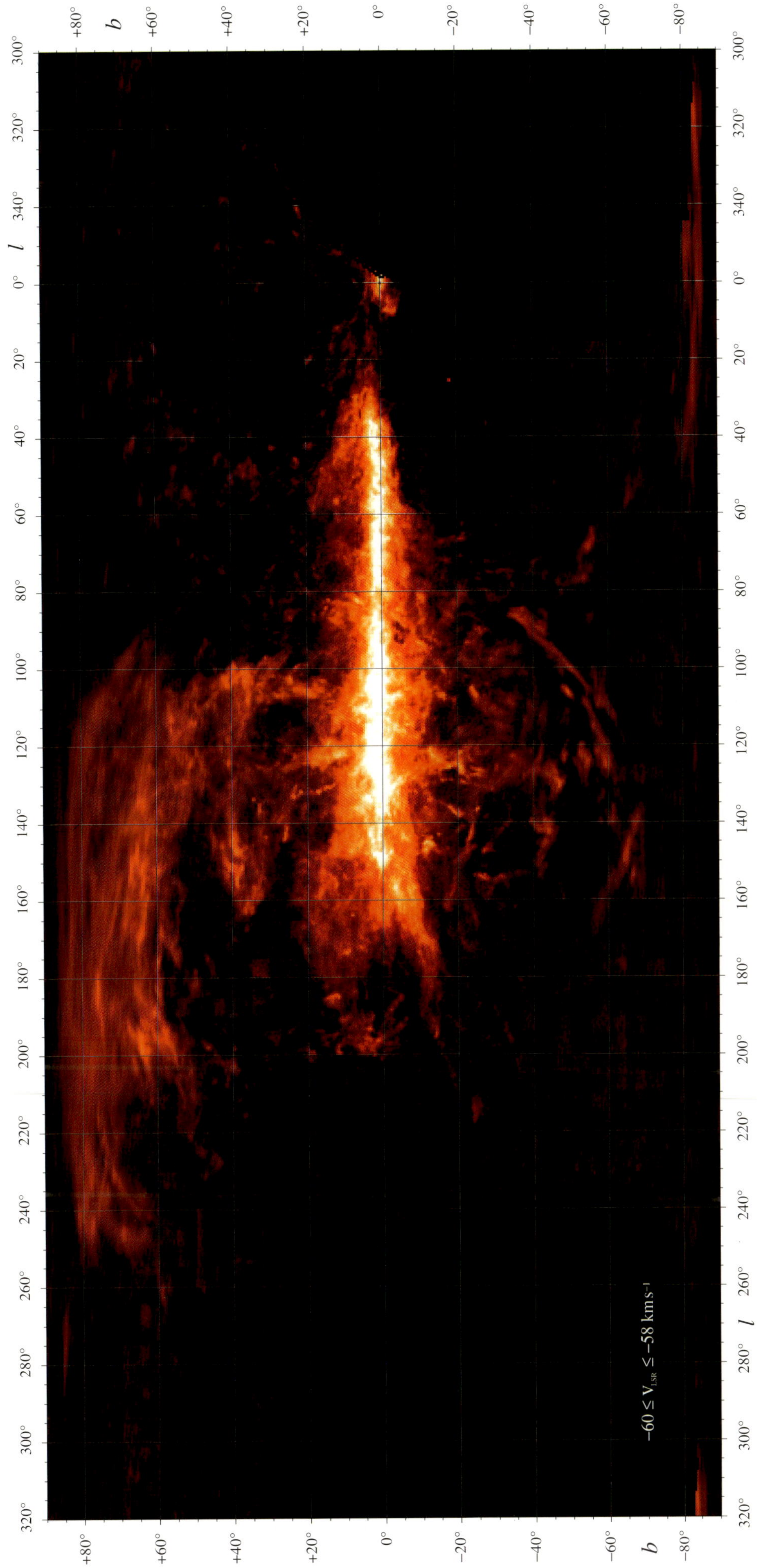

−60 ≤ V_LSR ≤ −58 km s⁻¹
l
b

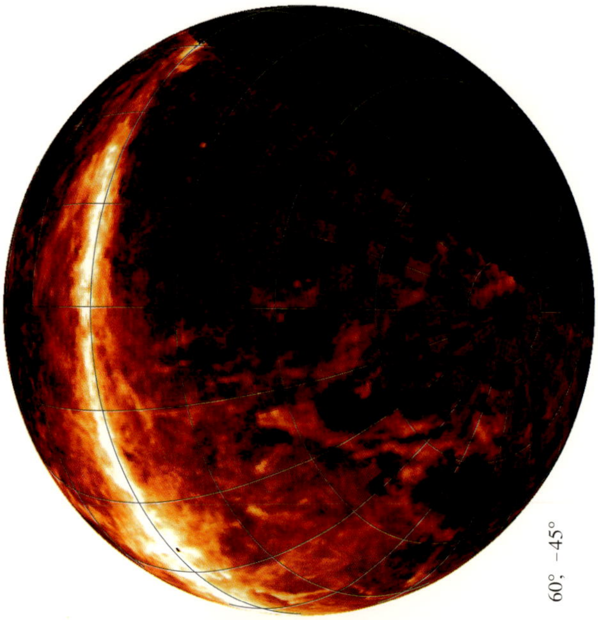

60°, −45°

180°, −45°

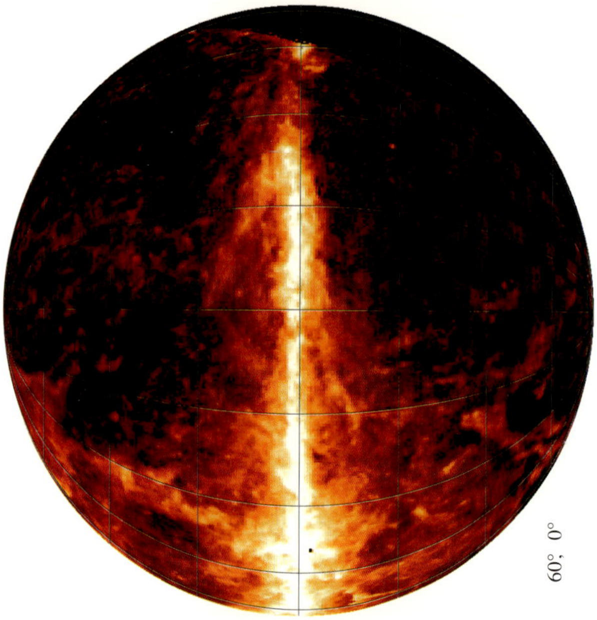

60°, 0°

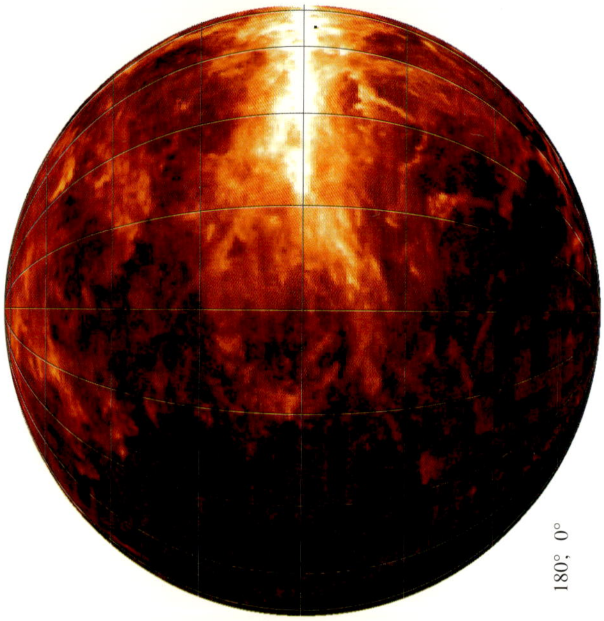

180°, 0°

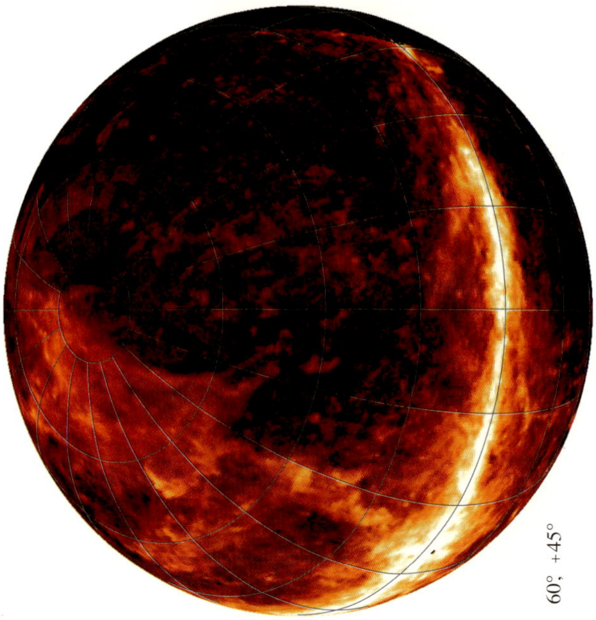

60°, +45°

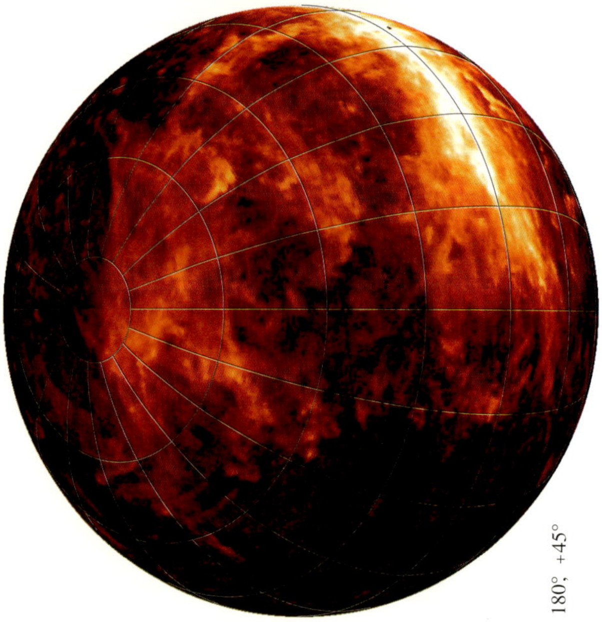

180°, +45°

$-50 \leq V_{LSR} \leq -48$ km s^{-1}

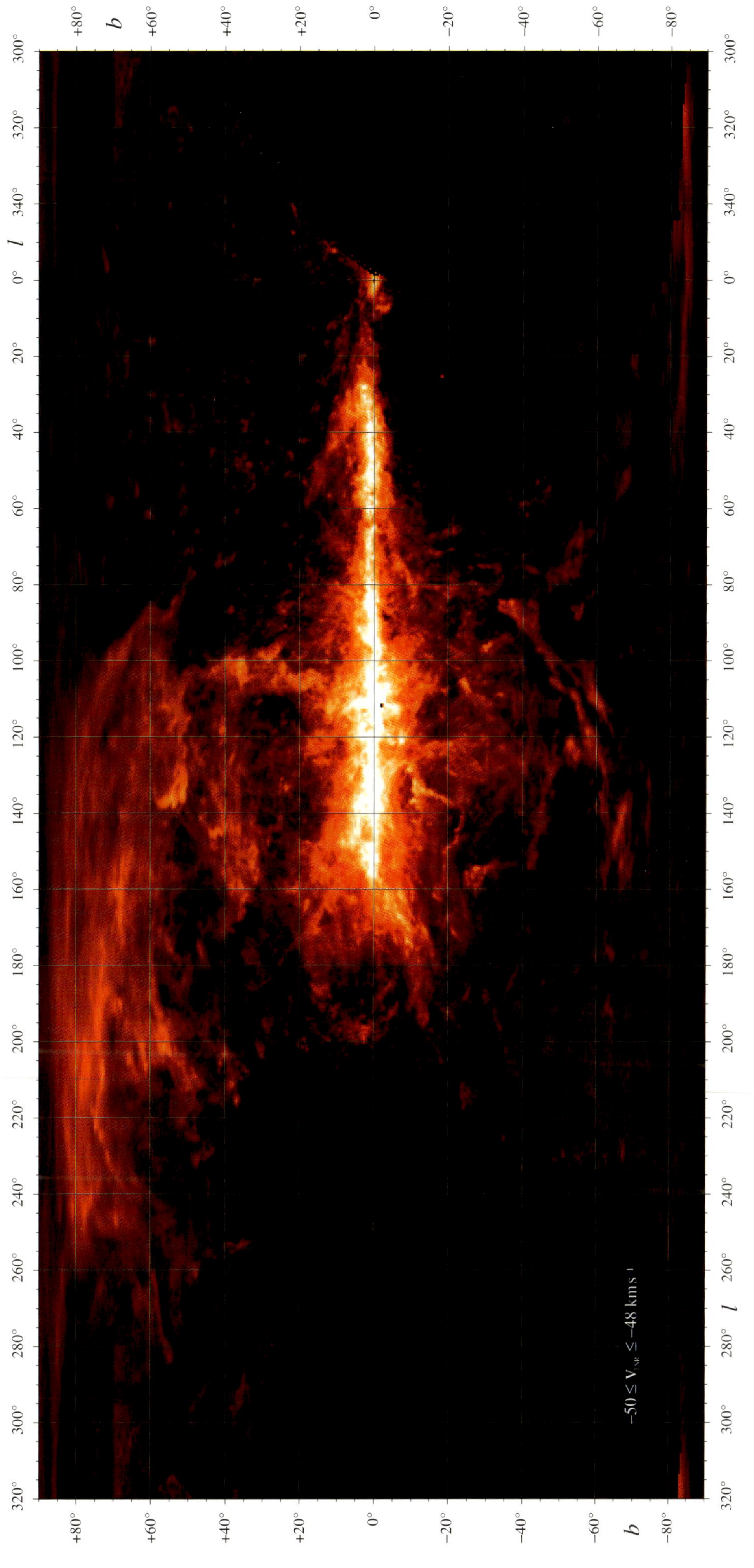
$-50 \leq V_{LSR} \leq -48$ km s^{-1}

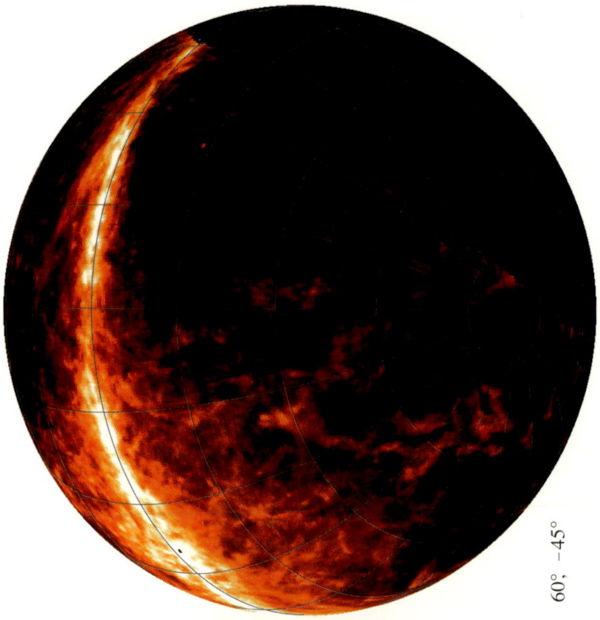

60°, −45°

180°, −45°

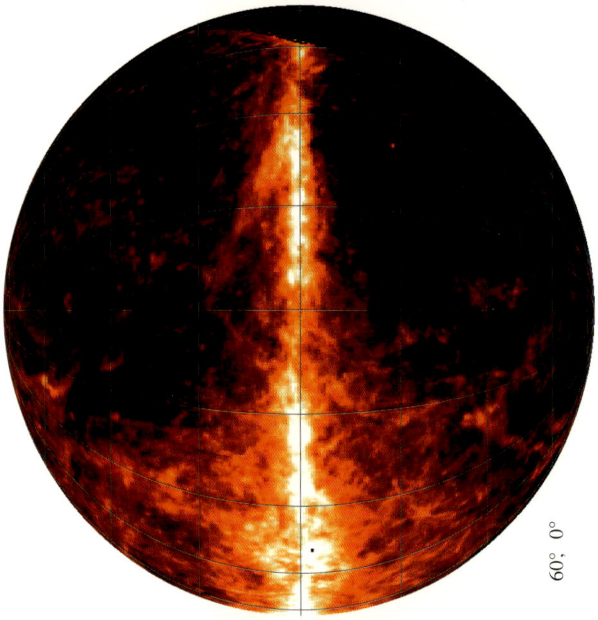

60°, 0°

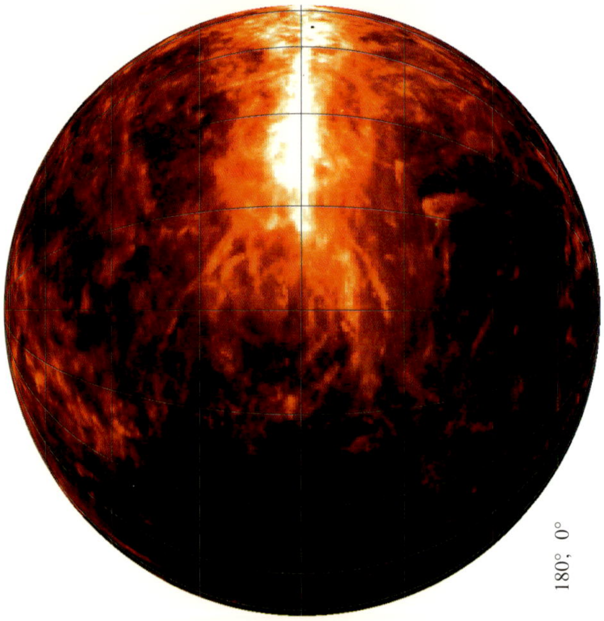

180°, 0°

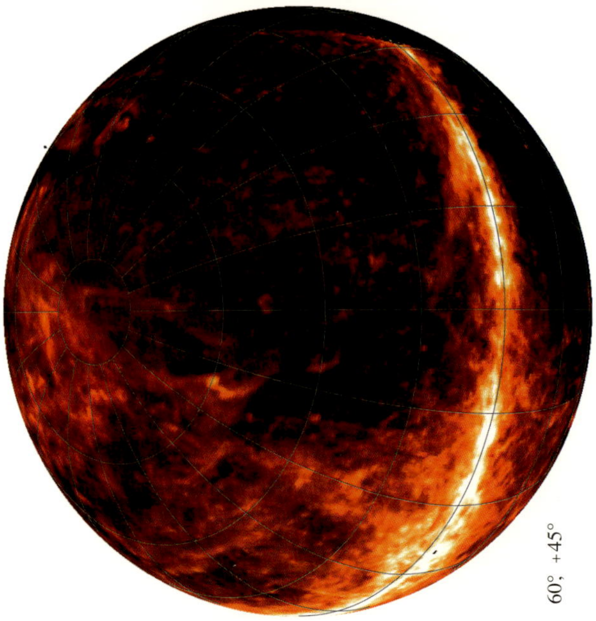

60°, +45°

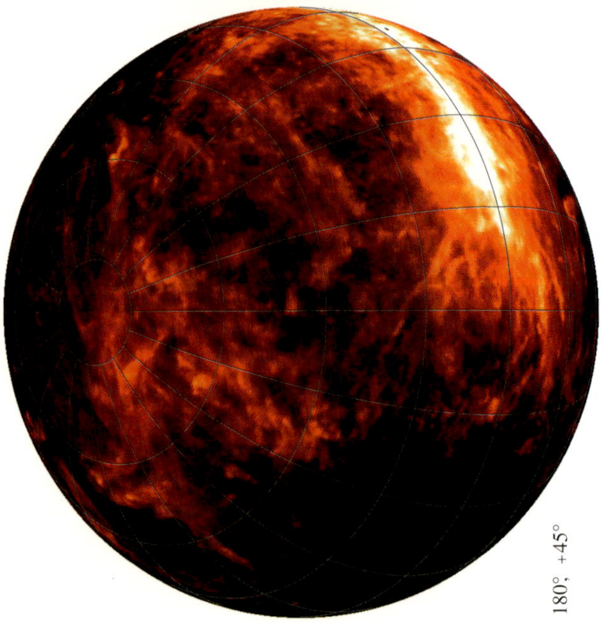

180°, +45°

$-40 \leq V_{LSR} \leq -38$ km s^{-1}

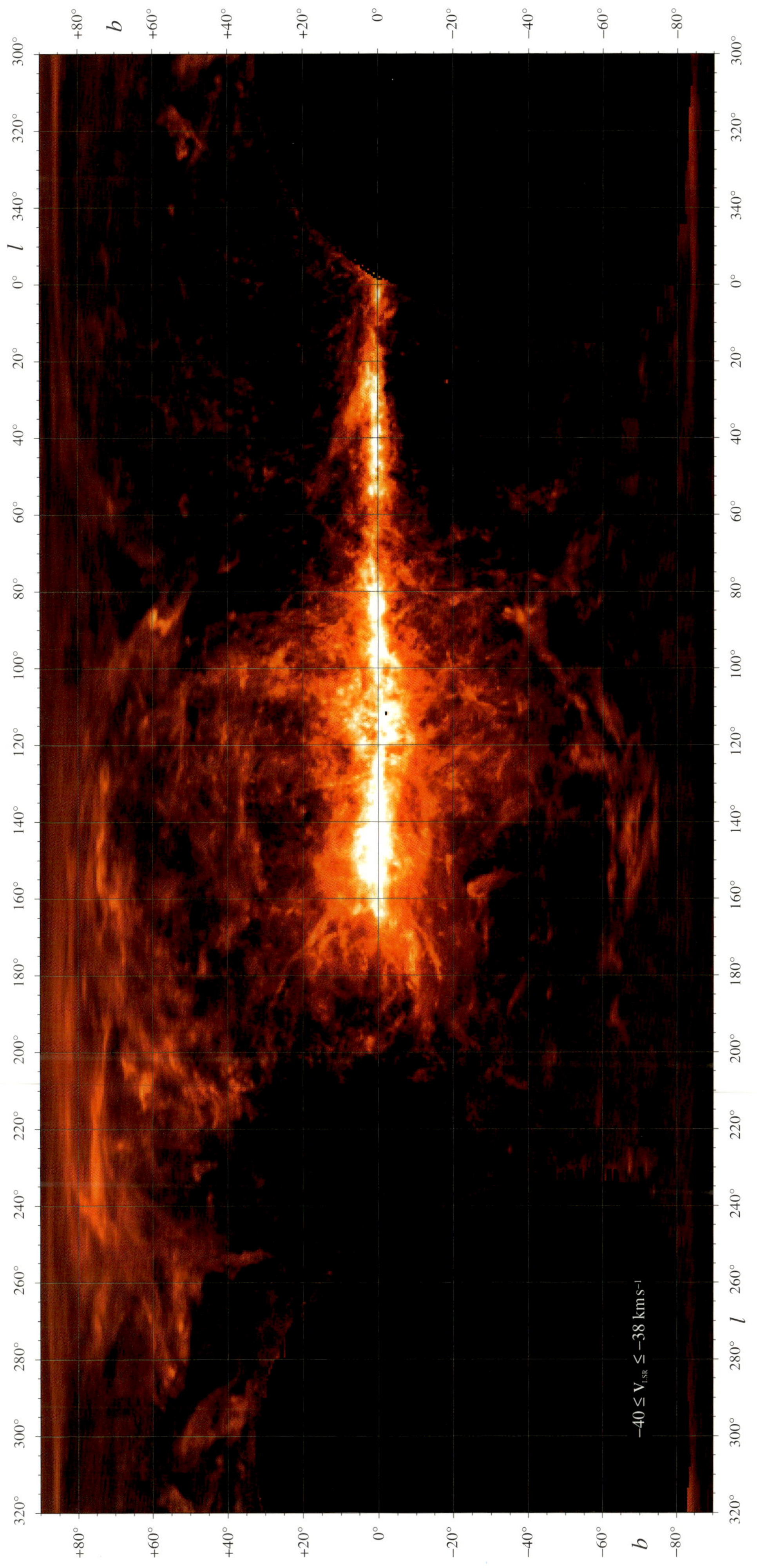

l
b
$-40 \le V_{LSR} \le -38$ km s^{-1}

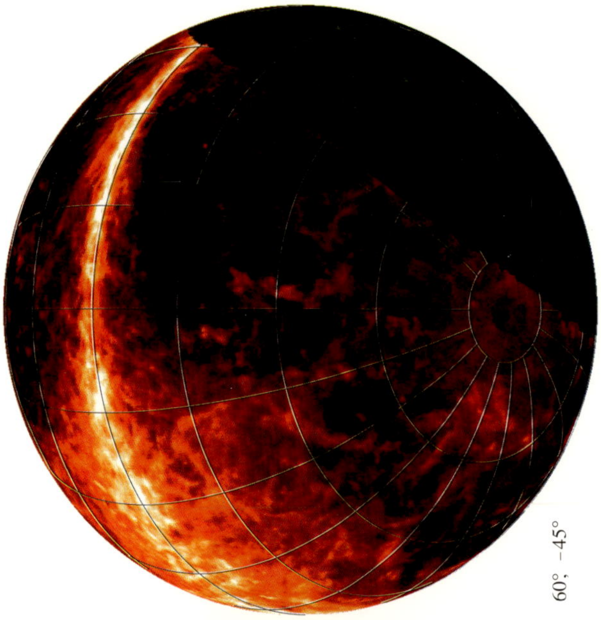

60°, −45°

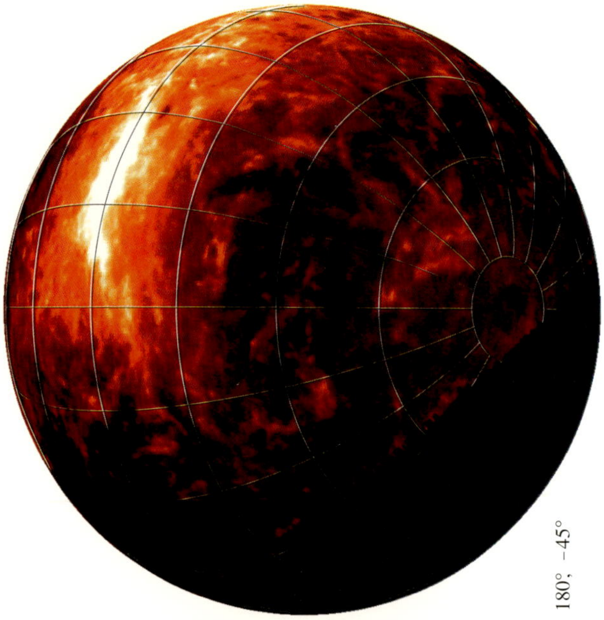

180°, −45°

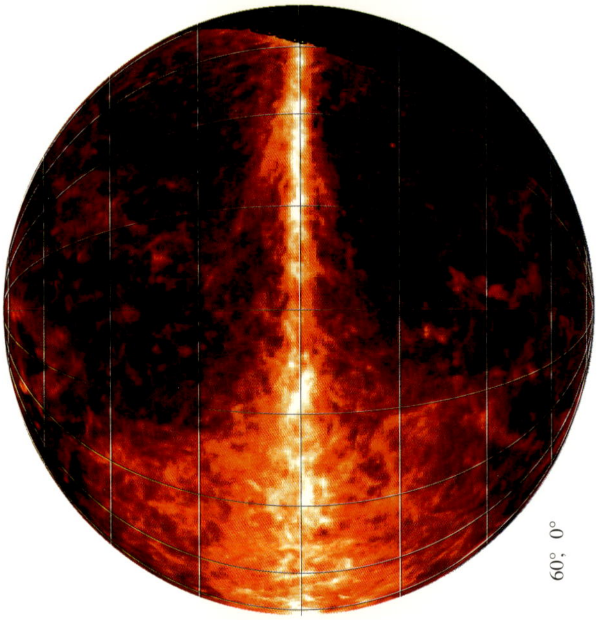

60°, 0°

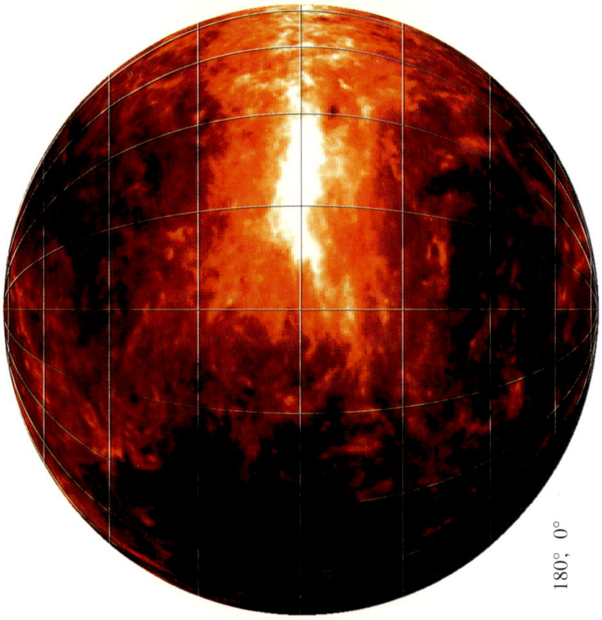

180°, 0°

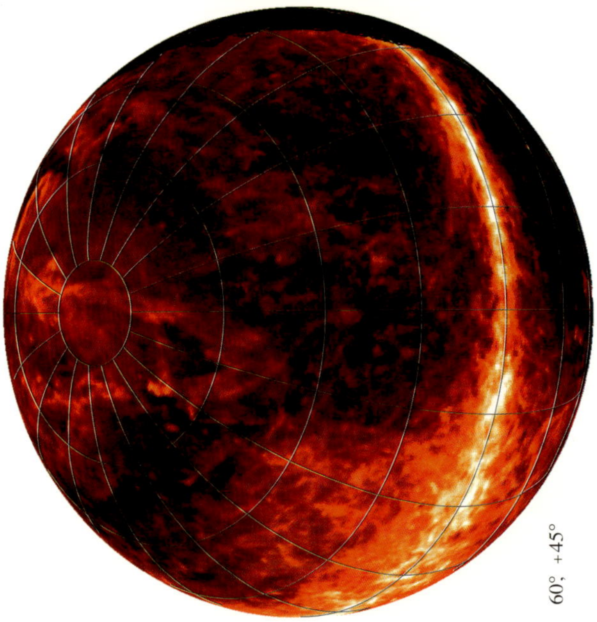

60°, +45°

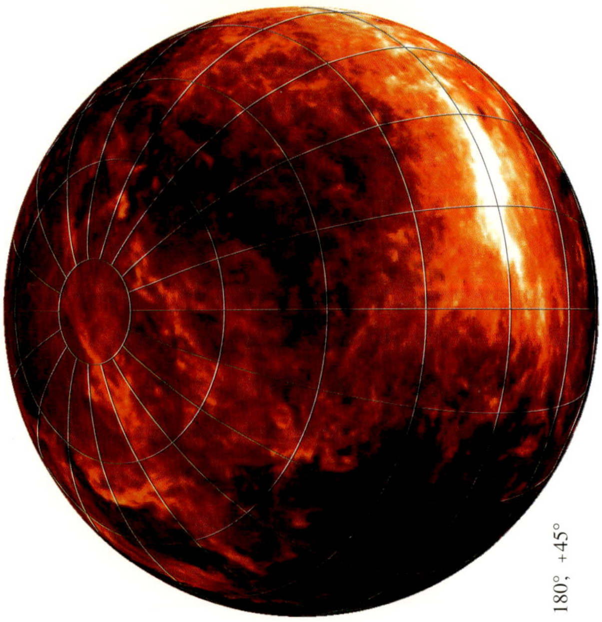

180°, +45°

$-30 \leq V_{LSR} \leq -28$ km s^{-1}

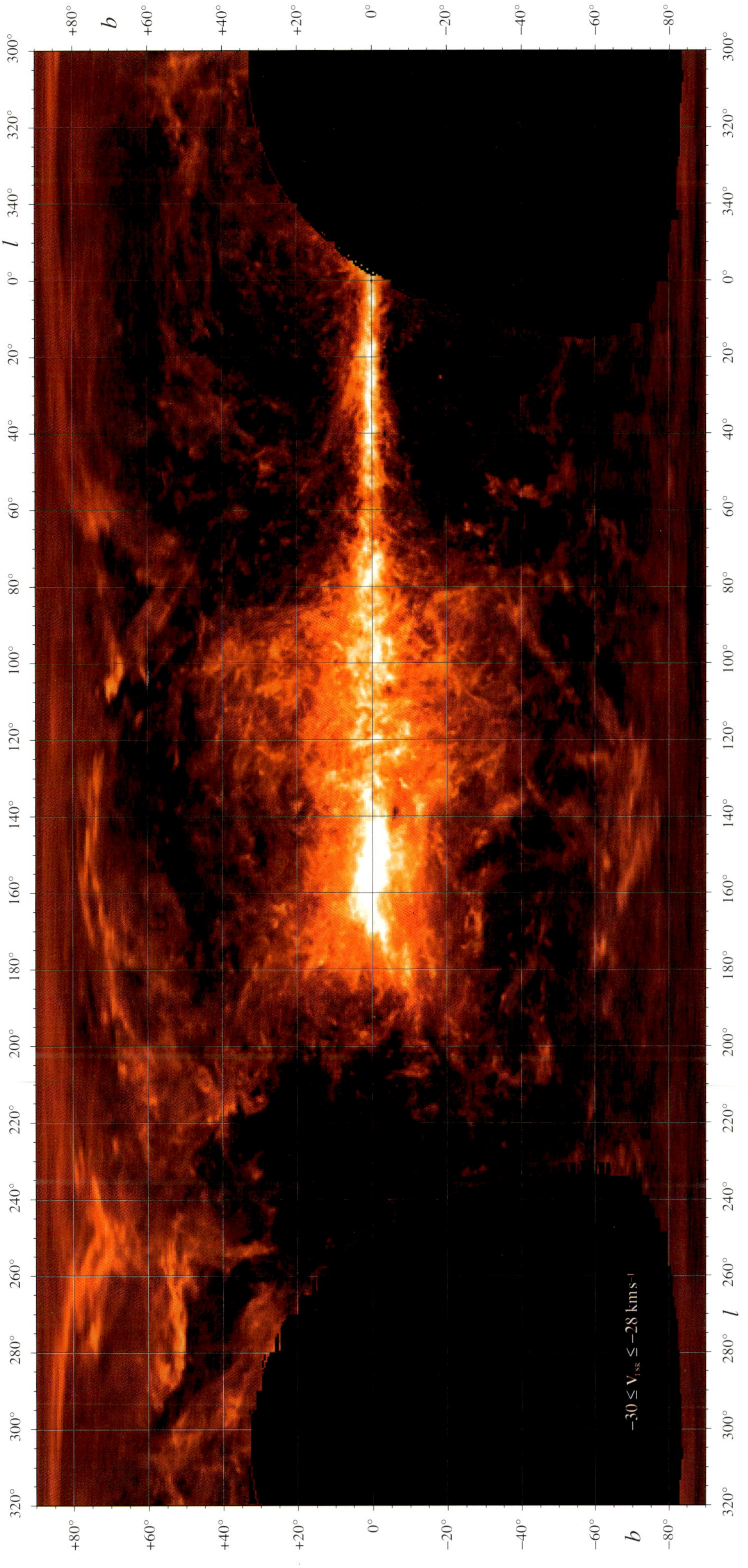
$-30 \leq V_{LSR} \leq -28$ km s^{-1}

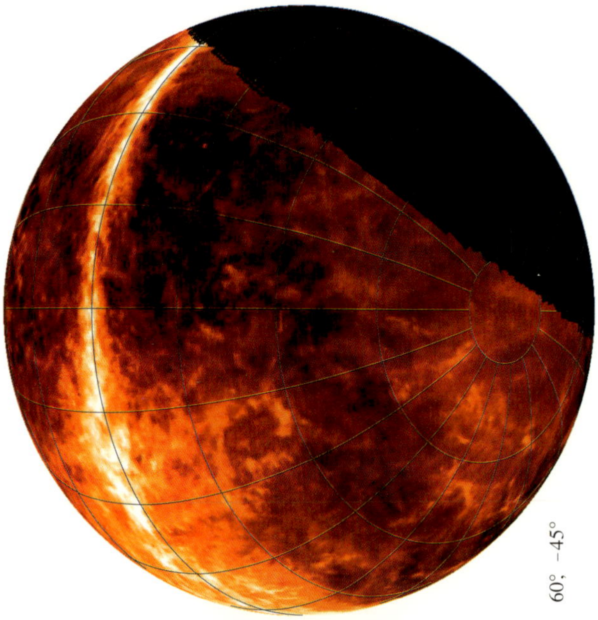
60°, −45°

180°, −45°

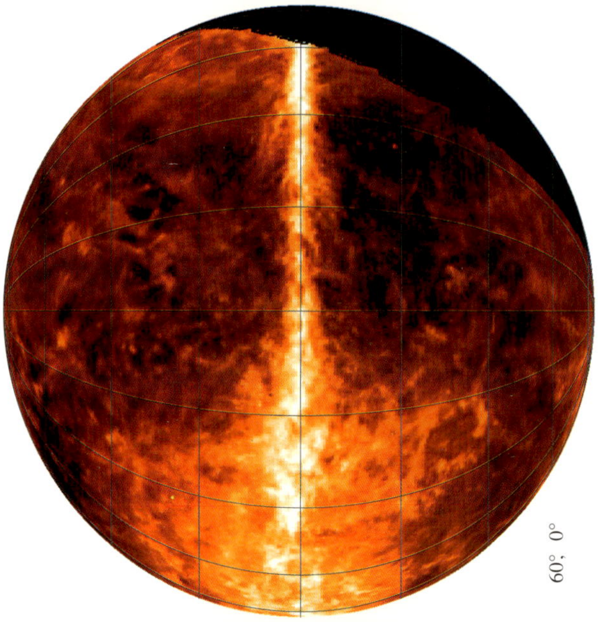
60°, 0°

180°, 0°

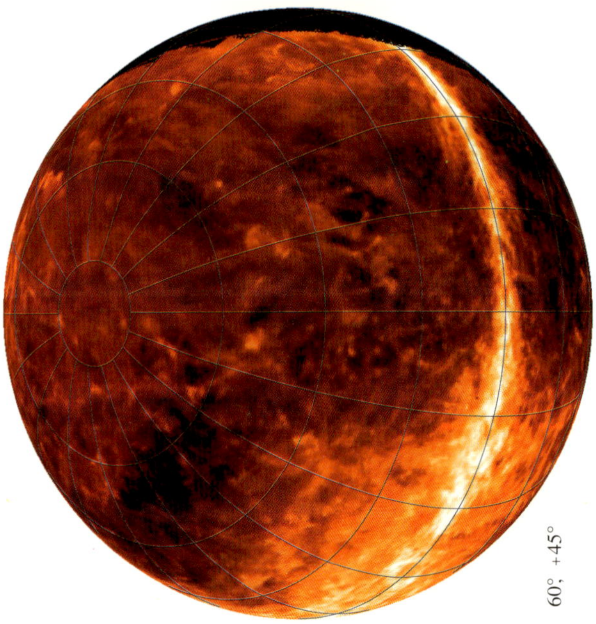
60°, +45°

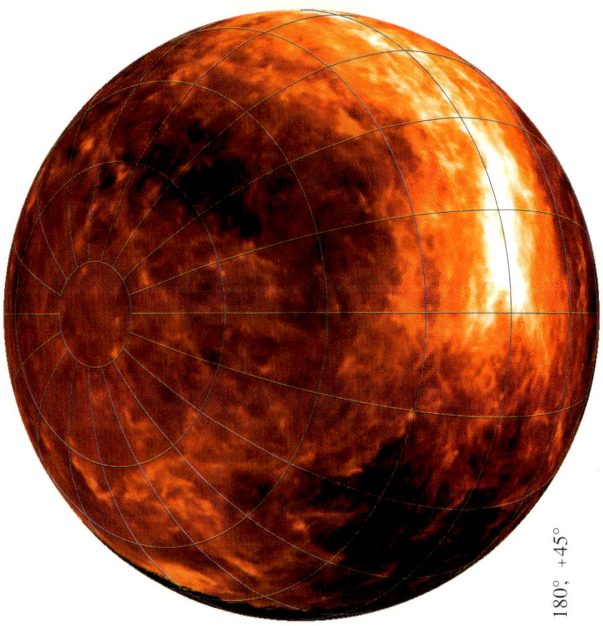
180°, +45°

$-20 \leq V_{LSR} \leq -18$ km s^{-1}

l
b
−20 ≤ V_LSR ≤ −18 km s−1

$-10 \leq V_{LSR} \leq -8$ km s^{-1}

−10 ≤ V_LSR ≤ −8 km s⁻¹
l
b
300° 320° 340° 0° 20° 40° 60° 80° 100° 120° 140° 160° 180° 200° 220° 240° 260° 280° 300° 320°
+80° +60° +40° +20° 0° −20° −40° −60° −80°

60°, −45°

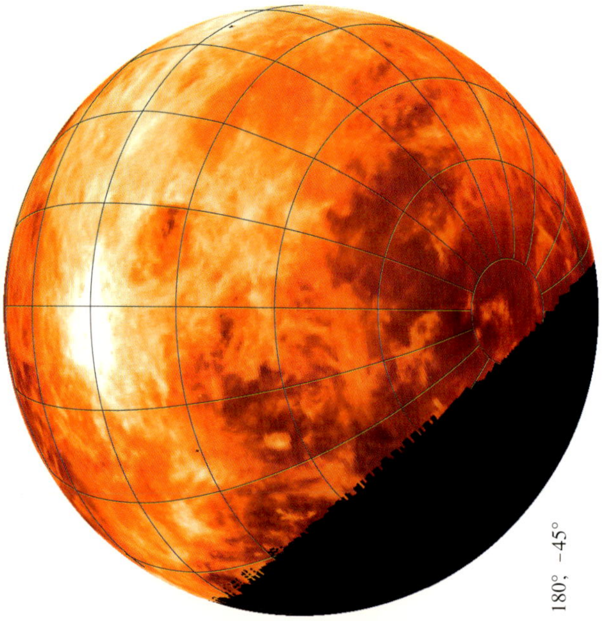

180°, −45°

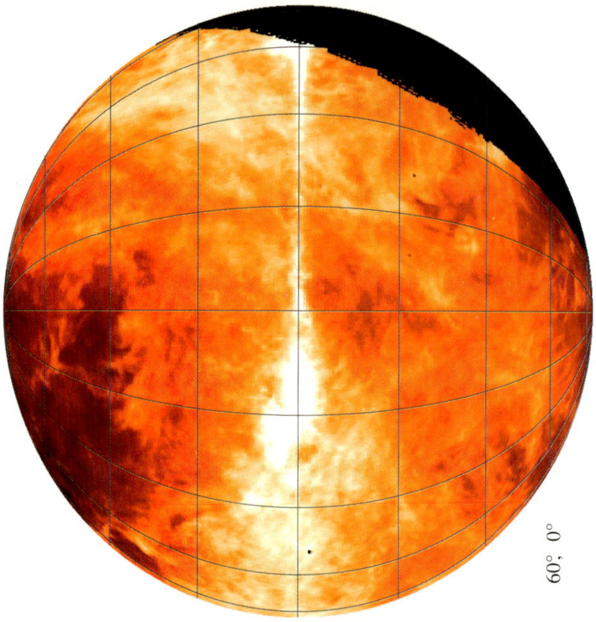

60°, 0°

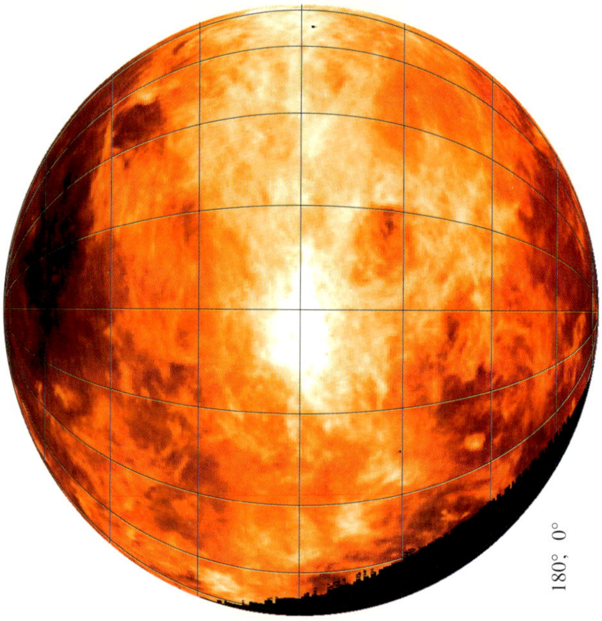

180°, 0°

60°, +45°

180°, +45°

$-1 \leq V_{LSR} \leq +1$ km s^{-1}

l
b
300° 320° 340° 0° 20° 40° 60° 80° 100° 120° 140° 160° 180° 200° 220° 240° 260° 280° 300° 320°
+80° +60° +40° +20° 0° −20° −40° −60° −80°
$-1 \le V_{LSR} \le +1$ km s^{-1}

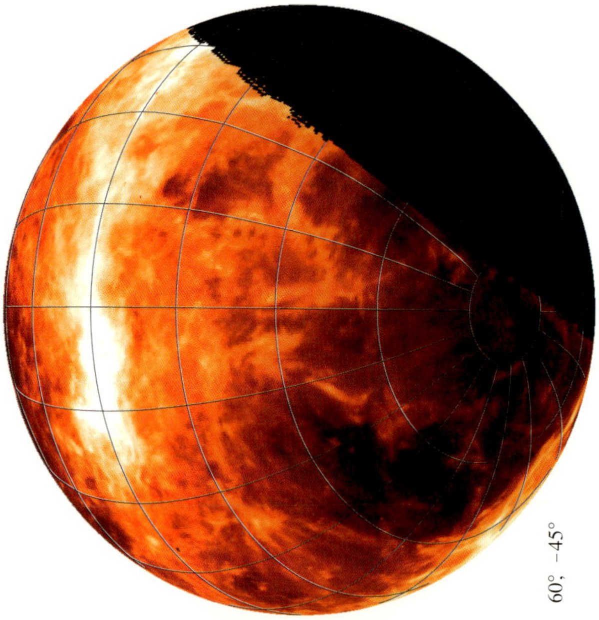

60°, −45°

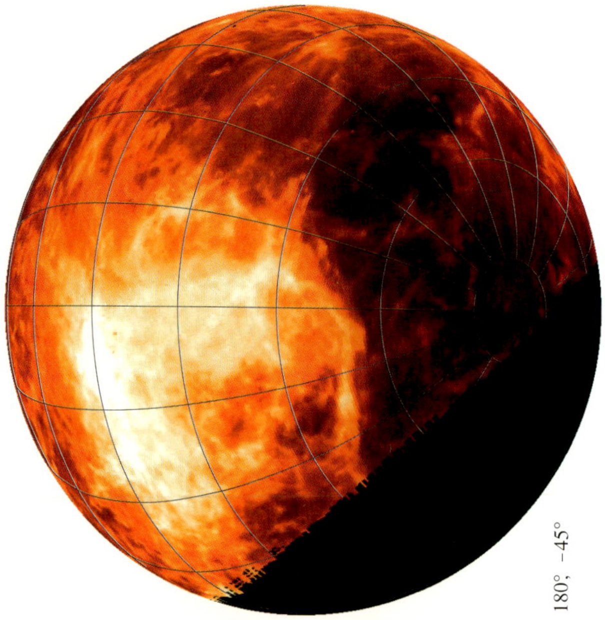

180°, −45°

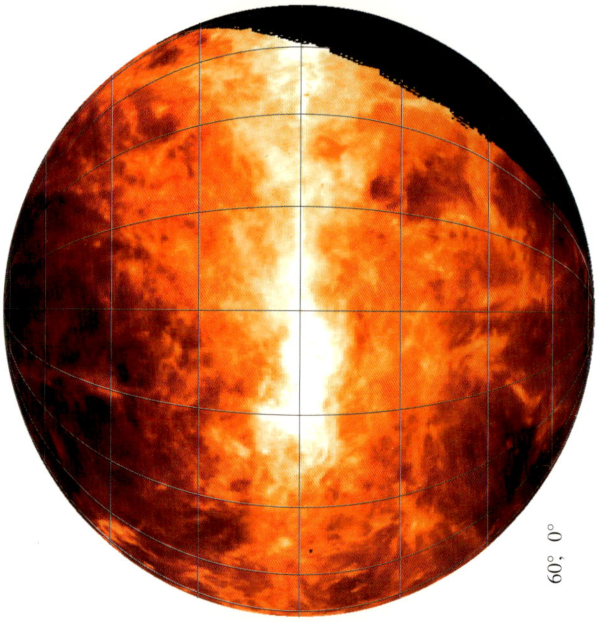

60°, 0°

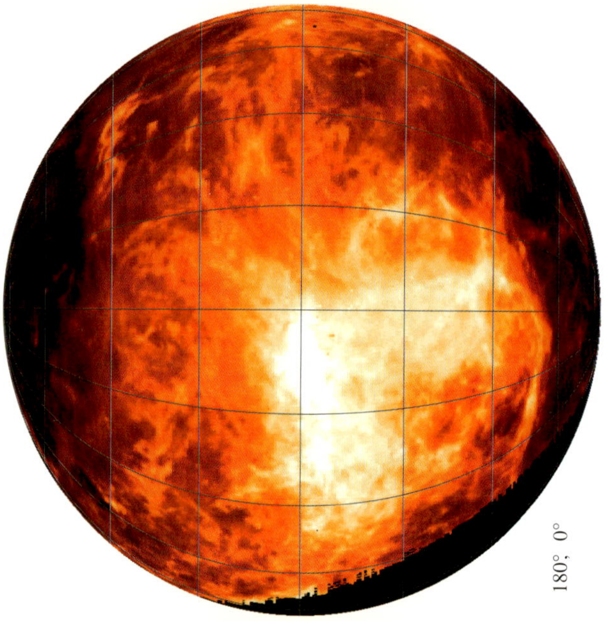

180°, 0°

60°, +45°

180°, +45°

$+8 \le V_{LSR} \le +10 \text{ km s}^{-1}$

$+8 \leq V_{LSR} \leq +10$ kms^{-1}

60°, −45°

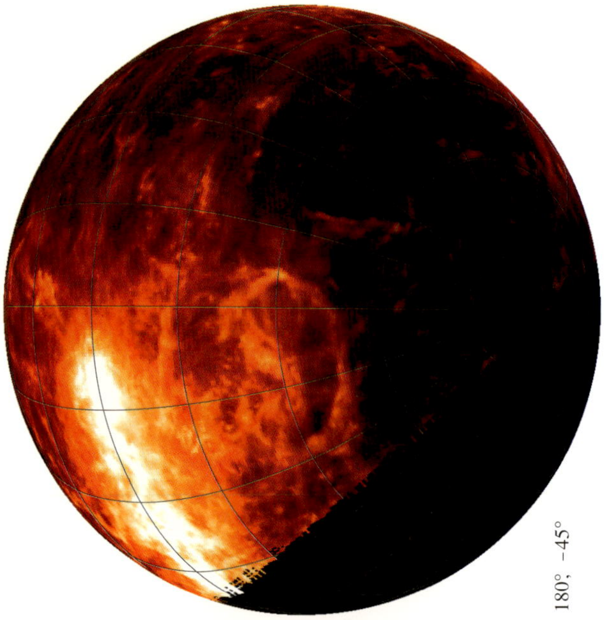

180°, −45°

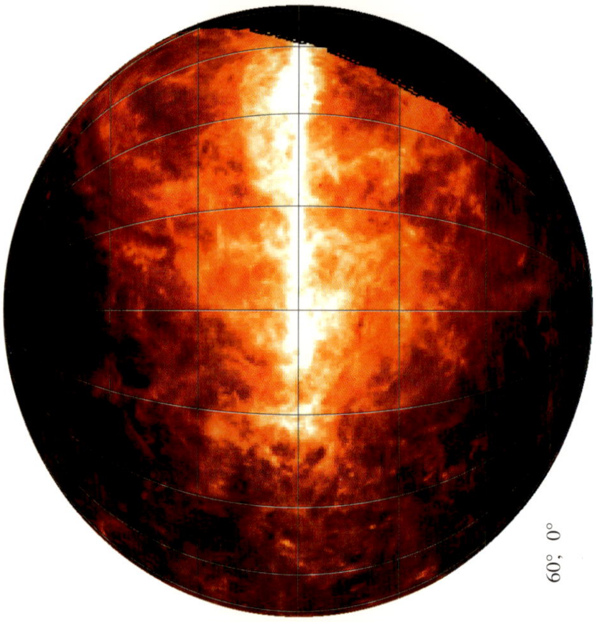

60°, 0°

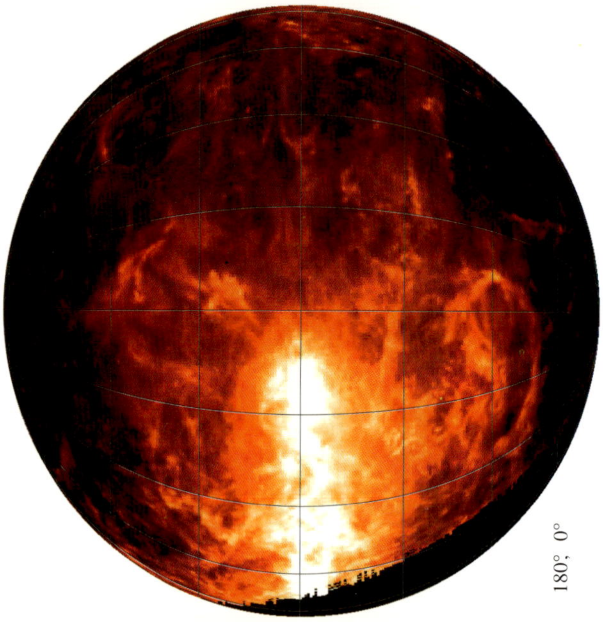

180°, 0°

60°, +45°

180°, +45°

$+18 \leq V_{LSR} \leq +20$ km s^{-1}

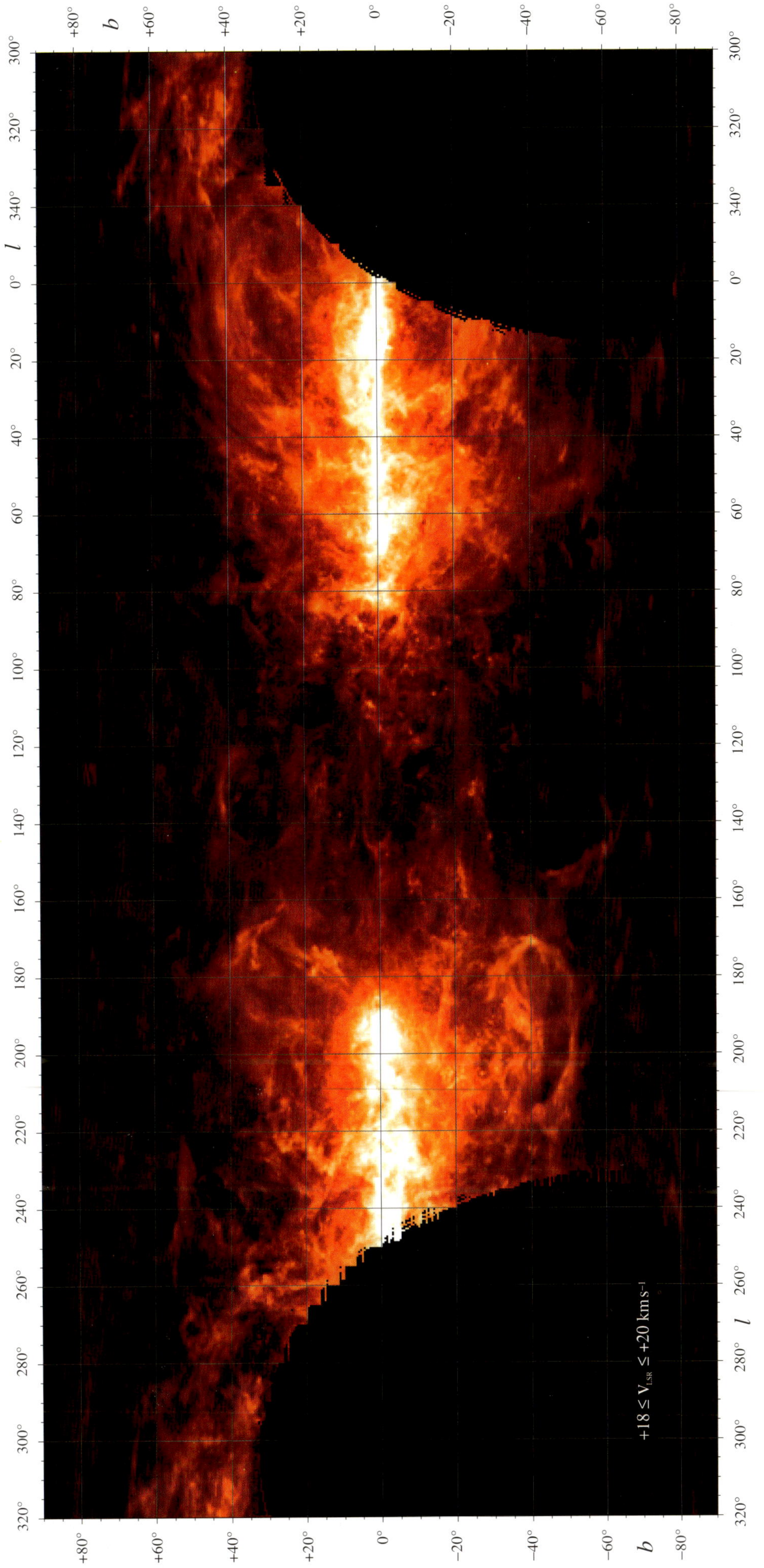
$+18 \le V_{LSR} \le +20$ km s^{-1}
l
b

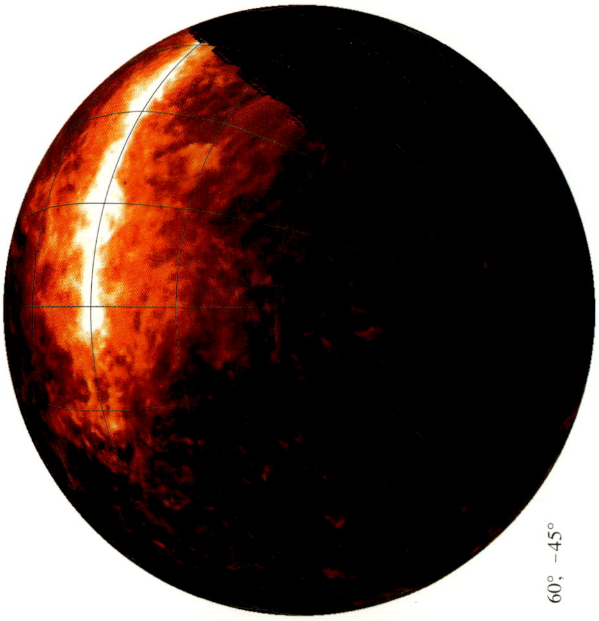

60°, −45°

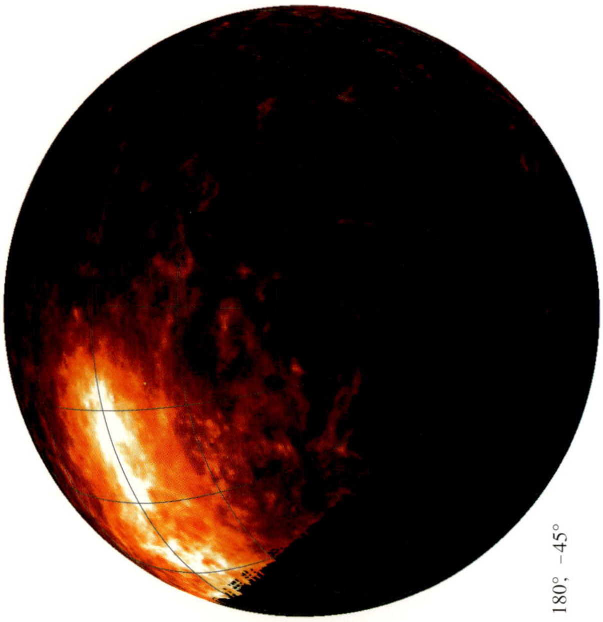

180°, −45°

60°, 0°

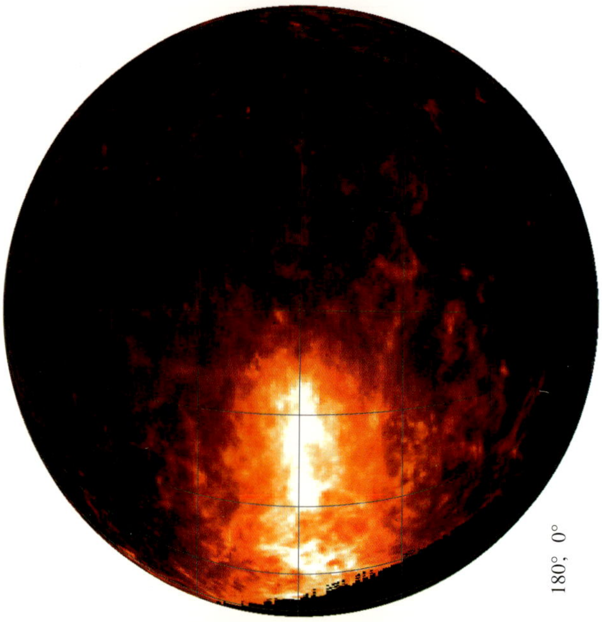

180°, 0°

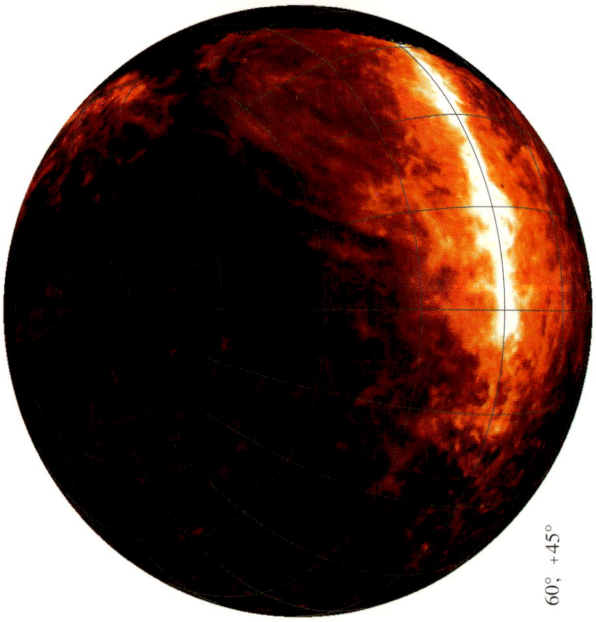

60°, +45°

180°, +45°

$+28 \leq V_{LSR} \leq +30$ km s^{-1}

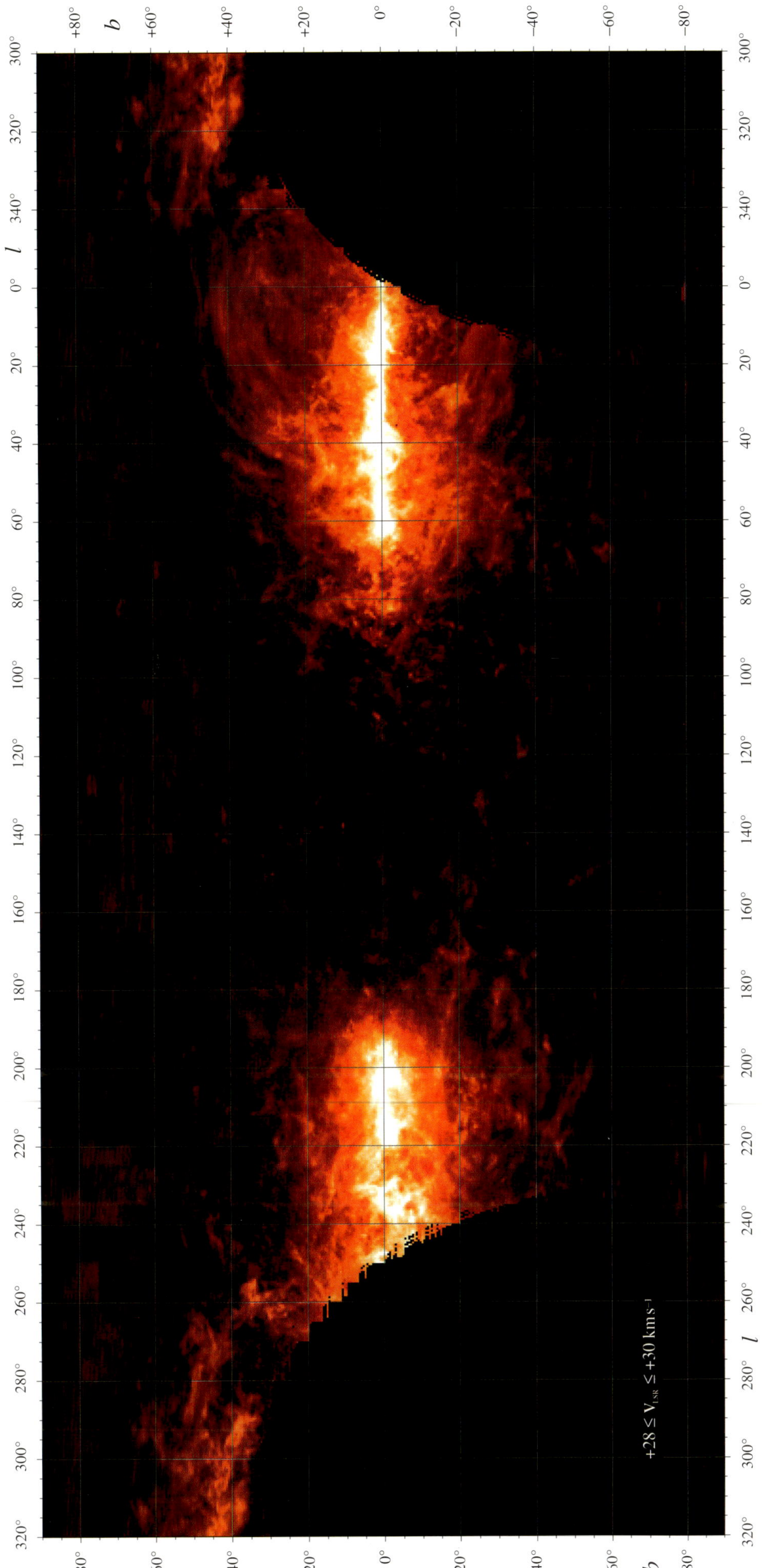
+28 ≤ V_{LSR} ≤ +30 km s^{-1}

60°, −45°

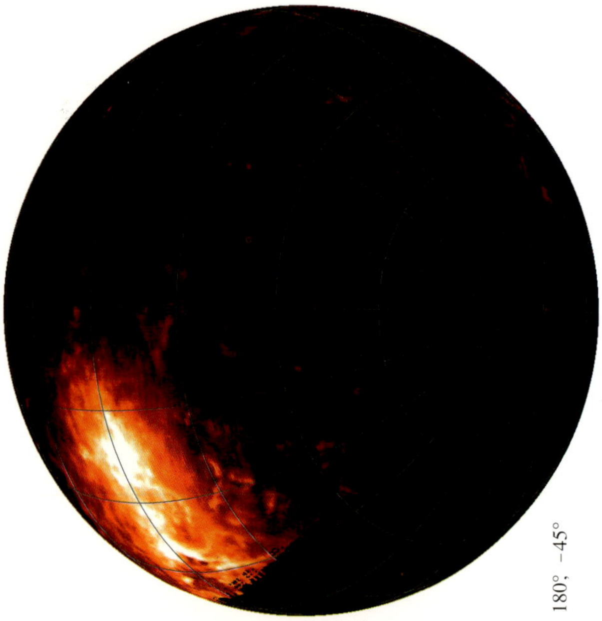

180°, −45°

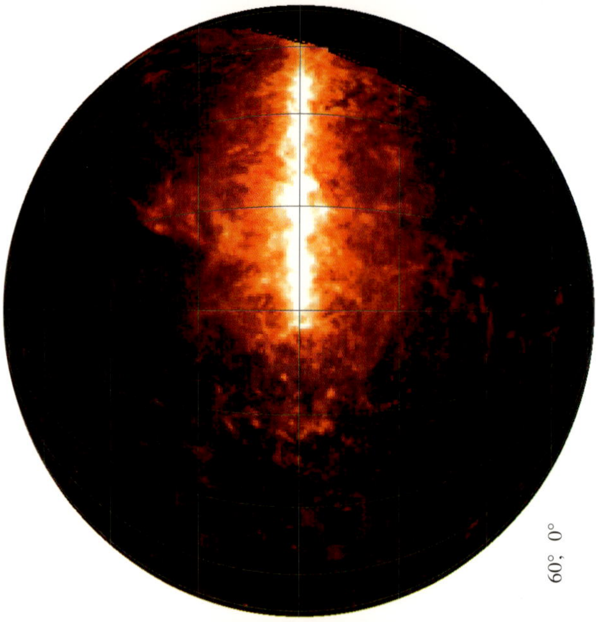

60°, 0°

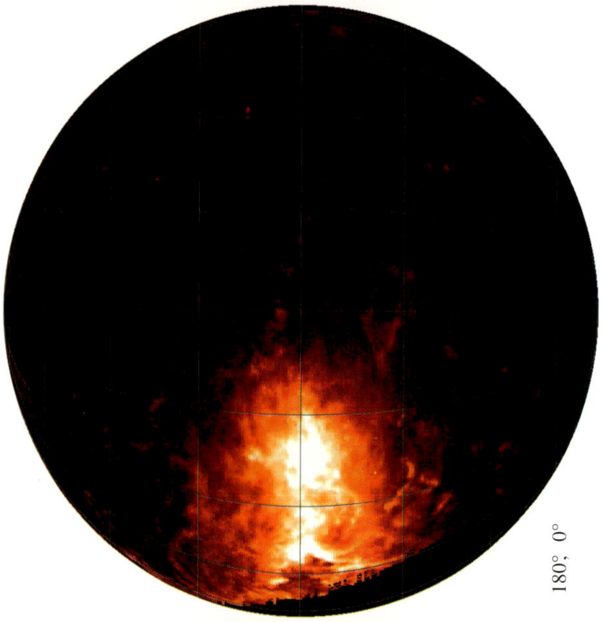

180°, 0°

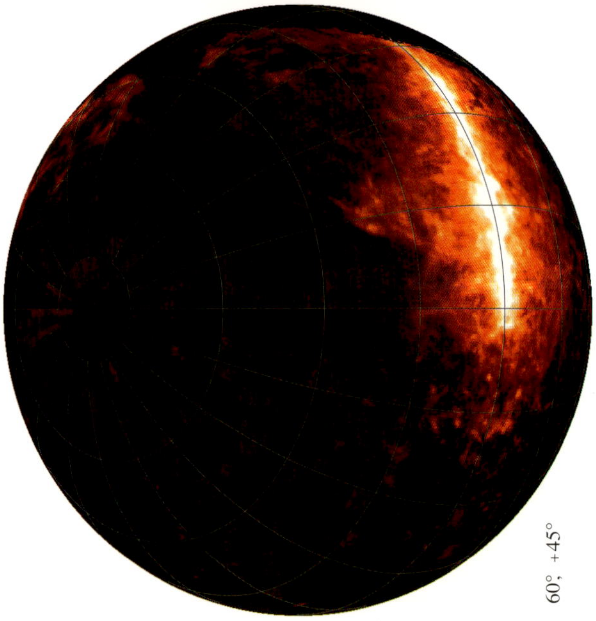

60°, +45°

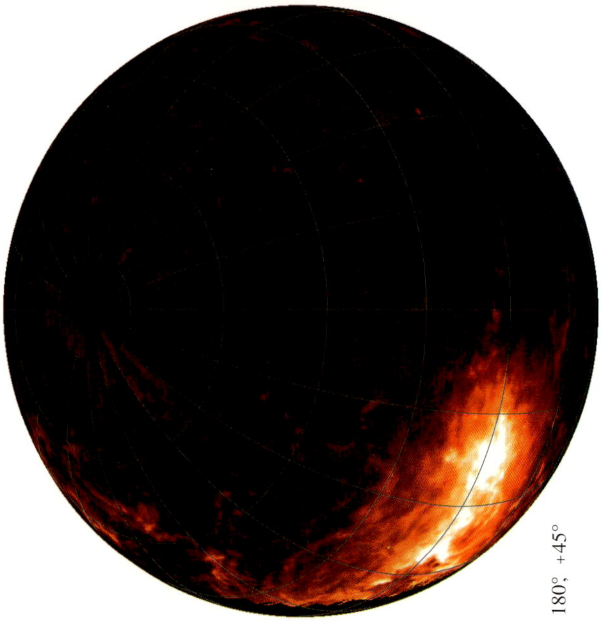

180°, +45°

$+38 \leq V_{LSR} \leq +40\ km\,s^{-1}$

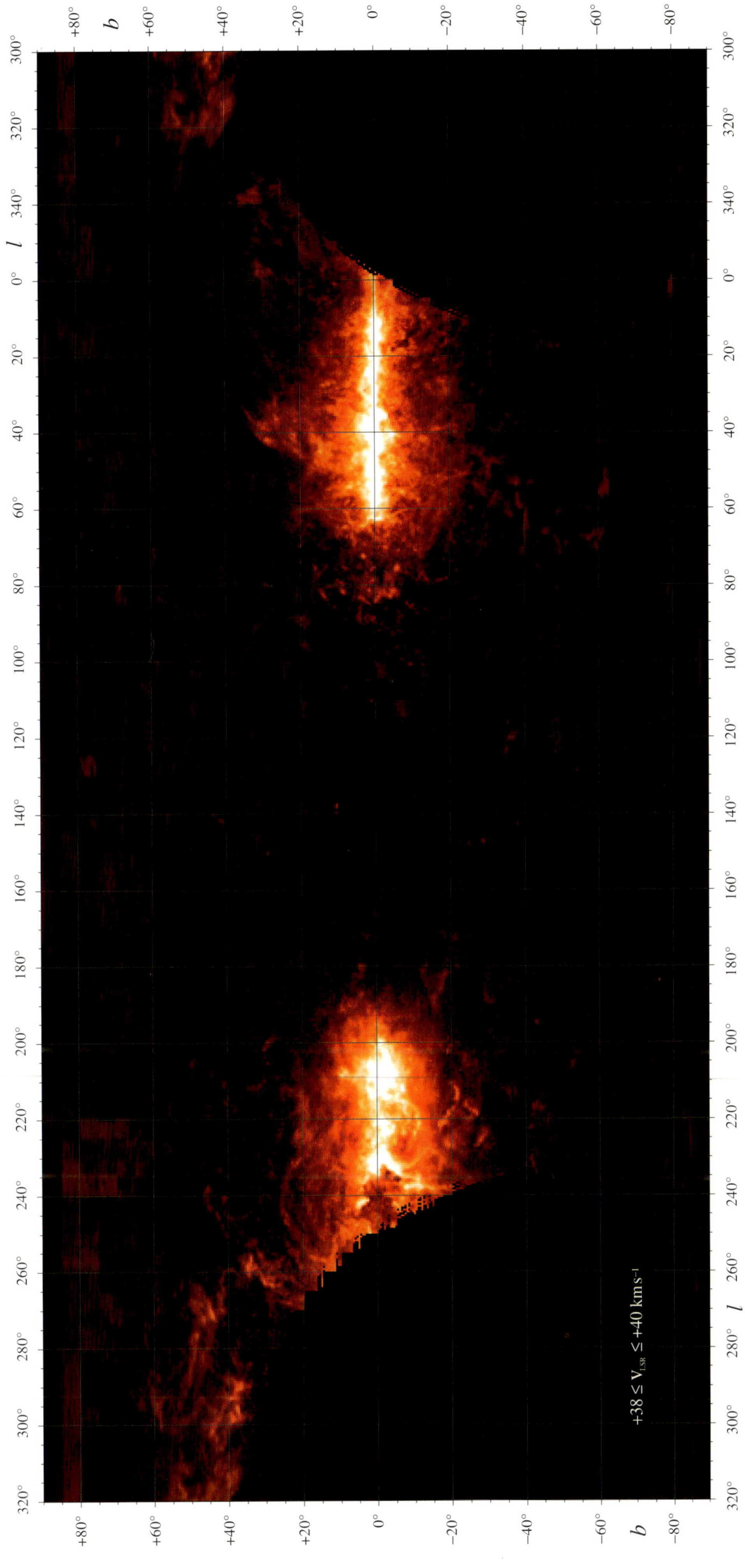
+38 ≤ V_LSR ≤ +40 km s⁻¹
l
b

Notulenboek IV
der N.A.C.
Vergadering 64 t/m 79

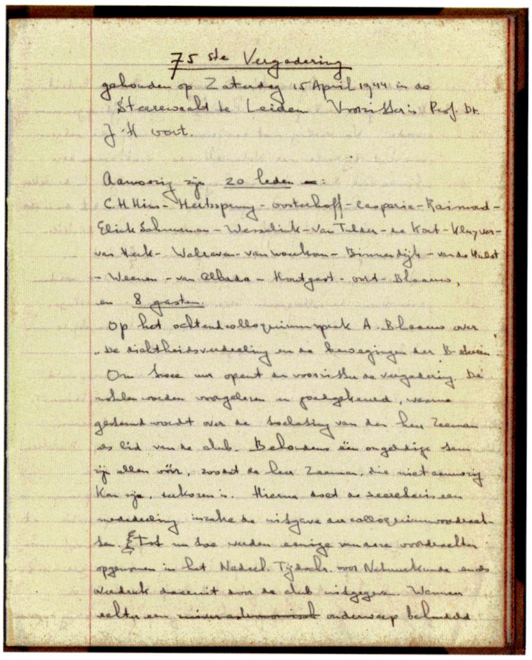
75 ste Vergadering
gehouden op Zaterdag 15 April 1944 in de Stadswald te Leiden. Voorzitter is Prof. Dr. J. H. Oort.
Aanwezig zijn 20 leden

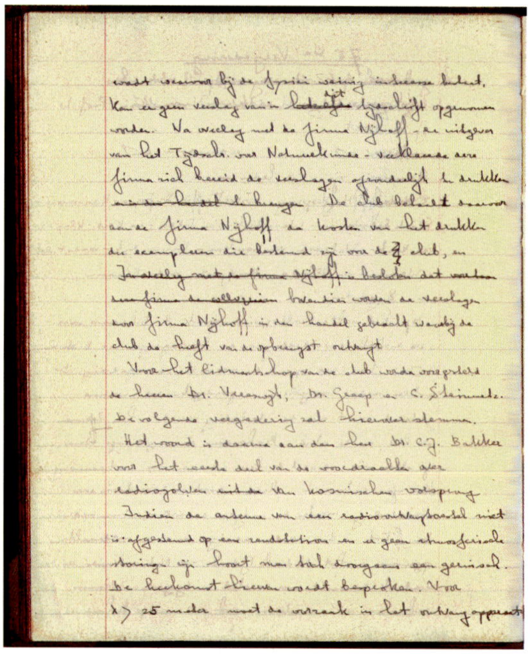

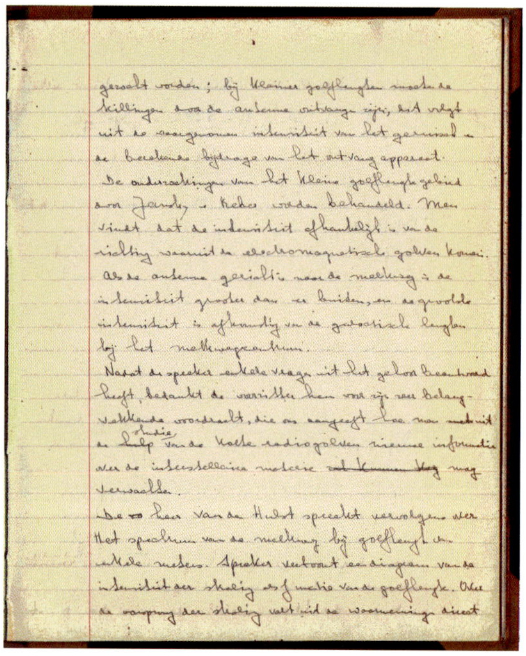

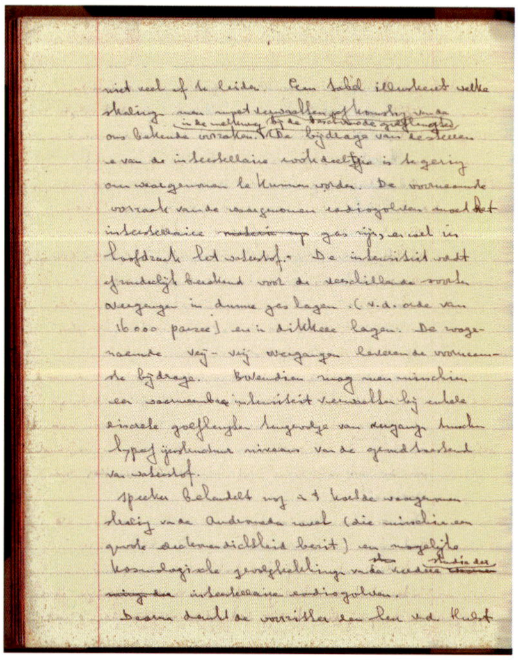

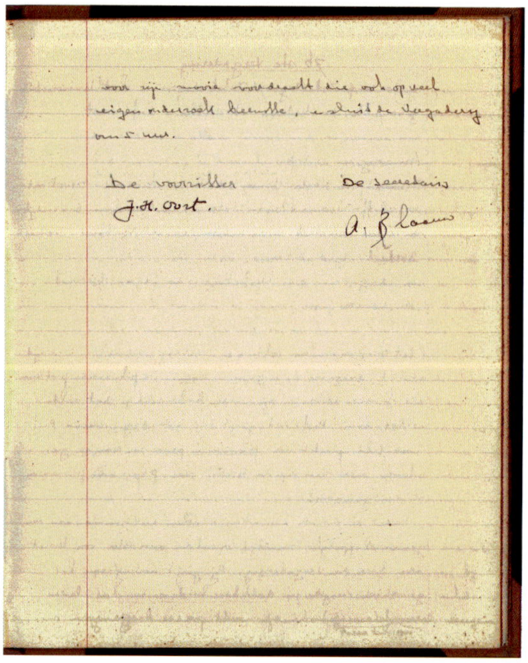
De voorzitter
J. H. Oort.
De secretaris
A. Blaauw

Minutes of the 75th meeting of the NAC

H.C. van de Hulst published his prediction that the 21-cm line would be observable from interstellar space in 1945, in the *Nederlandsch Tijdschrift voor Natuurkunde* (Netherlands Journal for Physics), Vol. 11, p. 210. The paper is written in Dutch; an English translation can be found in W.T. Sullivan III (ed.) *Classics in Radio Astronomy*, Reidel, 1982, paper 34. Van de Hulst's article was based on work which he first announced the previous year, at a meeting of the Nederlandse Astronomen Club (NAC) held in Leiden. A translation of the hand-written minutes (reproduced in facsimile on the previous page) of the 75th meeting of the Netherlands Astronomers' Club follows here.

75th Meeting

held on Saturday 15 April 1944 at the Observatory in Leiden.
Chairman Prof. Dr. J.H. Oort.

Present are 20 members: C.H. Hin - Hertzsprung - Oosterhoff - Casparie - Raimond - Elink Schuurman - Wesselink - van Tulder - de Kort - Kluyver - van Herk - Walraven - van Woerkom - Binnendijk - van de Hulst - Weenen - van Albada - Houtgast - Oort - Blaauw, and 8 guests.

During the morning colloquium A. Blaauw spoke about the density distribution and motions of B stars. At 2 o'clock the chairman opens the meeting. The minutes are read out and approved, after which voting takes place on the issue of admitting Mr. Zeeman to the club as a member. Except for one invalid ballot all were yea's, so that Mr. Zeeman, who could not be present, is elected. After this the secretary makes a statement regarding publication of the colloquium presentations. Until now some of these presentations have appeared in the "Nederl. Tijdschr. voor Natuurkunde" and have been distributed as reprints by the club. However, when a subject is treated for which there is little interest amongst physicists, then no report can appear in that journal. After discussions with the Nijhoff company — the publisher of the Tijdschr. voor Natuurkunde — the company stated that they were prepared to print the reports separately. The club pays Nijhoff for the costs of printing the copies intended for the club, and furthermore the reports will be offered for sale by Nijhoff whereby the club receives half of the proceeds.

Proposed for membership of the club are Dr. Vreeswijk, Dr. Greep, and C. Steinmetz. The next meeting will vote on this issue.

The word is then given to Dr. C.J. Bakker for the first part of the presentations on radio waves of cosmic origin. If the antenna of a radio receiver is not tuned to a transmitting station and if there are no atmospheric disturbances, one hears noise, nevertheless. The origin of this noise is discussed. For $\lambda > 25$ meters the origin must be sought in the receiving apparatus; in the case of shorter wavelengths the vibrations must have been received by the antenna; this follows from the observed intensity of the static and the known contribution of the receiving apparatus. Studies of the short wavelength region by Jansky and Reber are treated. It is found that the intensity depends on the direction from which the electromagnetic waves come. If the antenna is pointed toward the Milky Way, then the intensity is greater than otherwise, and the greatest intensity is contributed by the galactic longitude of the center of the Milky Way. After the speaker has answered several questions from the audience, the chairman thanks him for his very interesting presentation, which shows us how one may expect new information on the interstellar medium from the study of short radio waves.

Then Mr. van de Hulst speaks about the spectrum of the Milky Way at wavelengths of several meters. The speaker shows a diagram of the intensity of radiation as a function of wavelength. Not much about the origin of the radiation can be directly derived from the observations. A table illustrates which radiation one must expect, at the wavelengths considered, to originate from the causes known to us. In the Galaxy the contribution from stars and interstellar smoke particles is too weak to be observed. The principal cause of the observed radio waves must be the interstellar gas, and in fact mainly the hydrogen. The intensity is separately calculated for the various types of transitions in thin gas layers (on the order of 1600 parsec) and in thicker layers. The so-called free-free transitions provide the main contribution. Furthermore, one may perhaps expect an observable intensity at several discrete wavelengths as a consequence of transitions between hyperfine-structure levels of the ground state of hydrogen. The speaker next briefly treats the observed radiation from the Andromeda nebula (which perhaps possesses a high electron density) and possible cosmological inferences of further study of interstellar radio waves.

After this the chairman thanks Mr. van de Hulst for his fine presentation, which was also based on much of van de Hulst's own research, and closes the meeting at 5 o'clock.

The chairman	The secretary
J.H. Oort	A. Blaauw

$-150 \leq V_{LSR} \leq -140$ km s^{-1}

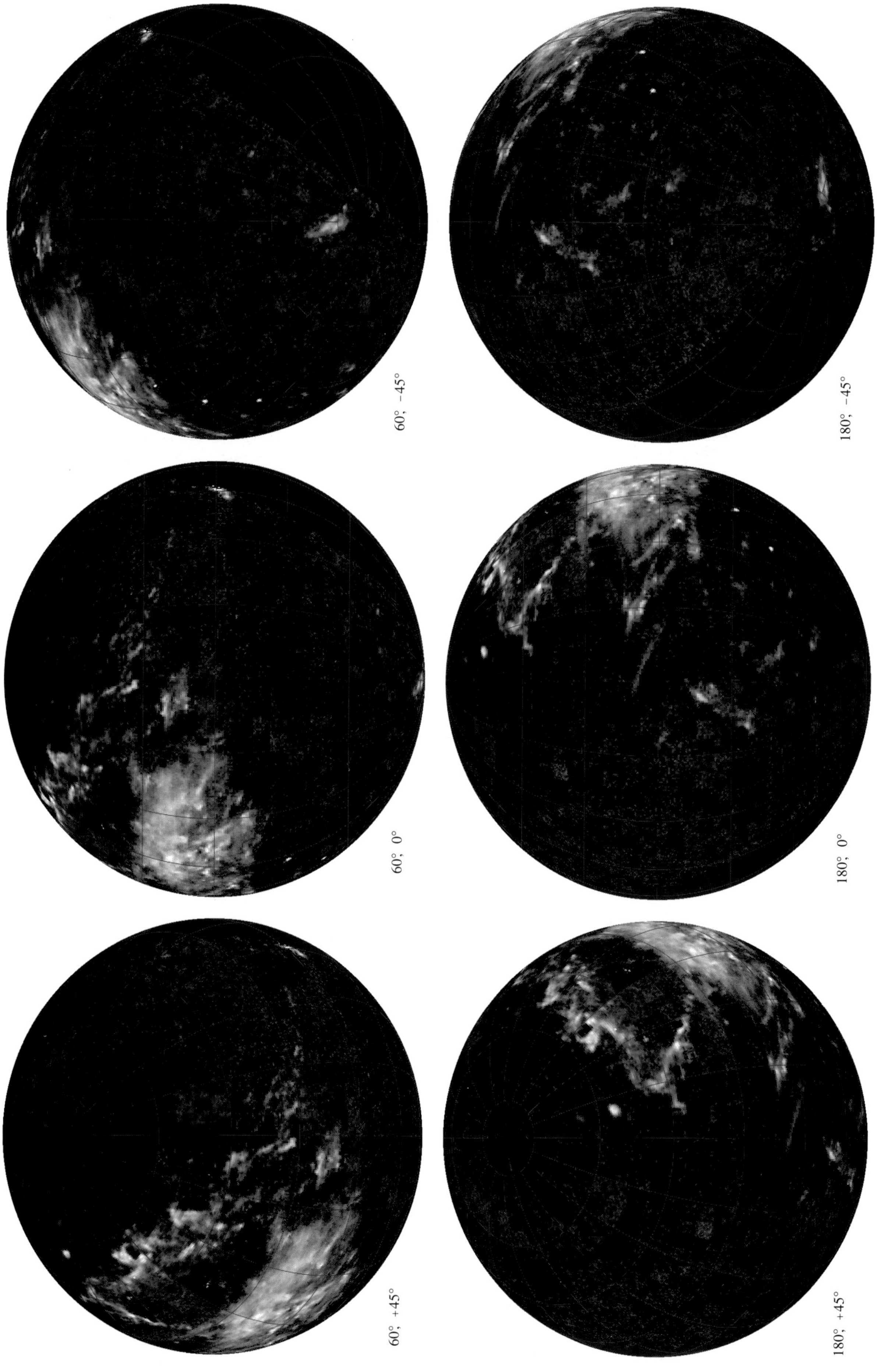

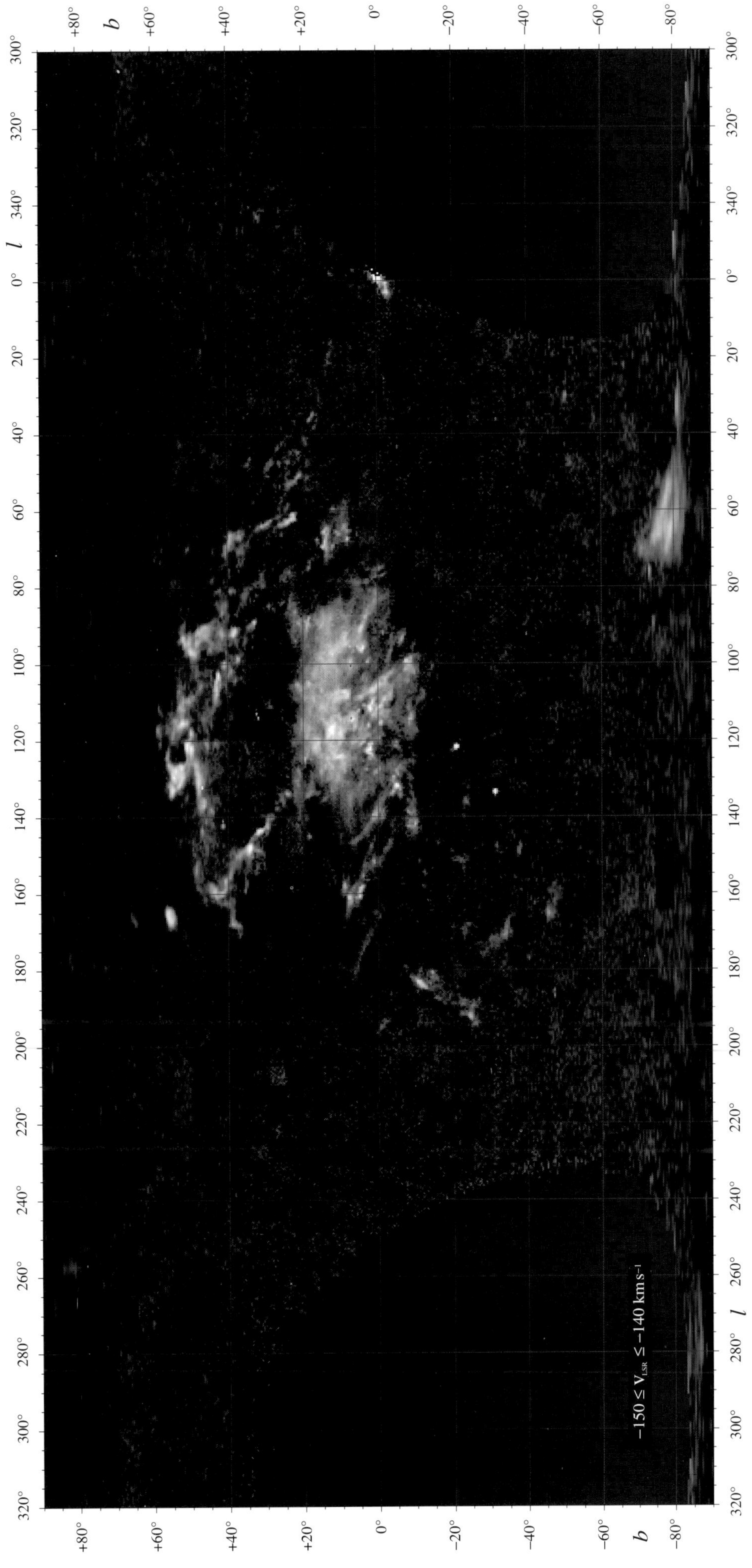

$-150 \leq V_{LSR} \leq -140$ km s^{-1}
l
b
300° 320° 340° 0° 20° 40° 60° 80° 100° 120° 140° 160° 180° 200° 220° 240° 260° 280° 300° 320°
+80° +60° +40° +20° 0° −20° −40° −60° −80°

$-140 \leq V_{LSR} \leq -130$ km s^{-1}

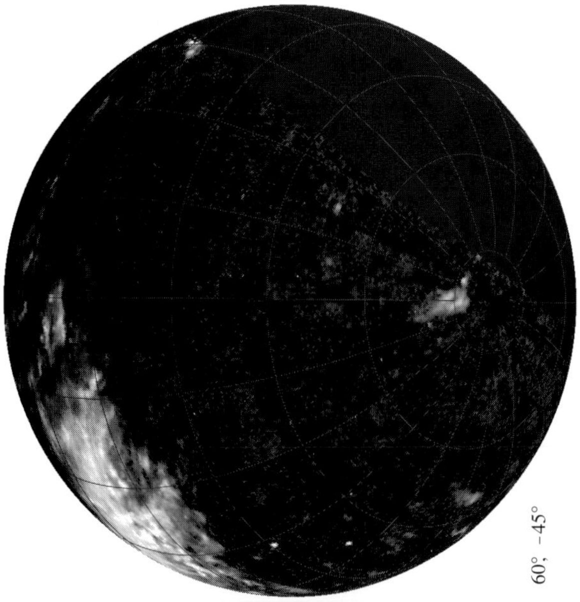

60°, −45°

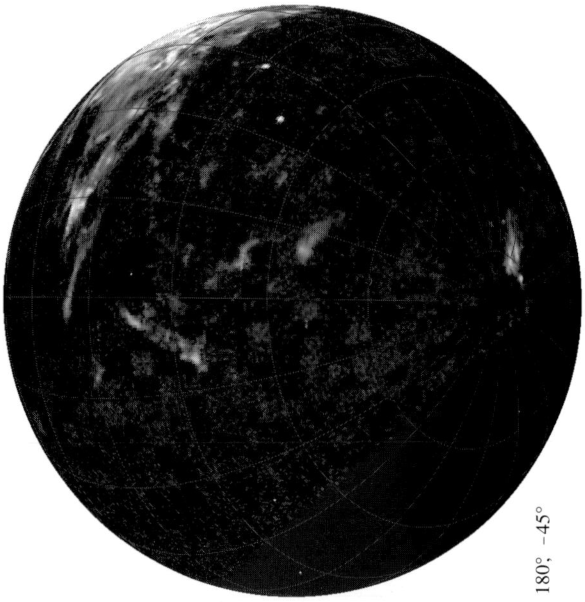

180°, −45°

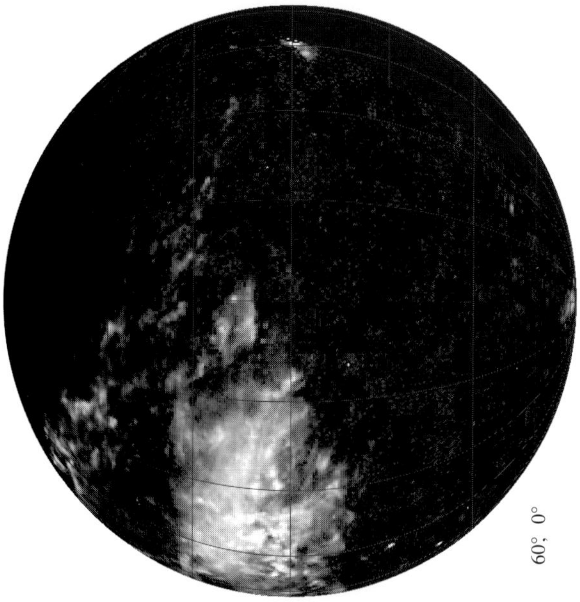

60°, 0°

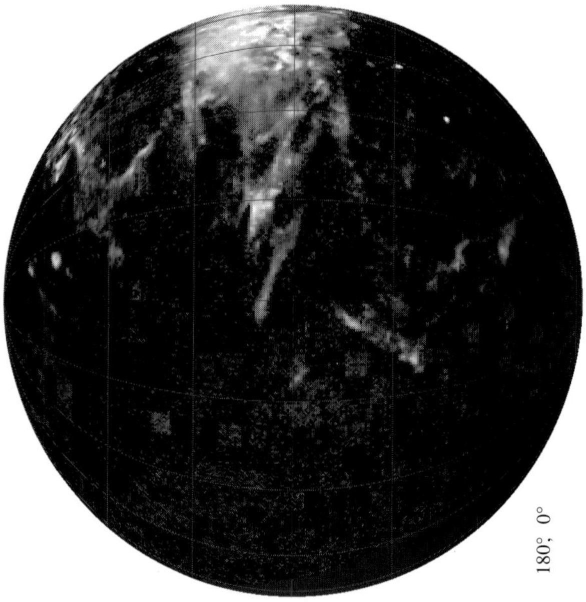

180°, 0°

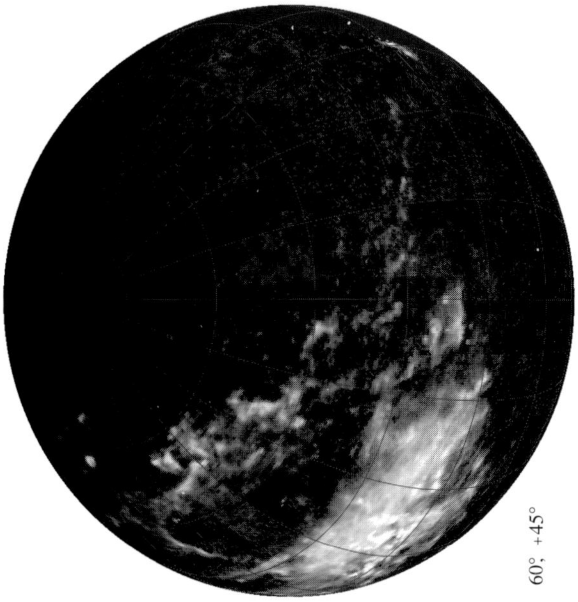

60°, +45°

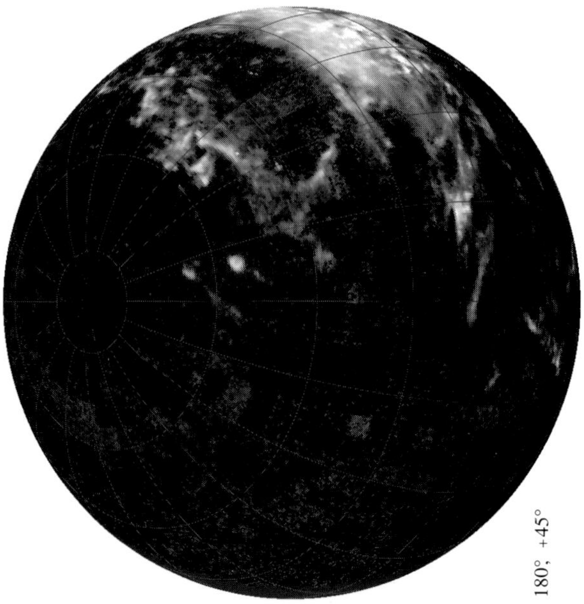

180°, +45°

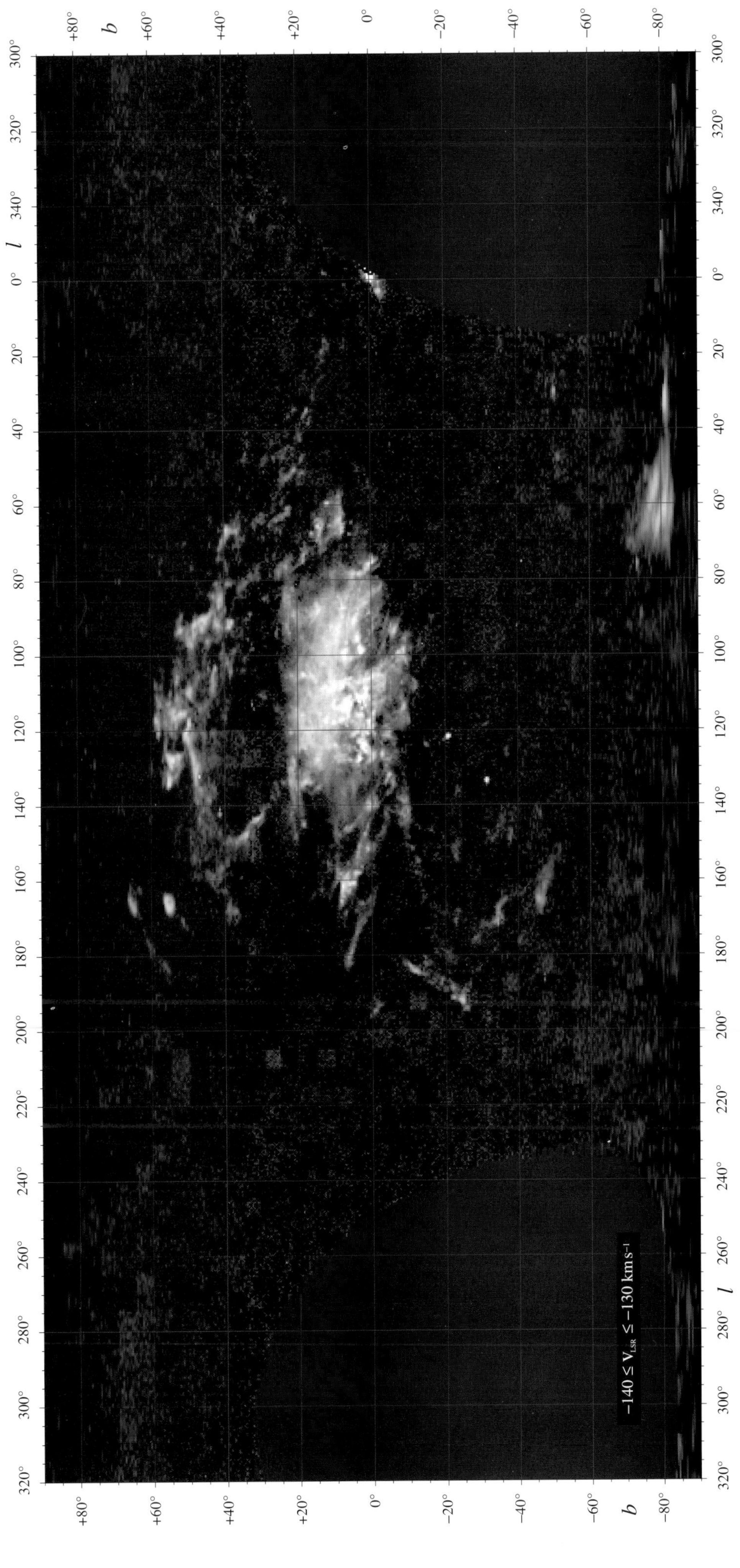
−140 ≤ V_LSR ≤ −130 kms⁻¹
l
b

$-130 \leq V_{LSR} \leq -120$ km s^{-1}

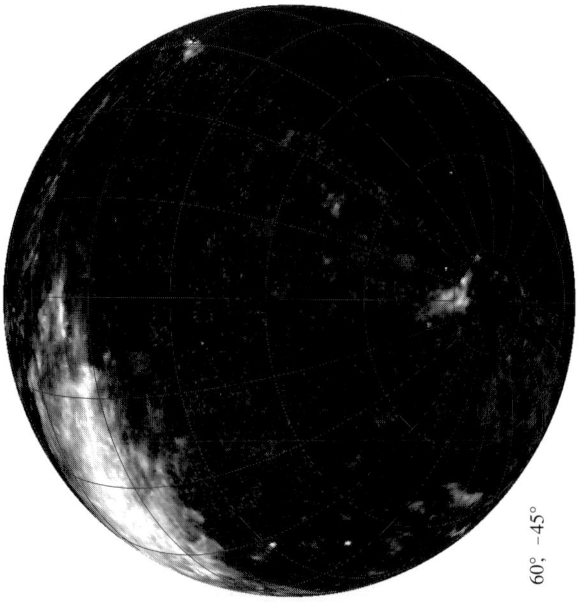

60°, −45°

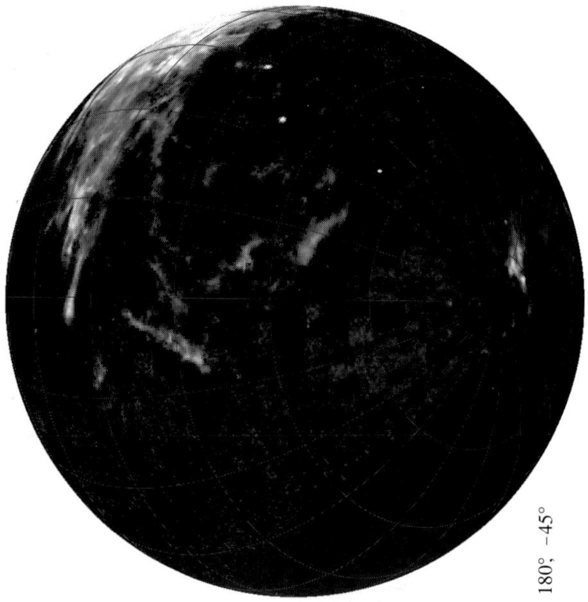

180°, −45°

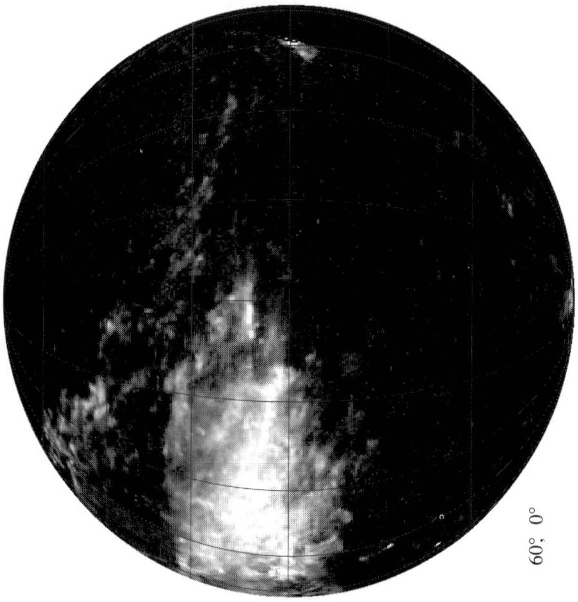

60°, 0°

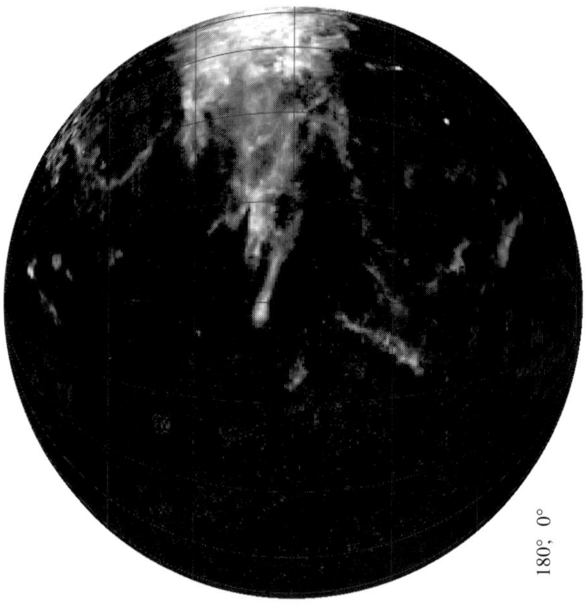

180°, 0°

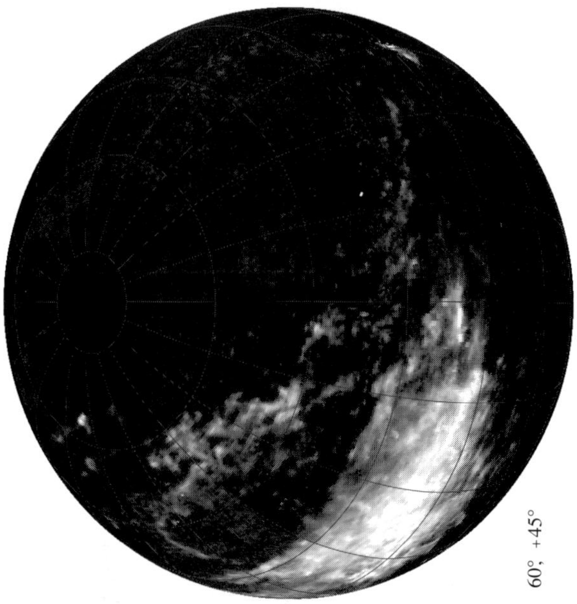

60°, +45°

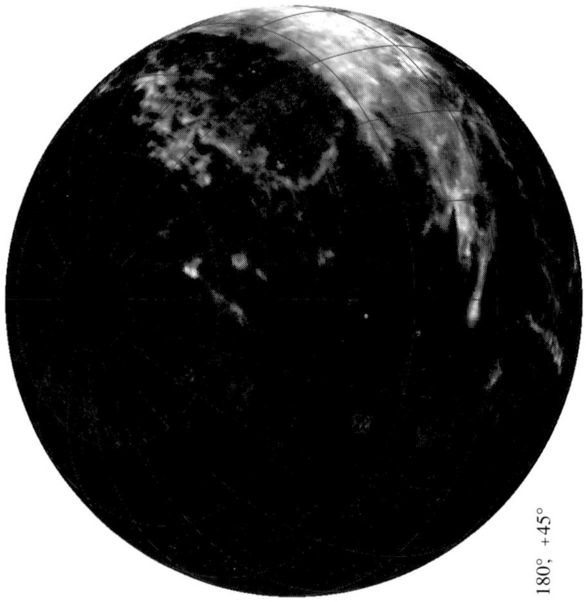

180°, +45°

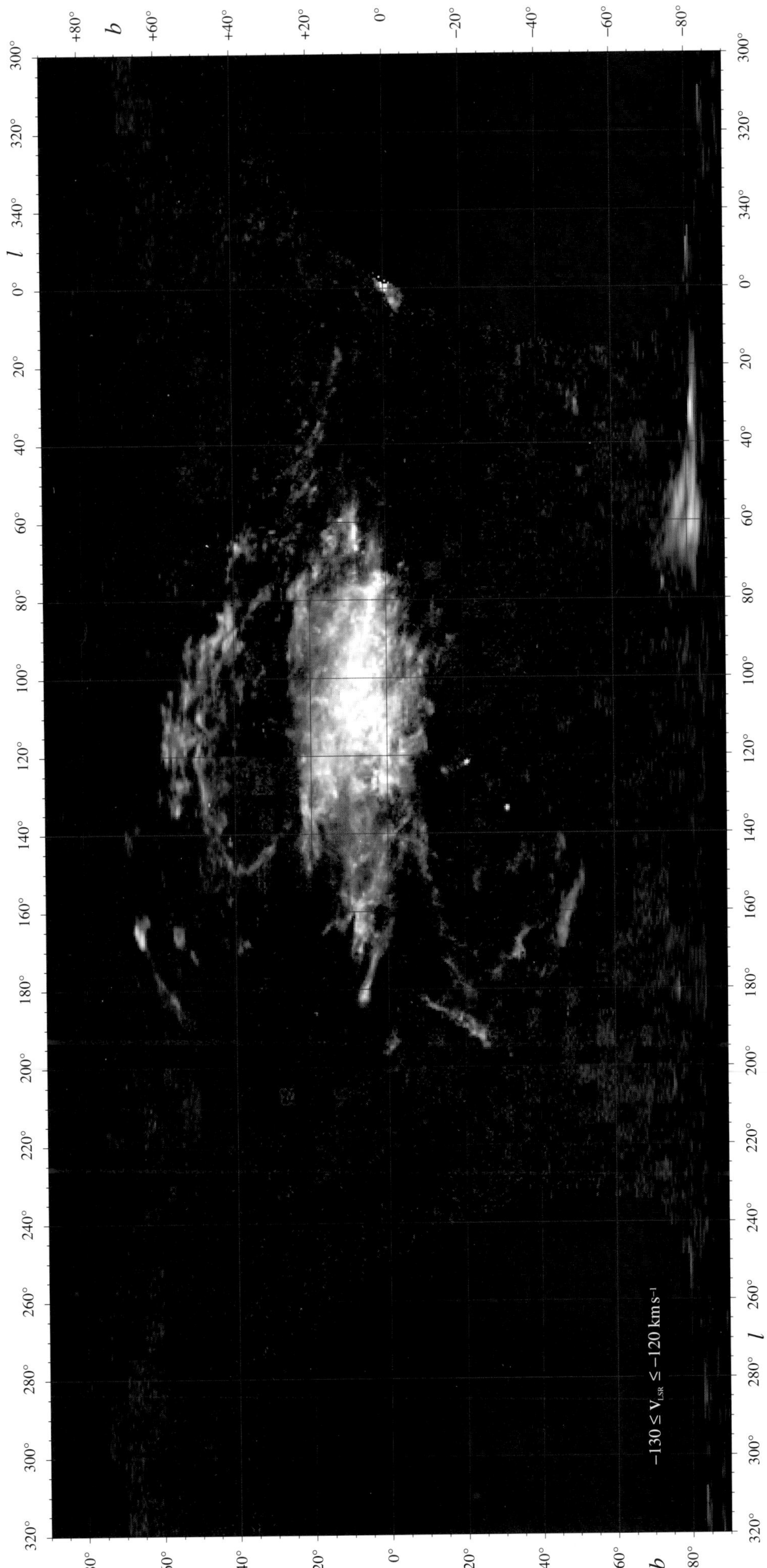
l
b
300° 320° 340° 0° 20° 40° 60° 80° 100° 120° 140° 160° 180° 200° 220° 240° 260° 280° 300° 320°
+80° +60° +40° +20° 0° −20° −40° −60° −80°
−130 ≤ V_LSR ≤ −120 km s−1

$-120 \le V_{LSR} \le -110$ km s^{-1}

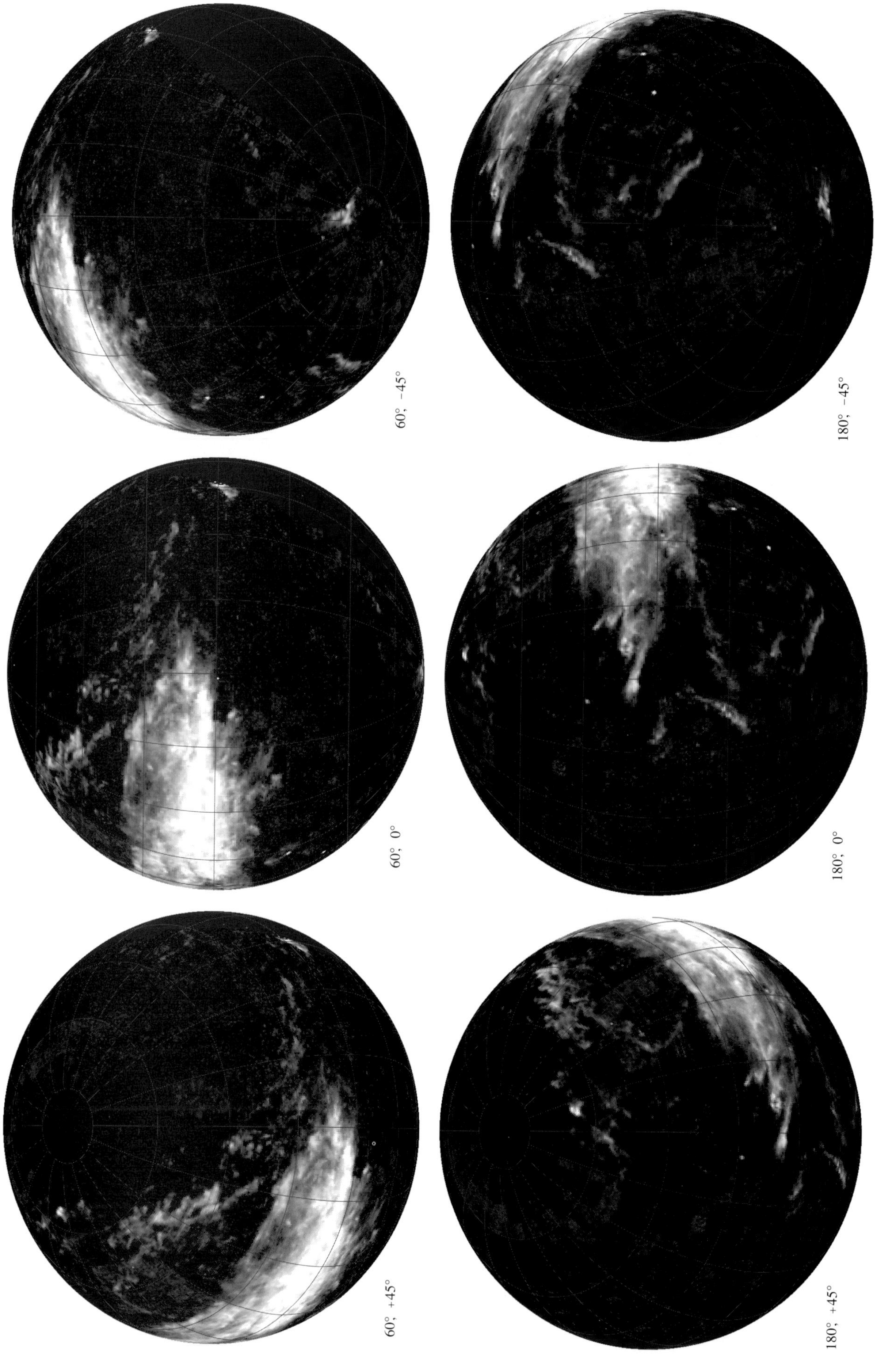

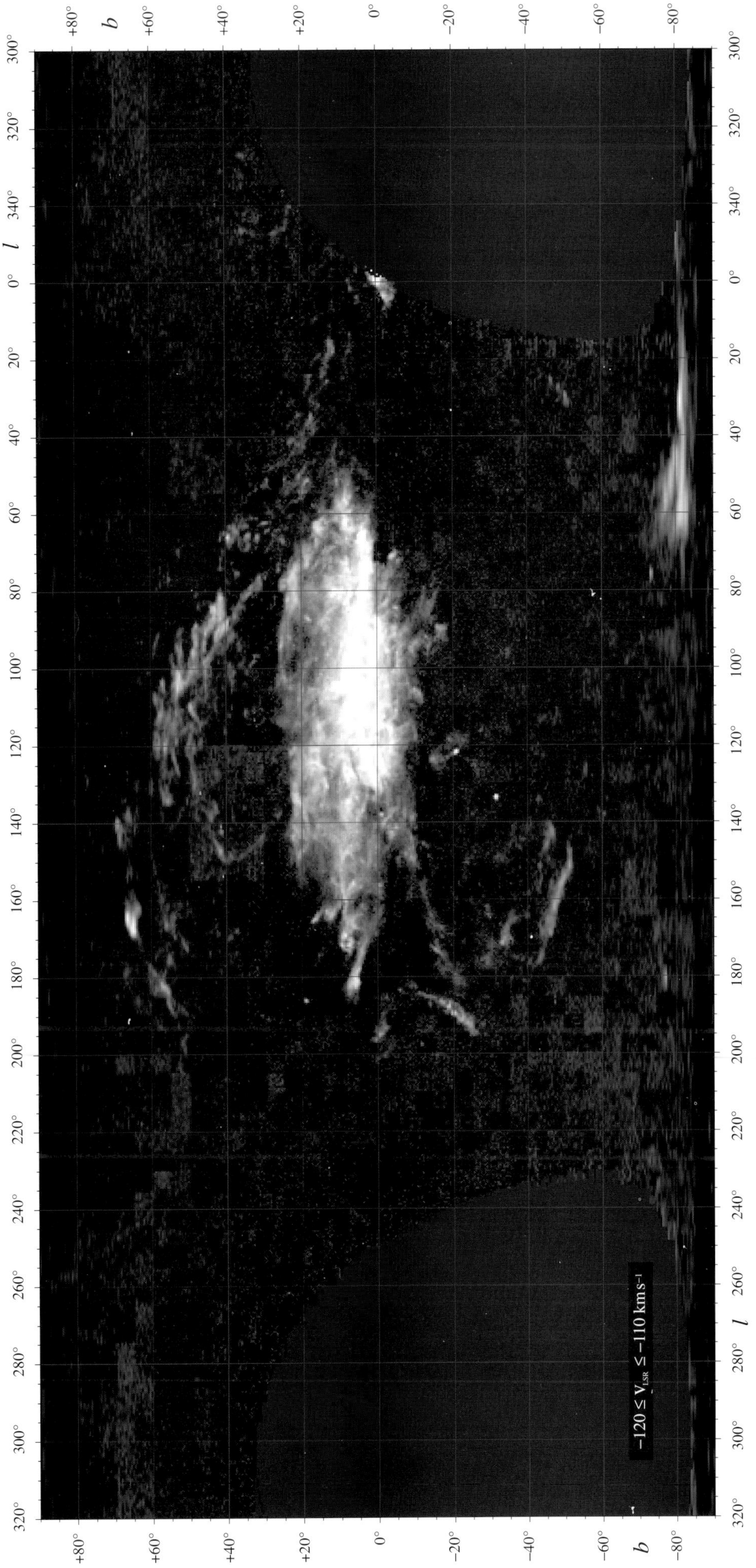

$-120 \le V_{LSR} \le -110$ km s^{-1}
l
b

$-110 \leq V_{LSR} \leq -100$ km s^{-1}

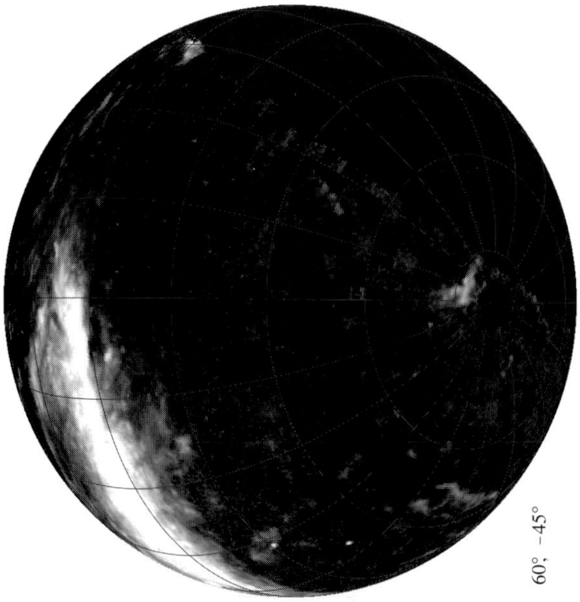

60°, −45°

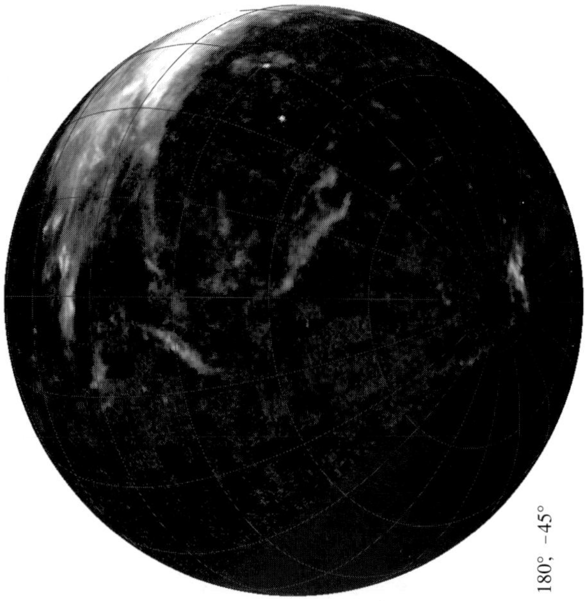

180°, −45°

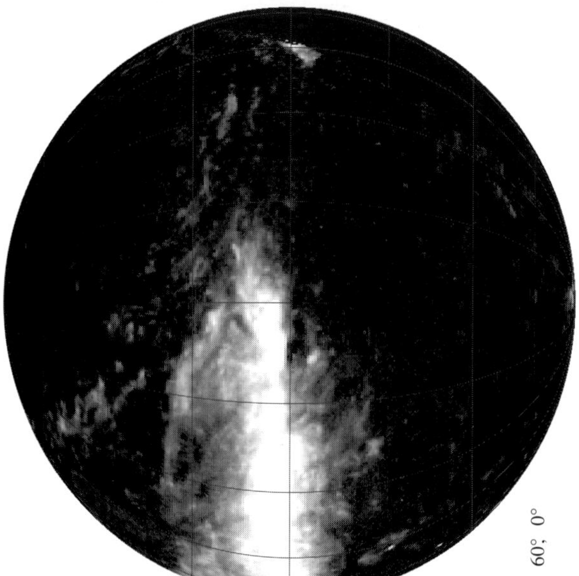

60°, 0°

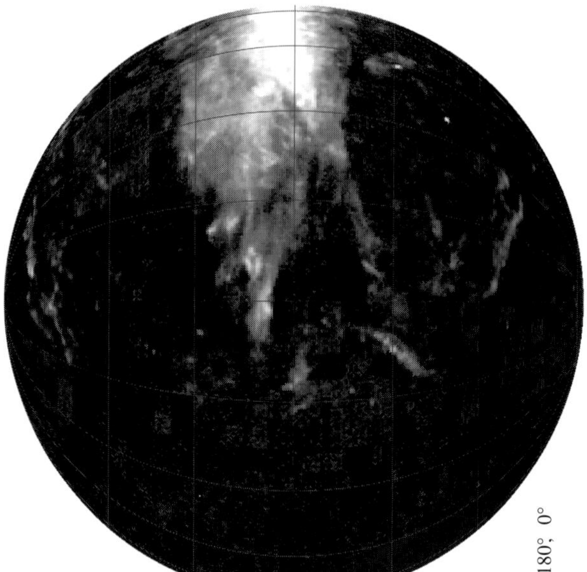

180°, 0°

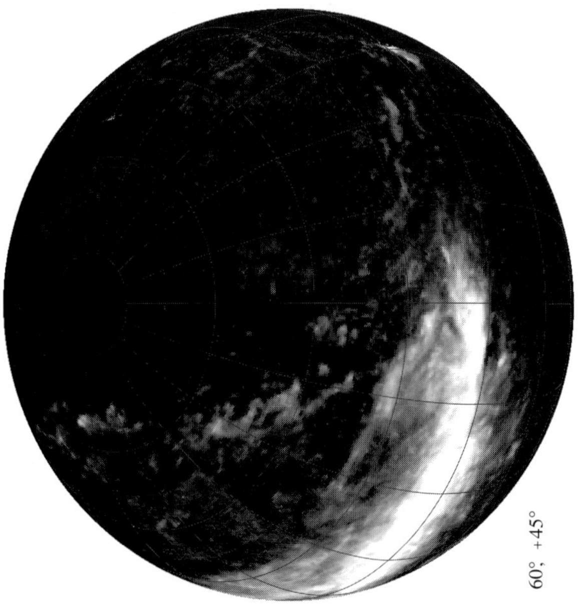

60°, +45°

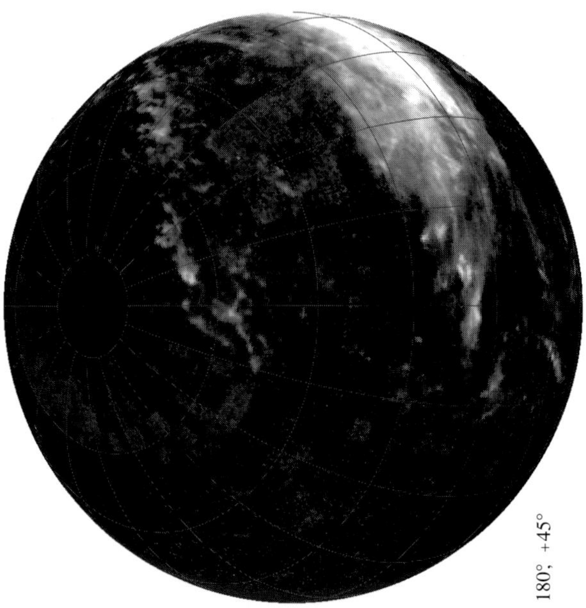

180°, +45°

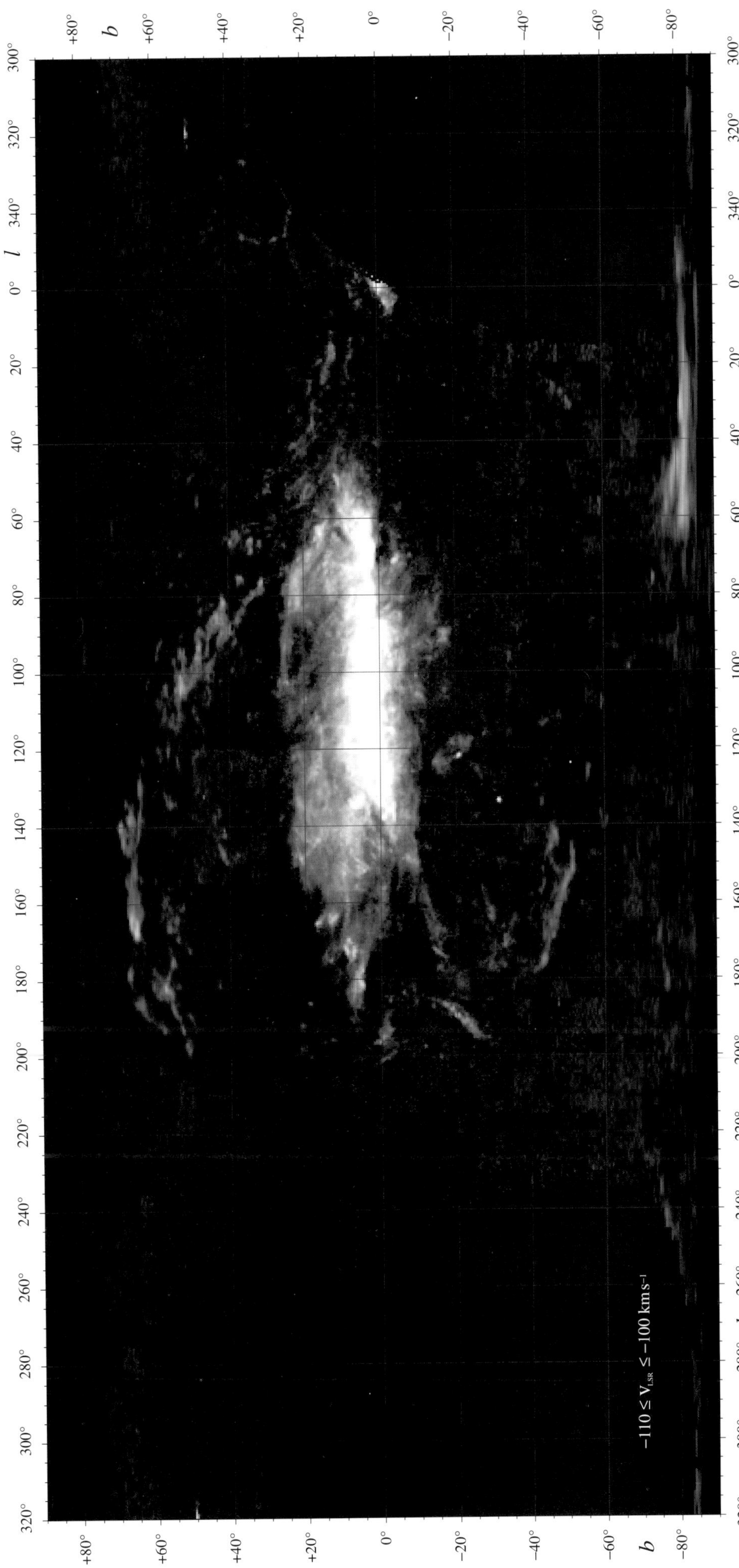
l
b
−110 ≤ V_LSR ≤ −100 kms⁻¹

$-100 \leq V_{LSR} \leq -90$ km s^{-1}

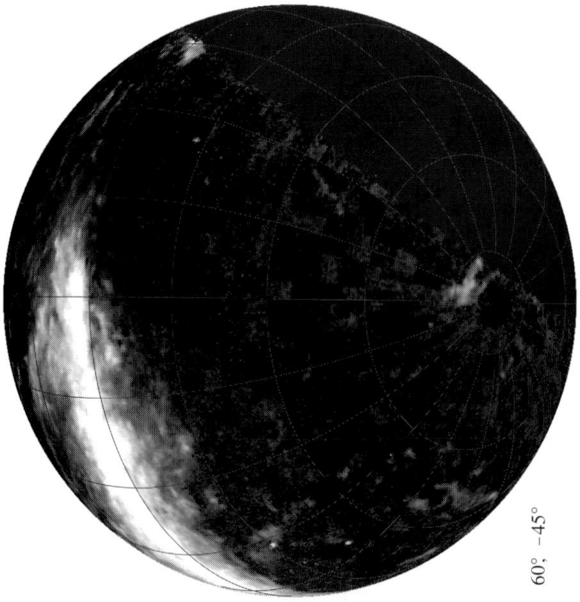

60°; −45°

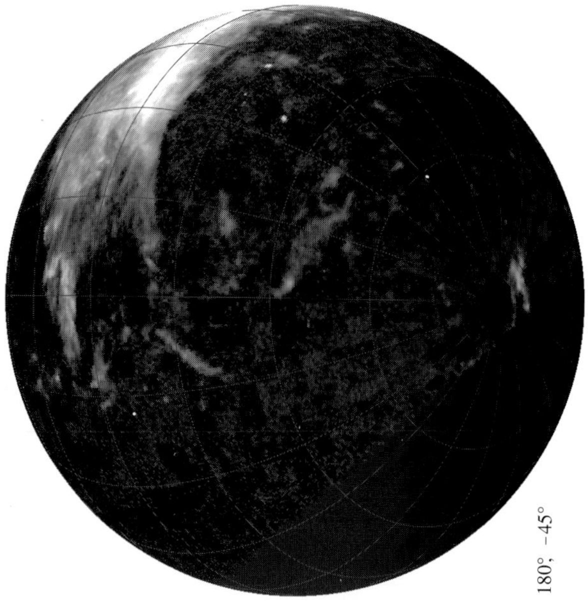

180°; −45°

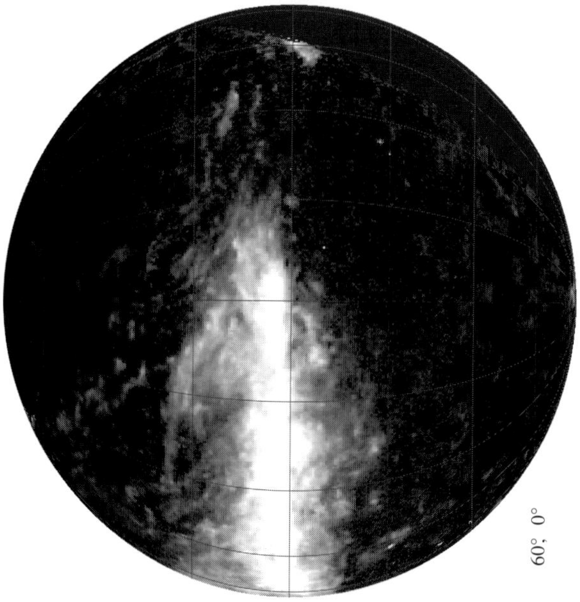

60°; 0°

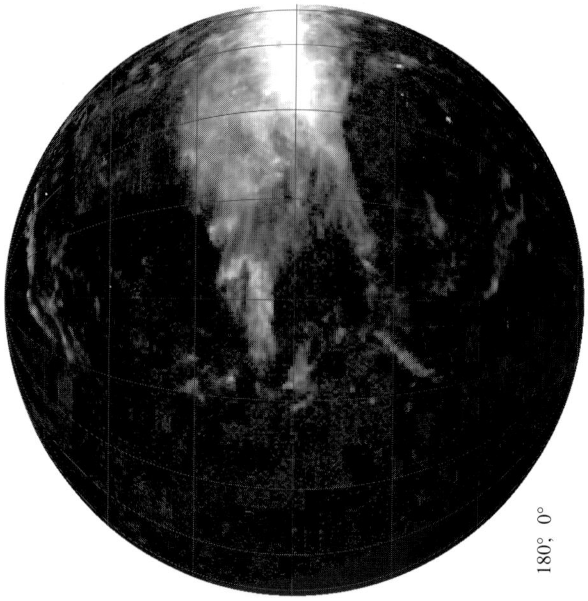

180°; 0°

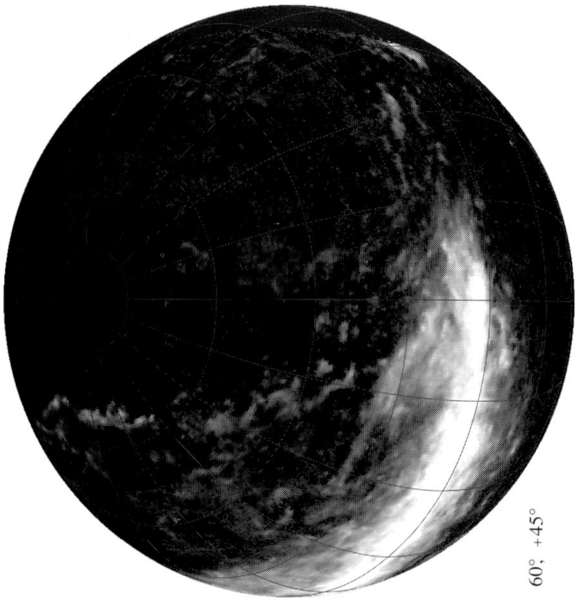

60°; +45°

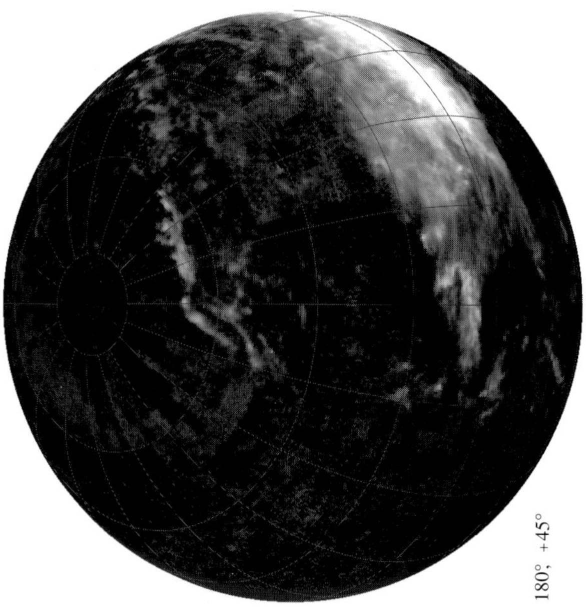

180°; +45°

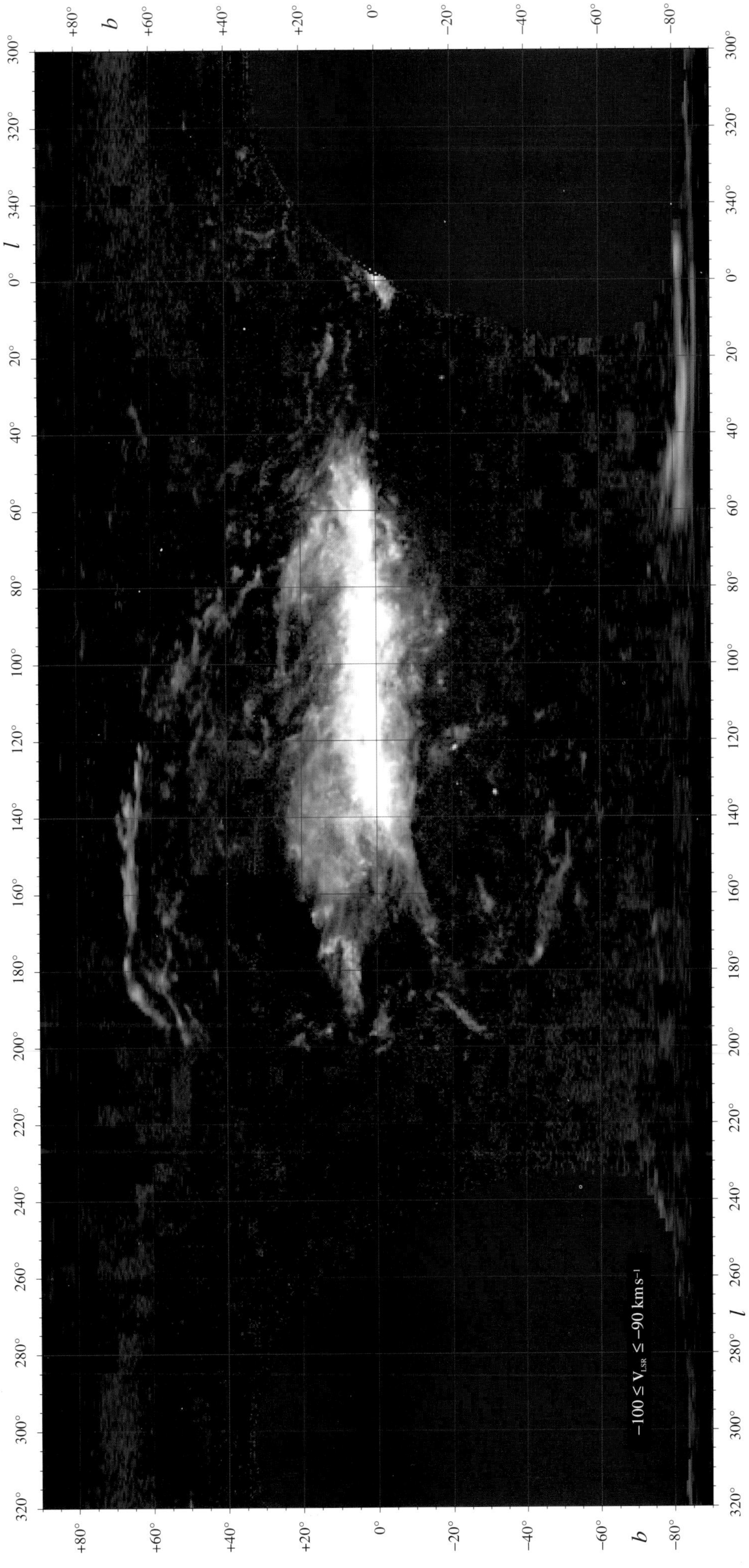

−100 ≤ V_LSR ≤ −90 km s−1
l
b

$-90 \leq V_{LSR} \leq -80$ km s^{-1}

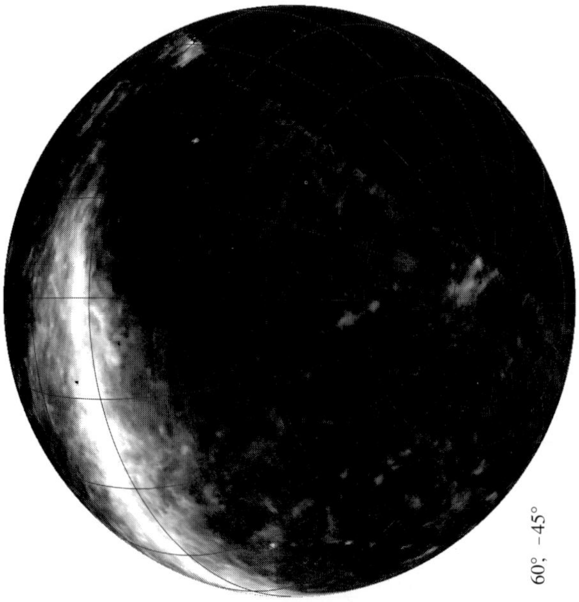

60°, −45°

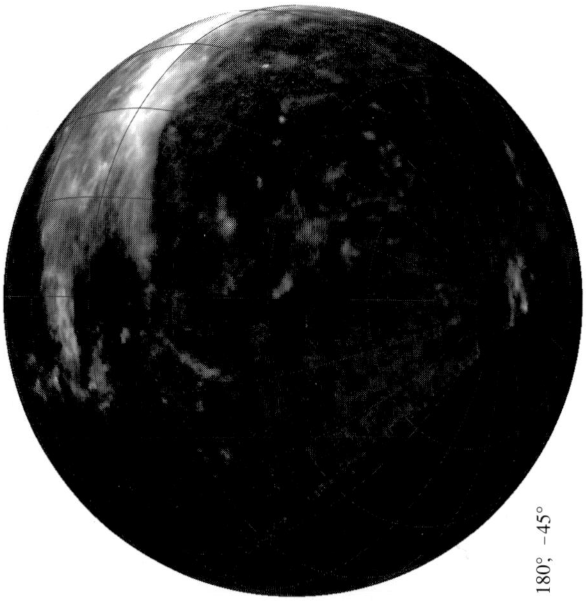

180°, −45°

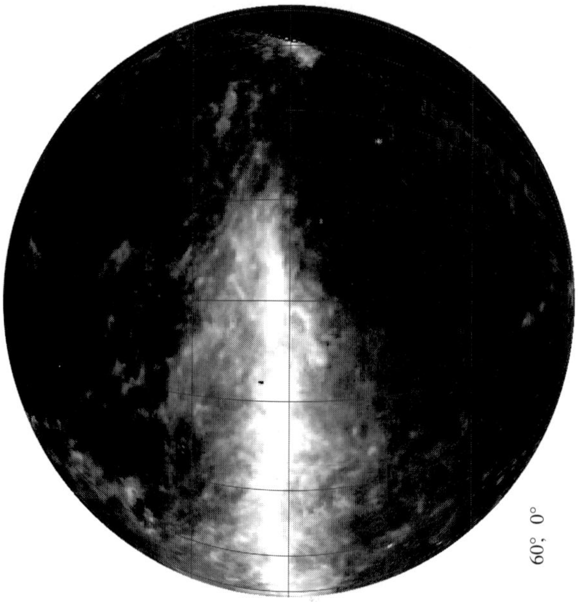

60°, 0°

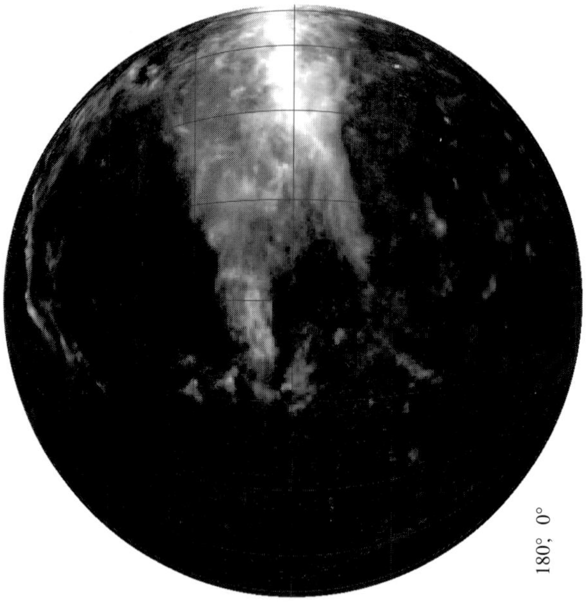

180°, 0°

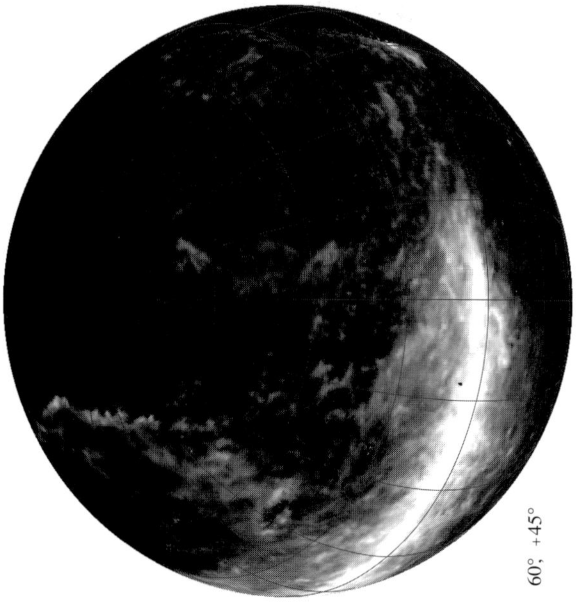

60°, +45°

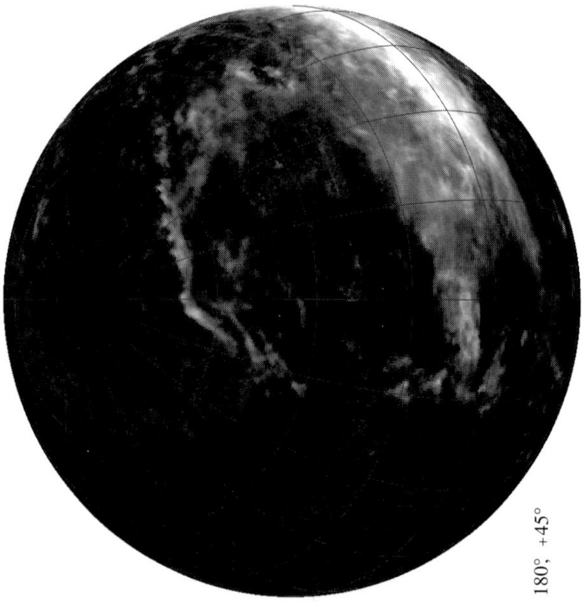

180°, +45°

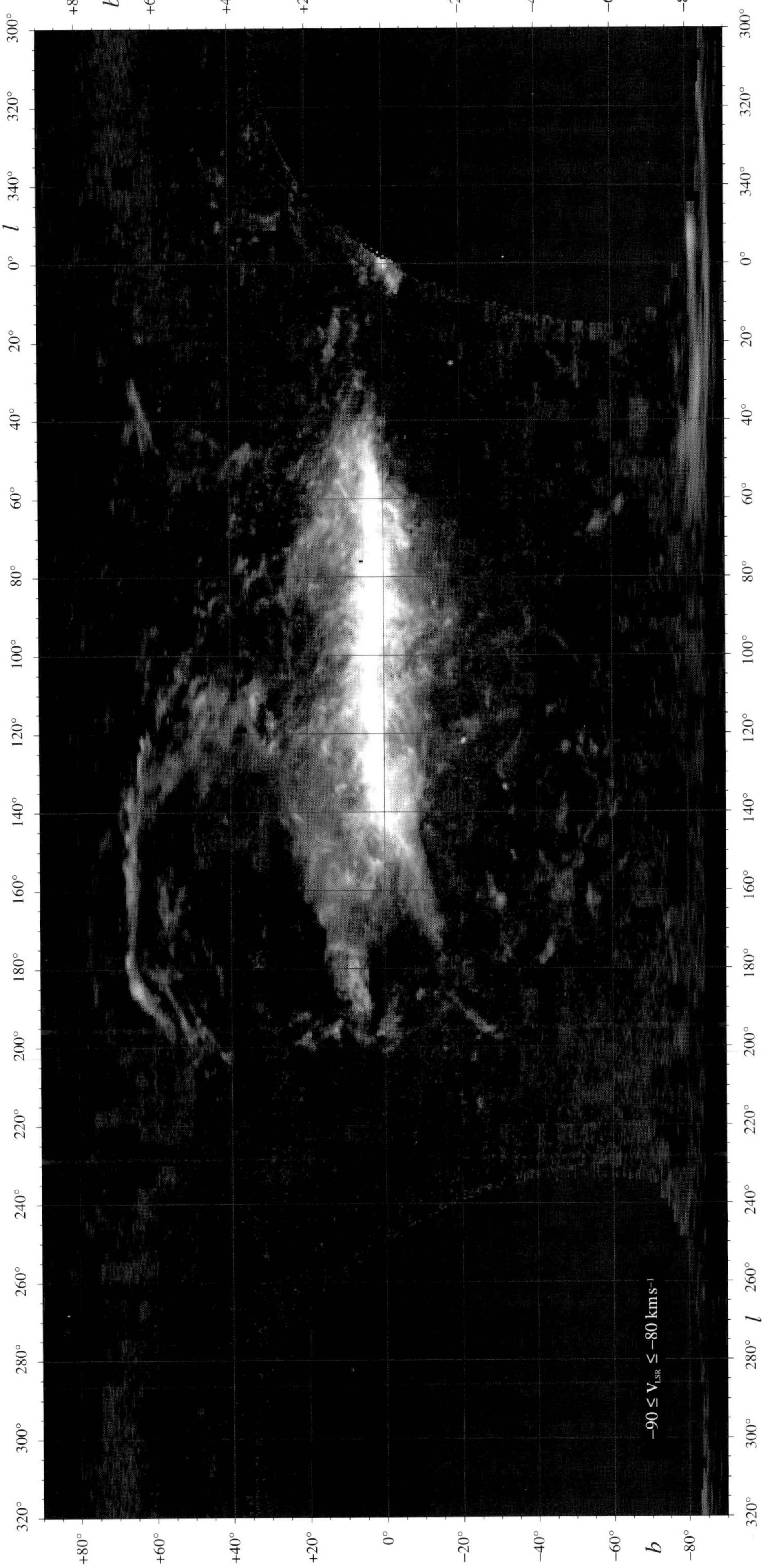
l
b
300° 320° 340° 0° 20° 40° 60° 80° 100° 120° 140° 160° 180° 200° 220° 240° 260° 280° 300° 320°
+80° +60° +40° +20° 0° −20° −40° −60° −80°
−90 ≤ V_LSR ≤ −80 kms−1

$-80 \leq V_{LSR} \leq -70$ km s^{-1}

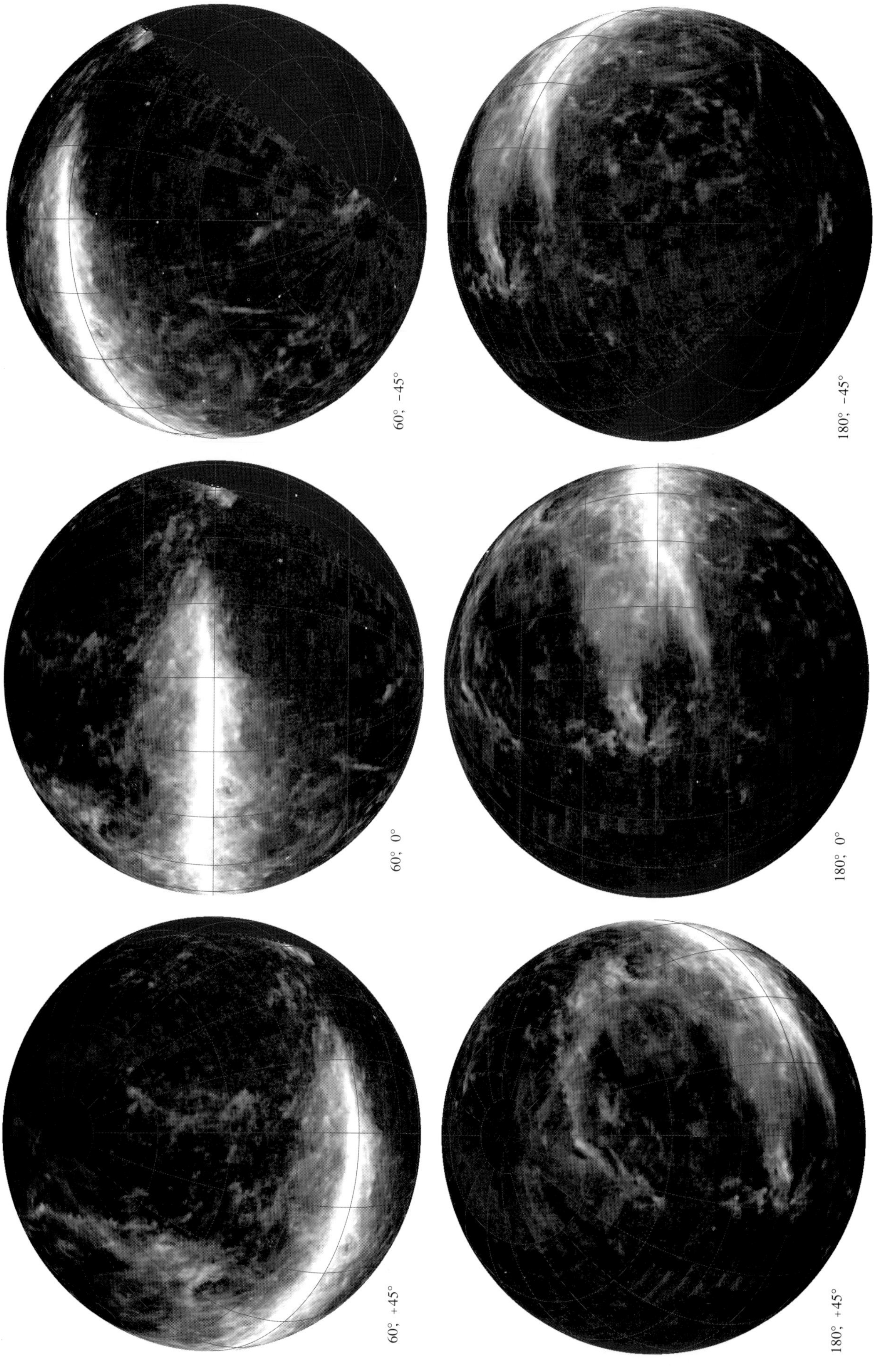

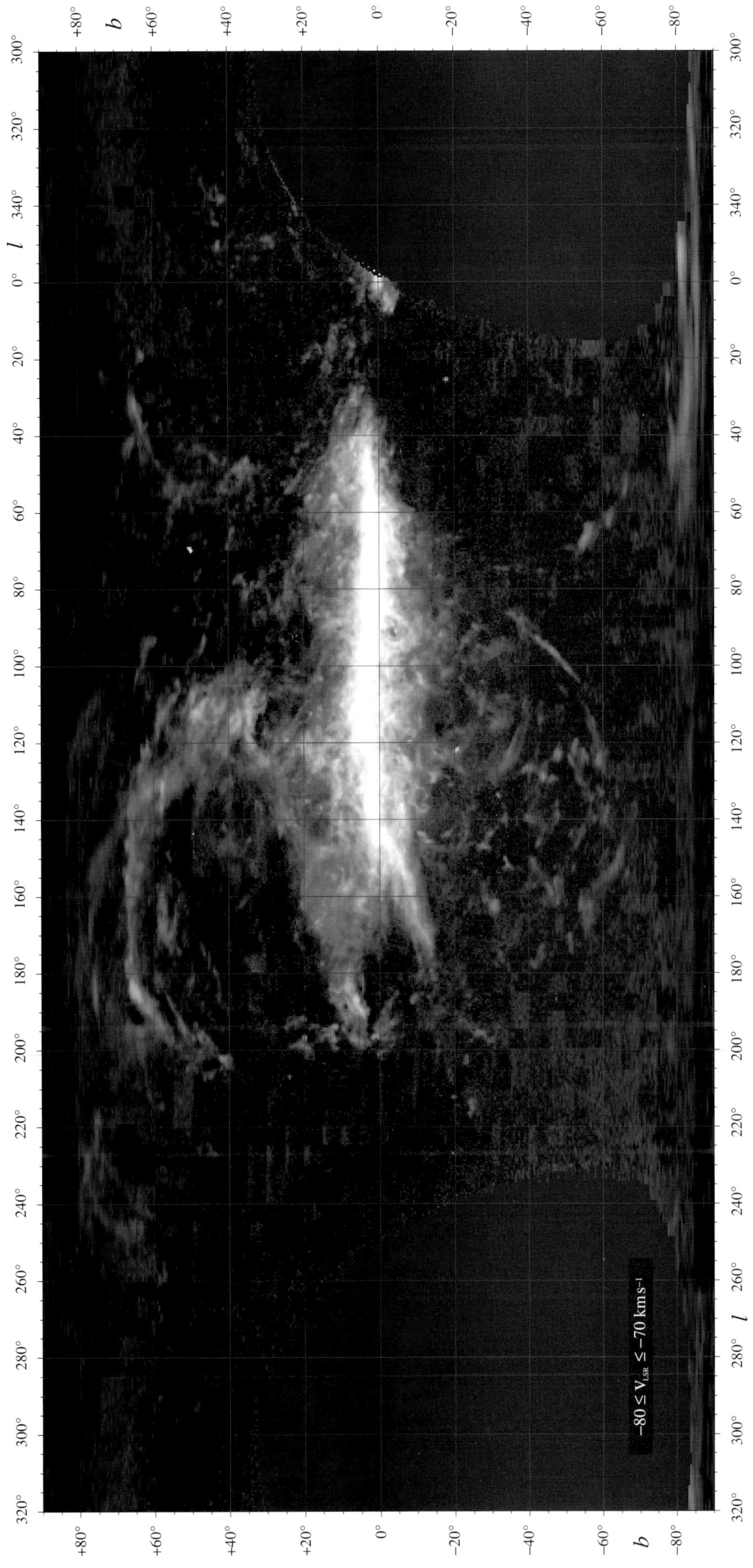

l
b
−80 ≤ V_LSR ≤ −70 kms−1
300° 320° 340° 0° 20° 40° 60° 80° 100° 120° 140° 160° 180° 200° 220° 240° 260° 280° 300° 320°
+80° +60° +40° +20° 0° −20° −40° −60° −80°

$-70 \le V_{LSR} \le -68$ km s^{-1}

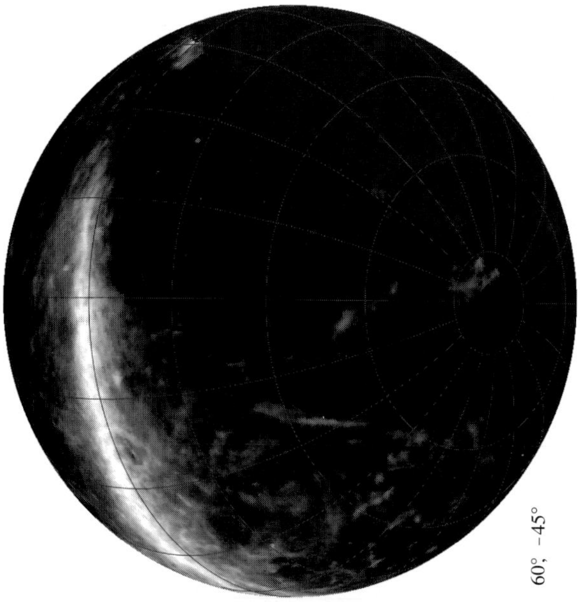

60°, −45°

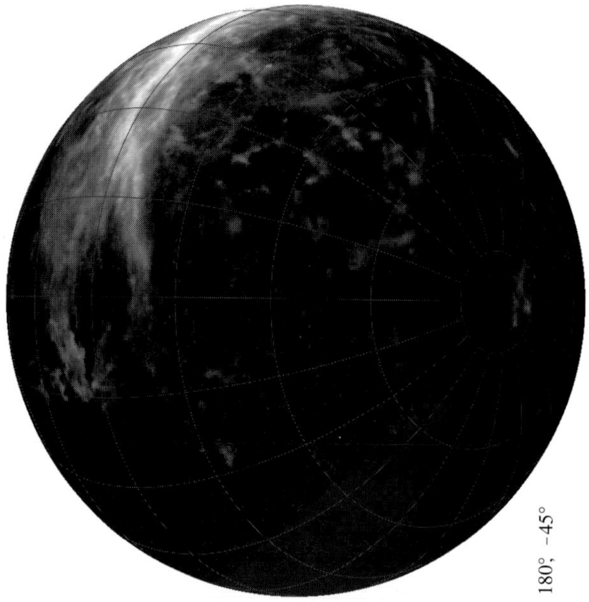

180°, −45°

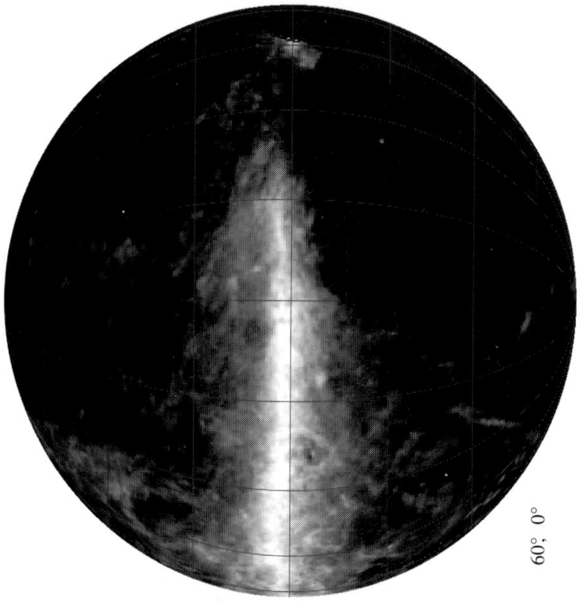

60°, 0°

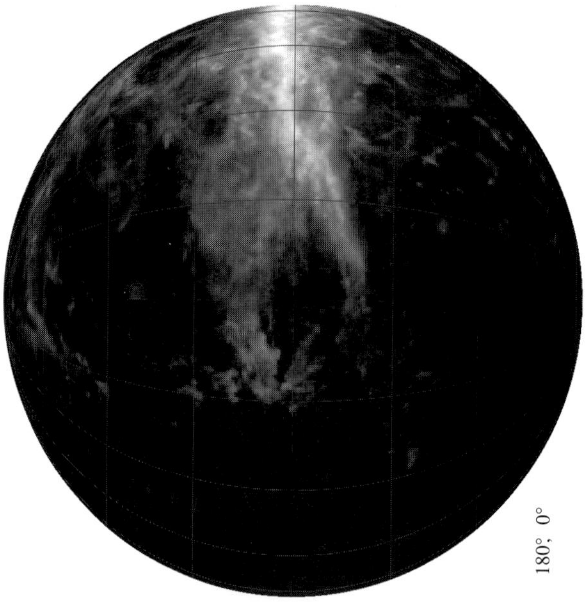

180°, 0°

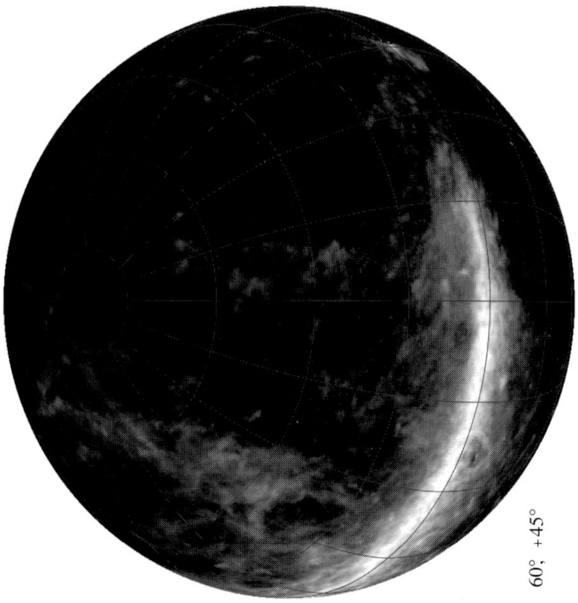

60°, +45°

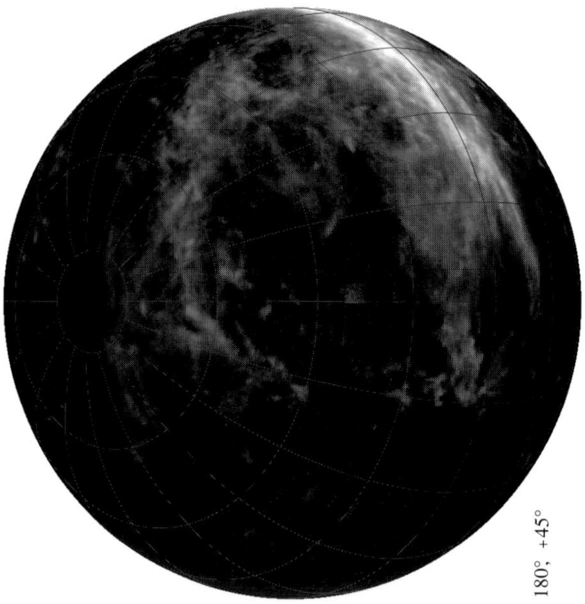

180°, +45°

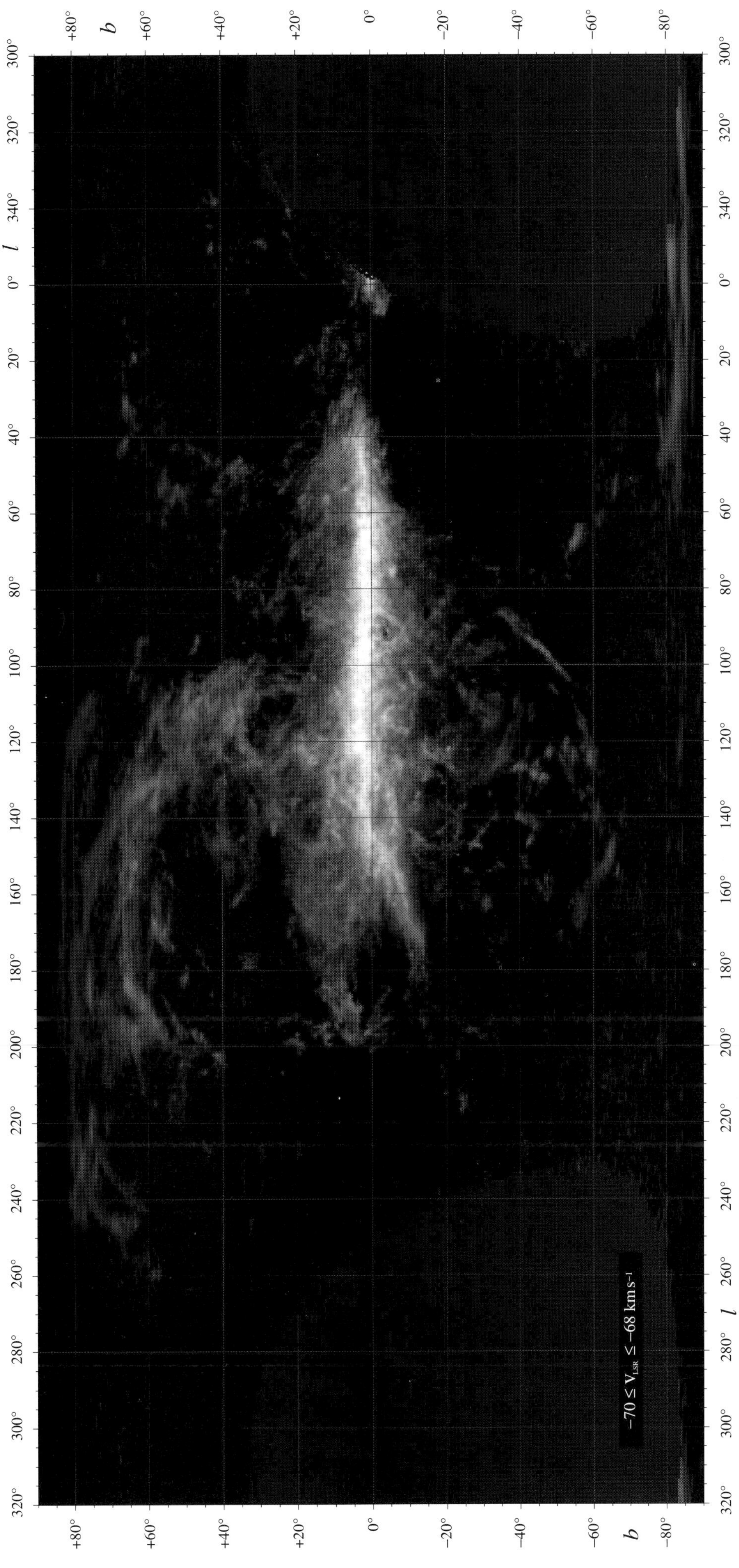

l
b
300° 320° 340° 0° 20° 40° 60° 80° 100° 120° 140° 160° 180° 200° 220° 240° 260° 280° 300° 320°
+80° +60° +40° +20° 0° −20° −40° −60° −80°
$-70 \le V_{LSR} \le -68$ kms−1

$-68 \leq V_{LSR} \leq -66$ km s^{-1}

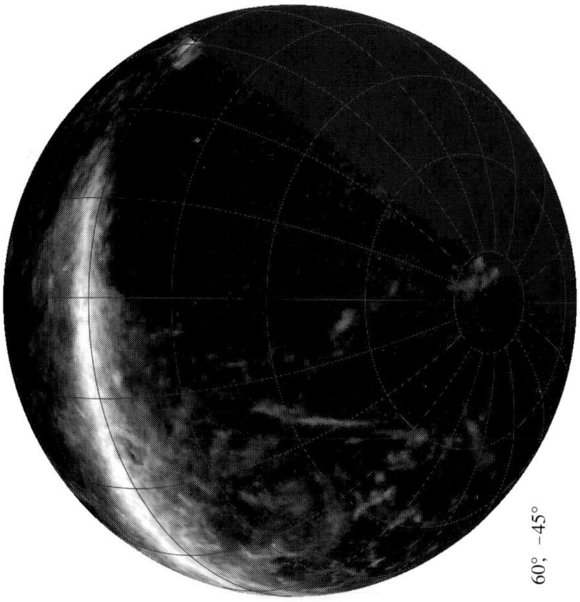

60°, −45°

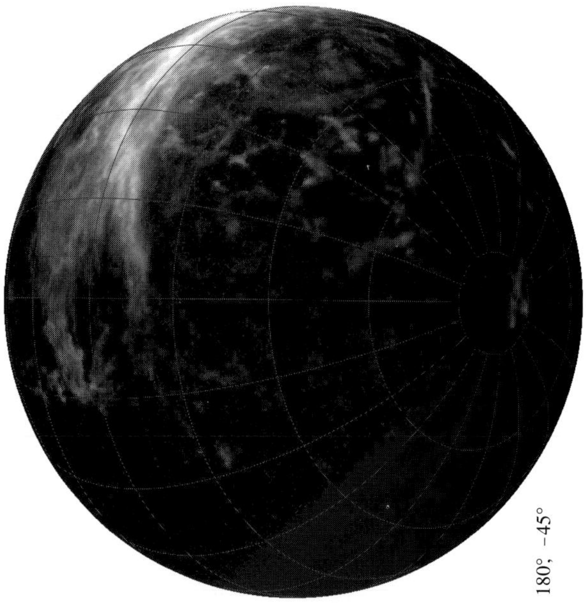

180°, −45°

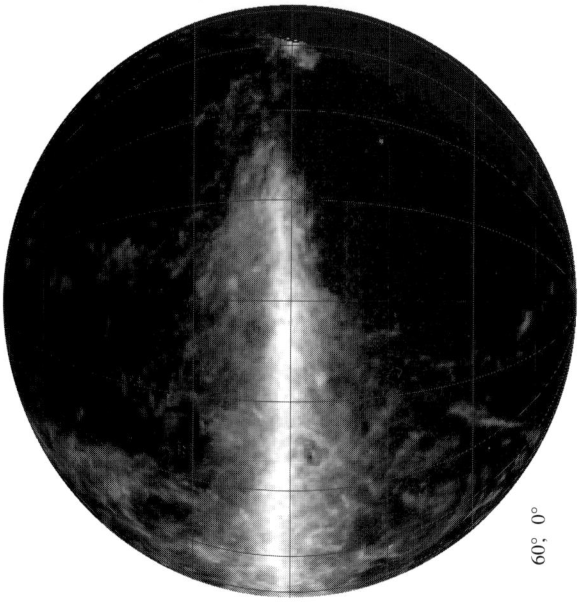

60°, 0°

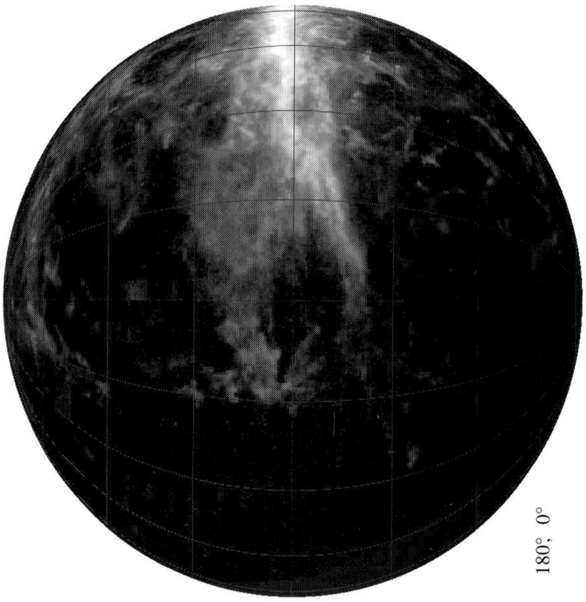

180°, 0°

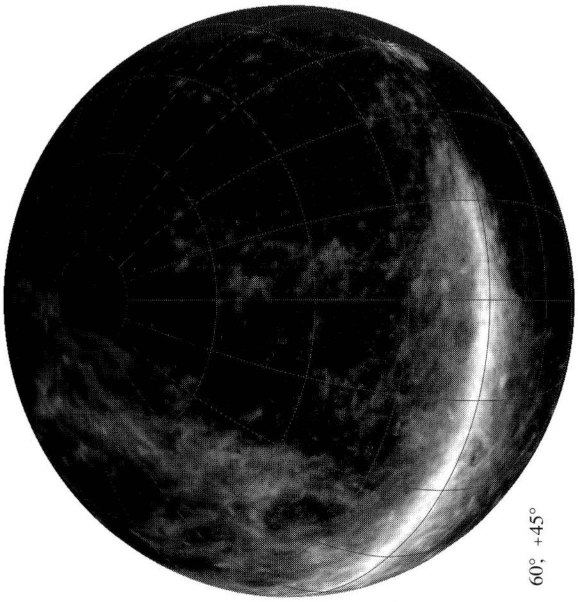

60°, +45°

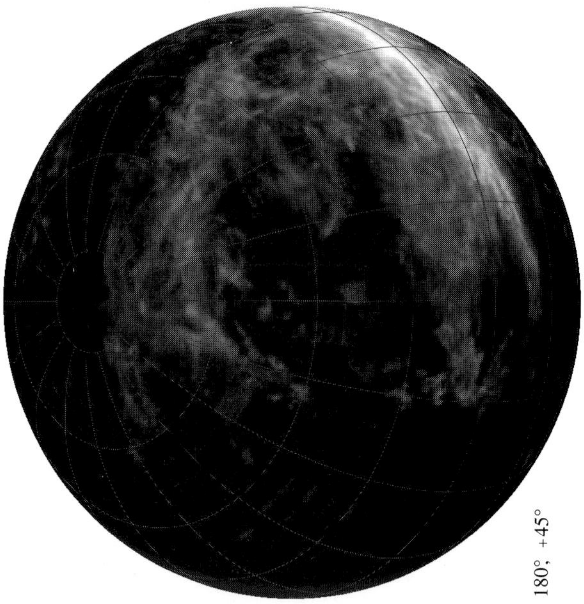

180°, +45°

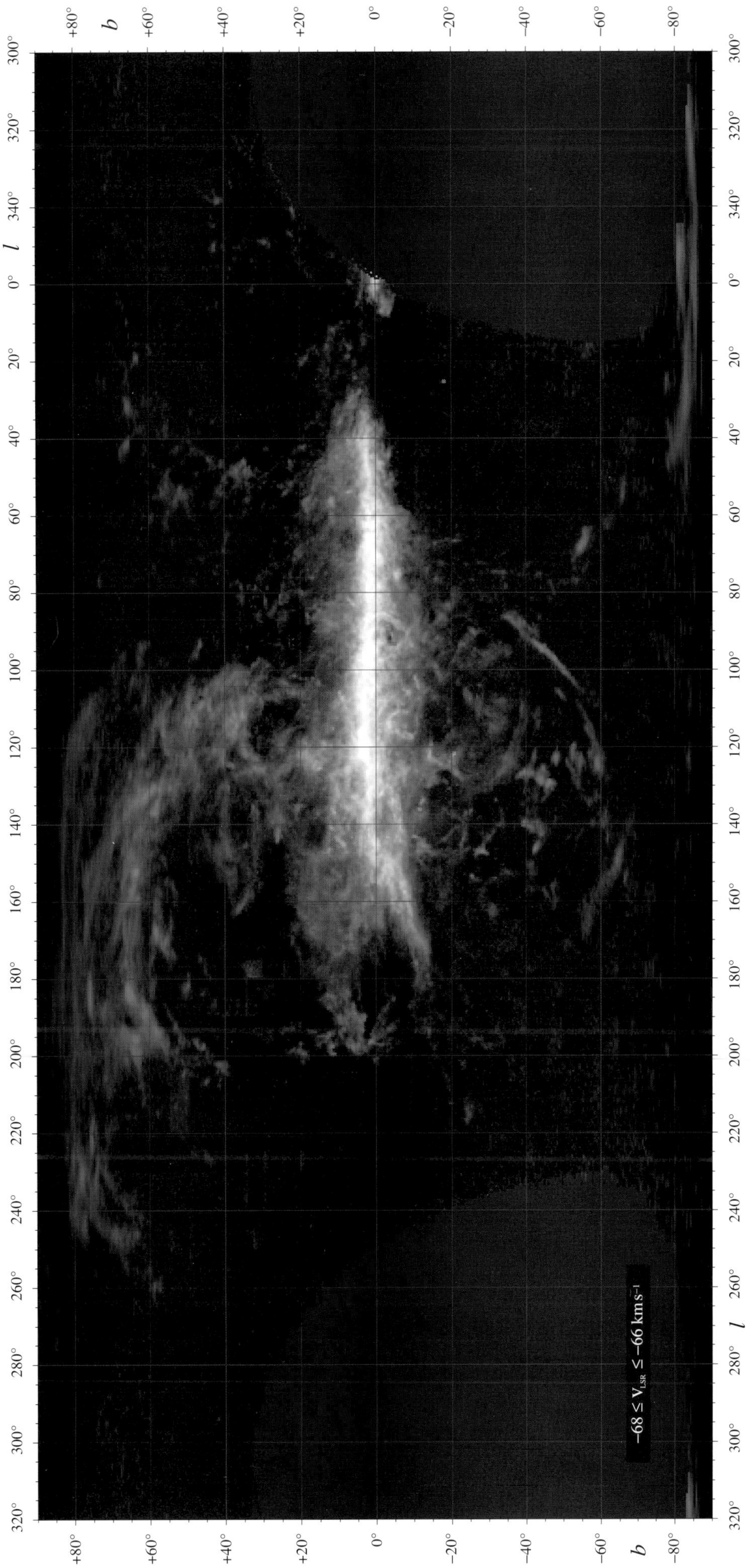
l
b
+80°
+60°
+40°
+20°
0°
−20°
−40°
−60°
−80°
300°
320°
340°
0°
20°
40°
60°
80°
100°
120°
140°
160°
180°
200°
220°
240°
260°
280°
$-68 \le V_{LSR} \le -66$ km s^{-1}

$-66 \le V_{LSR} \le -64$ km s^{-1}

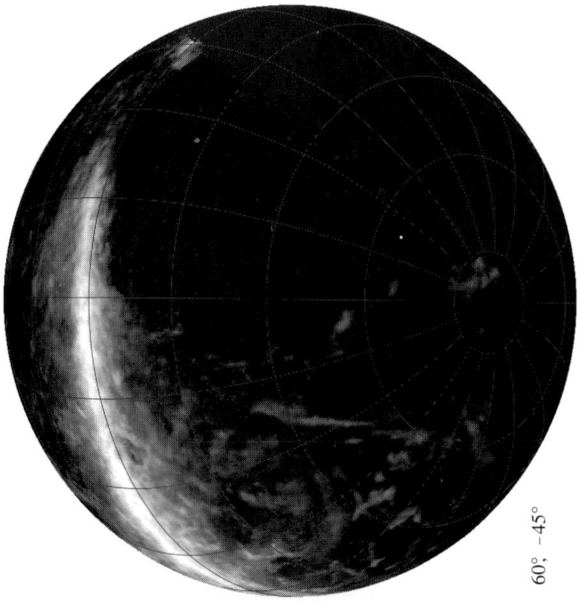

60°, −45°

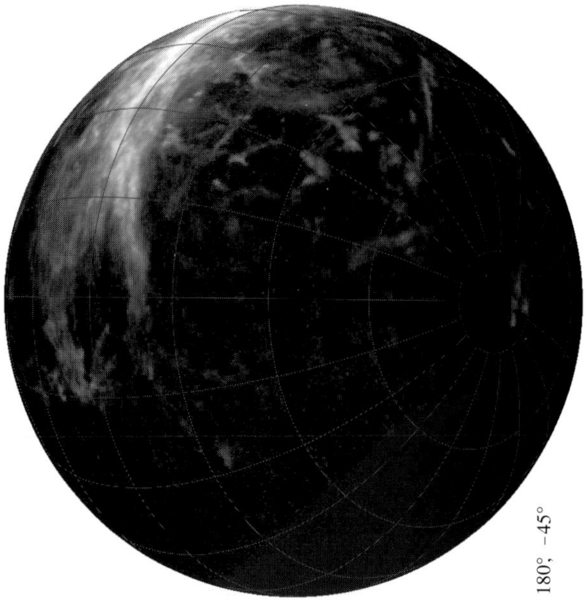

180°, −45°

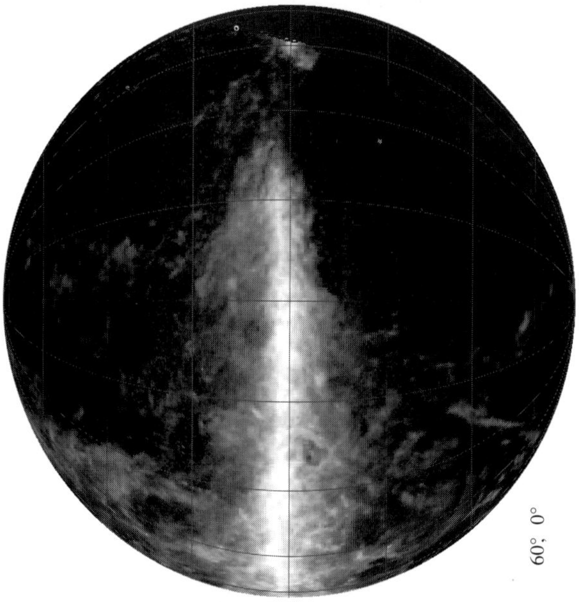

60°, 0°

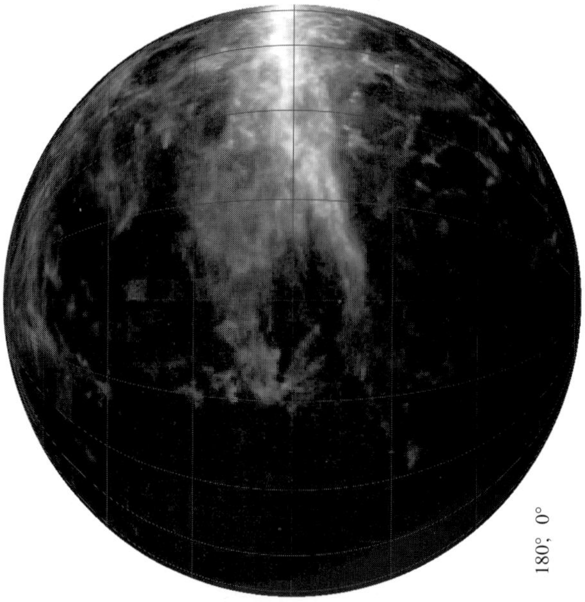

180°, 0°

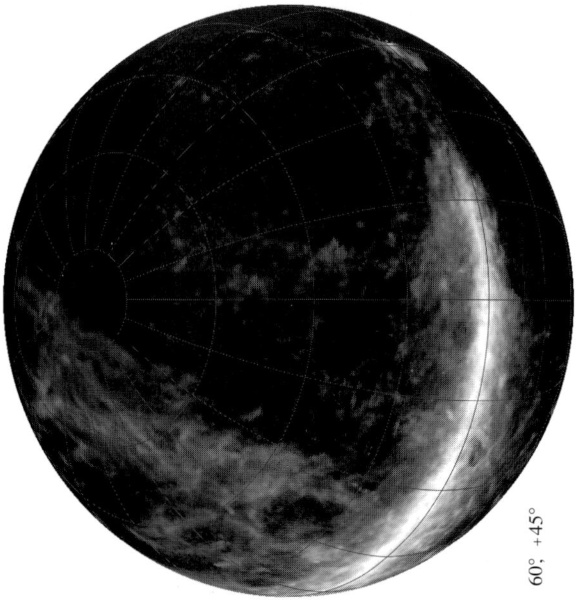

60°, +45°

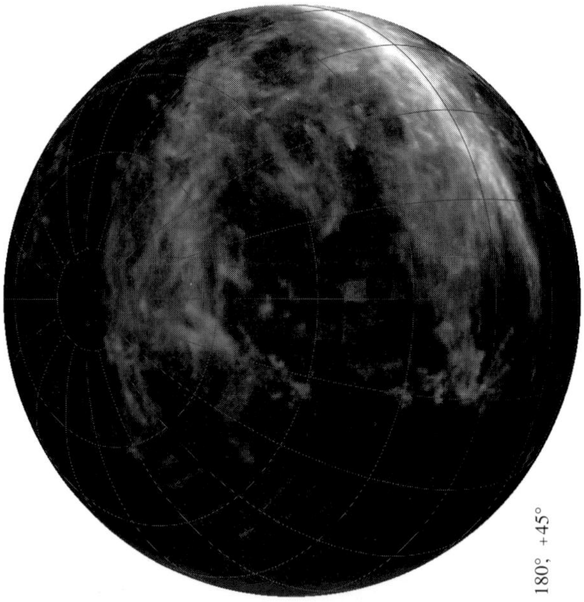

180°, +45°

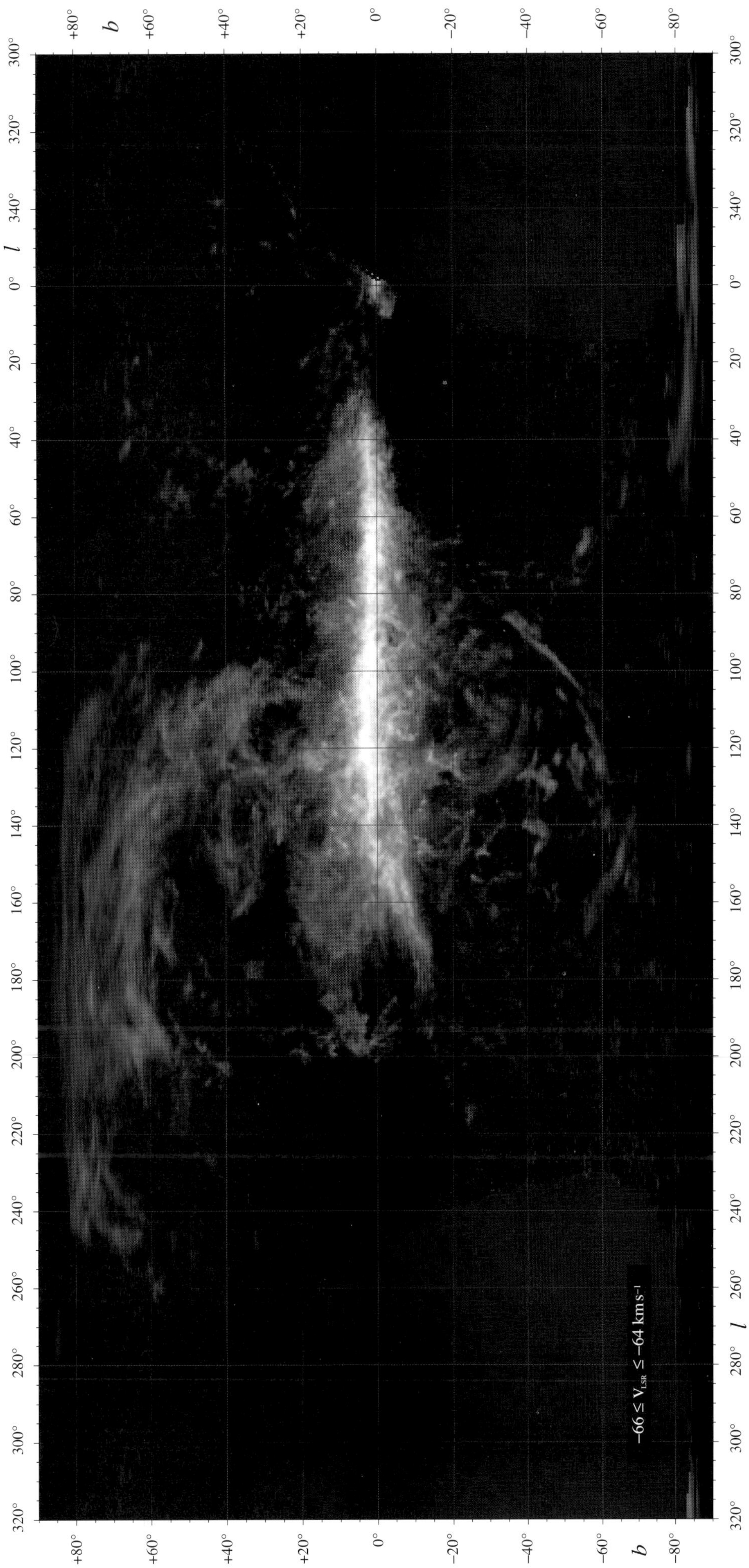
$-66 \le V_{LSR} \le -64$ kms^{-1}
l
b

$-64 \leq V_{LSR} \leq -62$ km s^{-1}

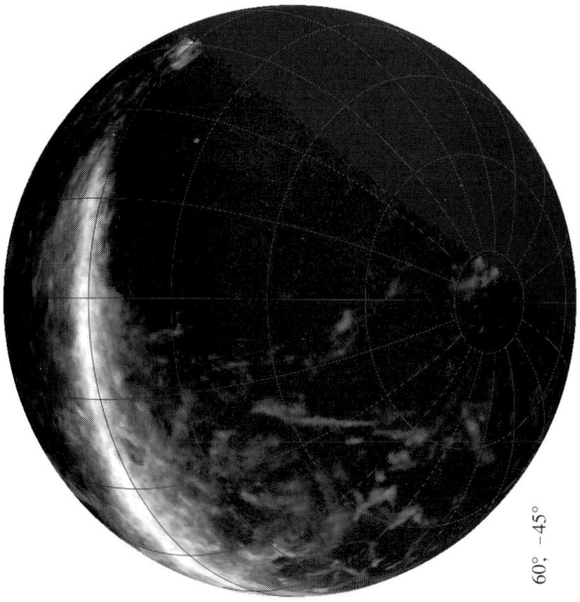

60°, −45°

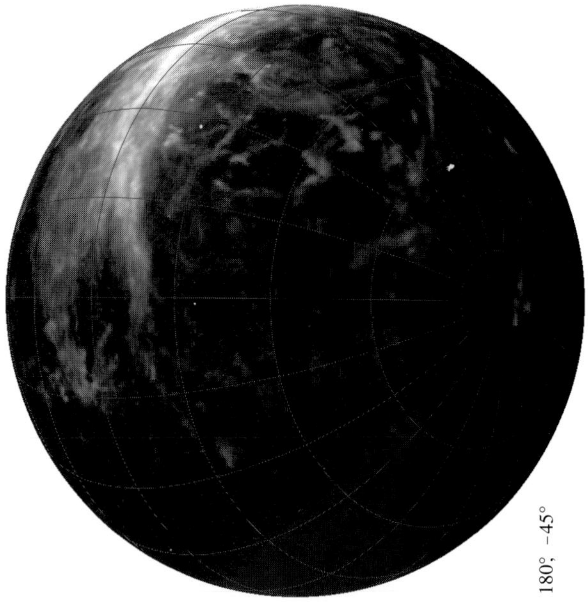

180°, −45°

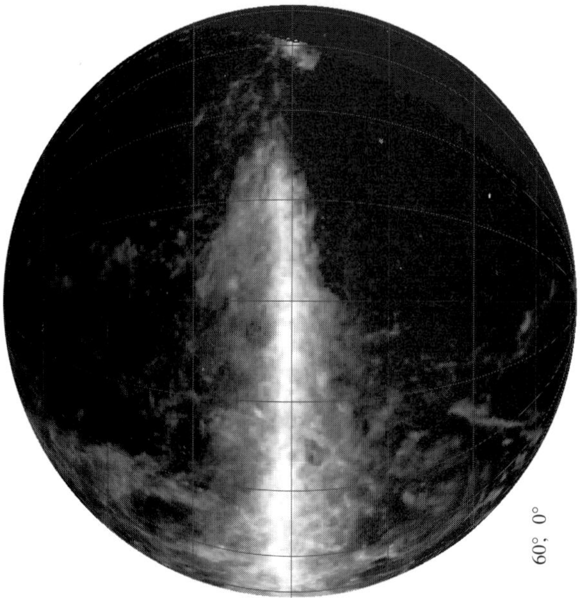

60°, 0°

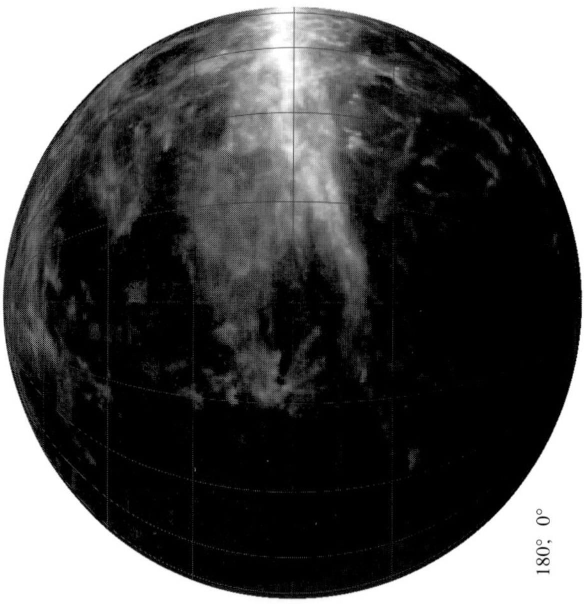

180°, 0°

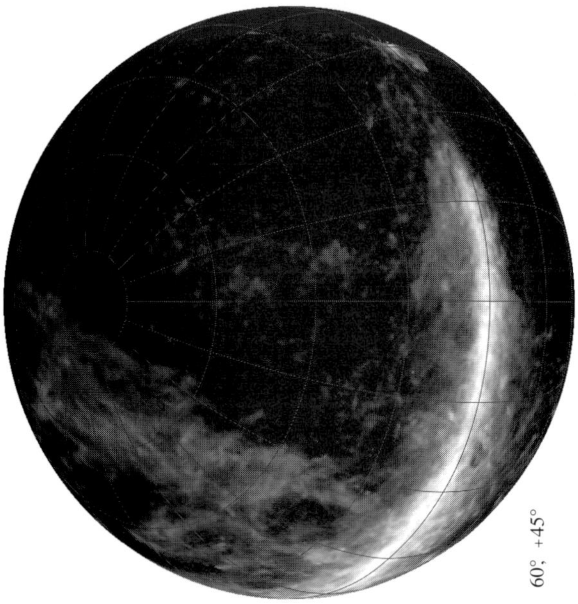

60°, +45°

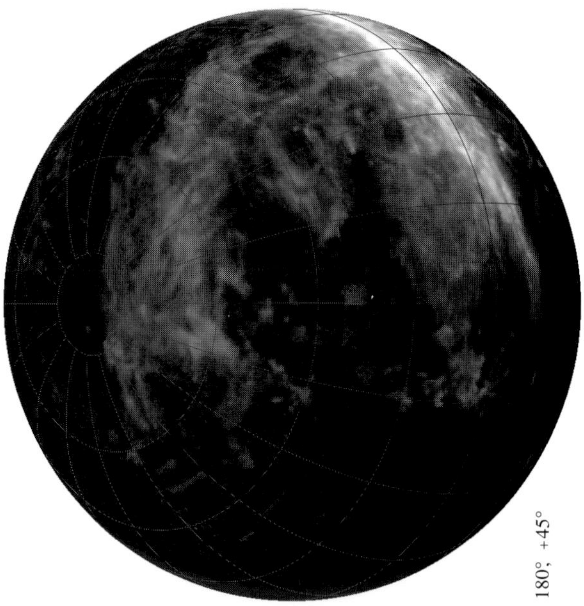

180°, +45°

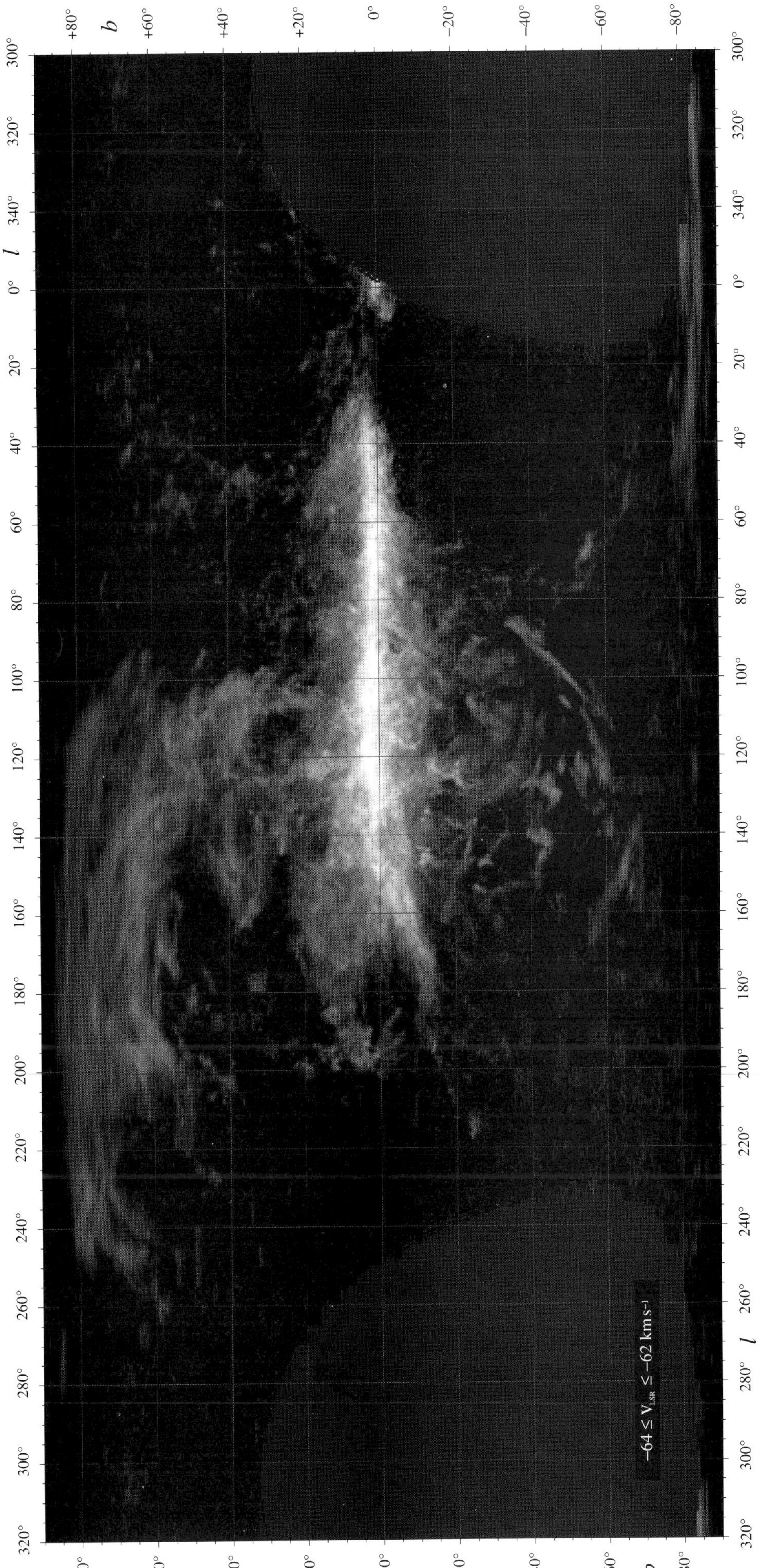

−64 ≤ V_LSR ≤ −62 km s−1
l
b

$-62 \le V_{LSR} \le -60$ km s^{-1}

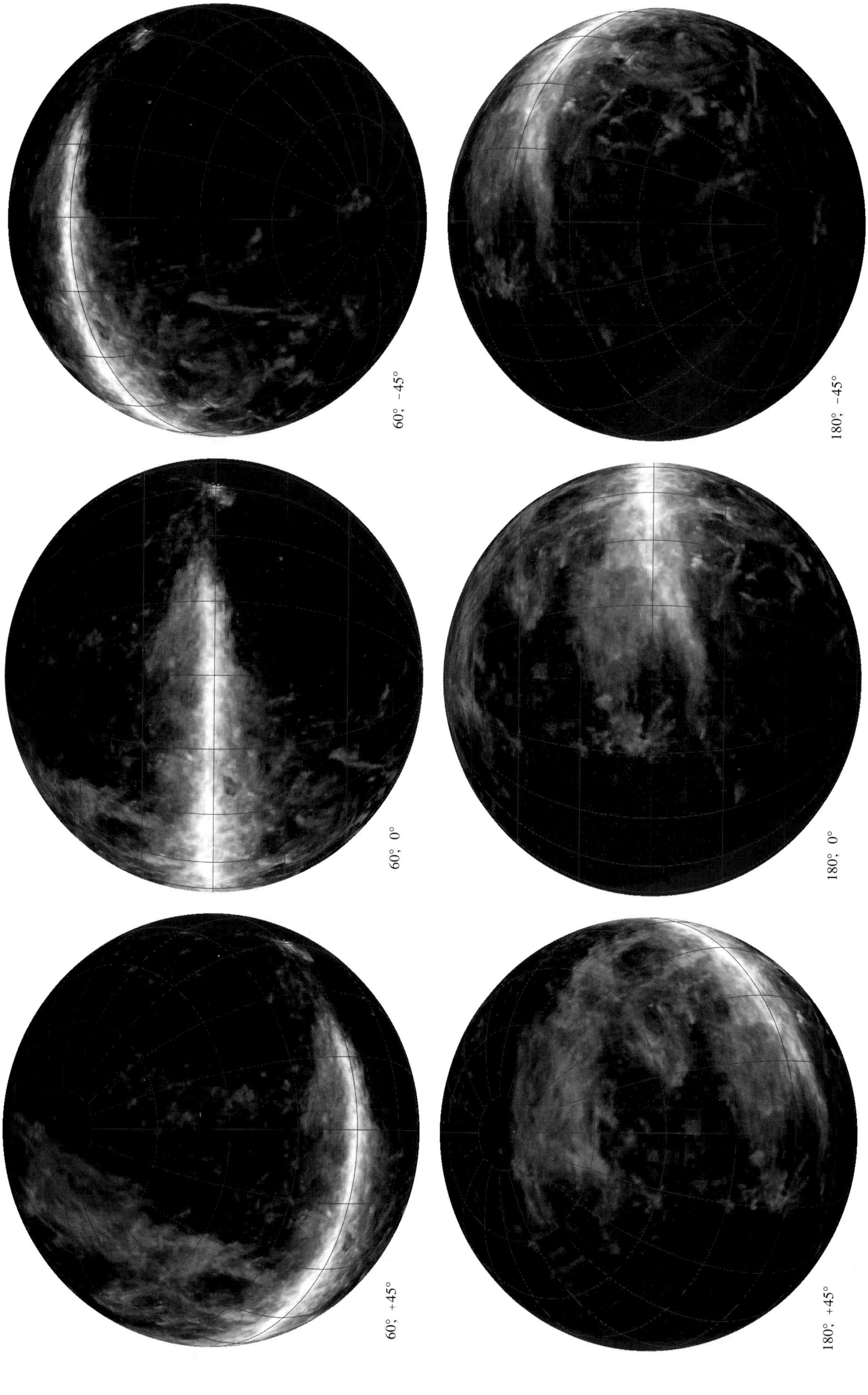

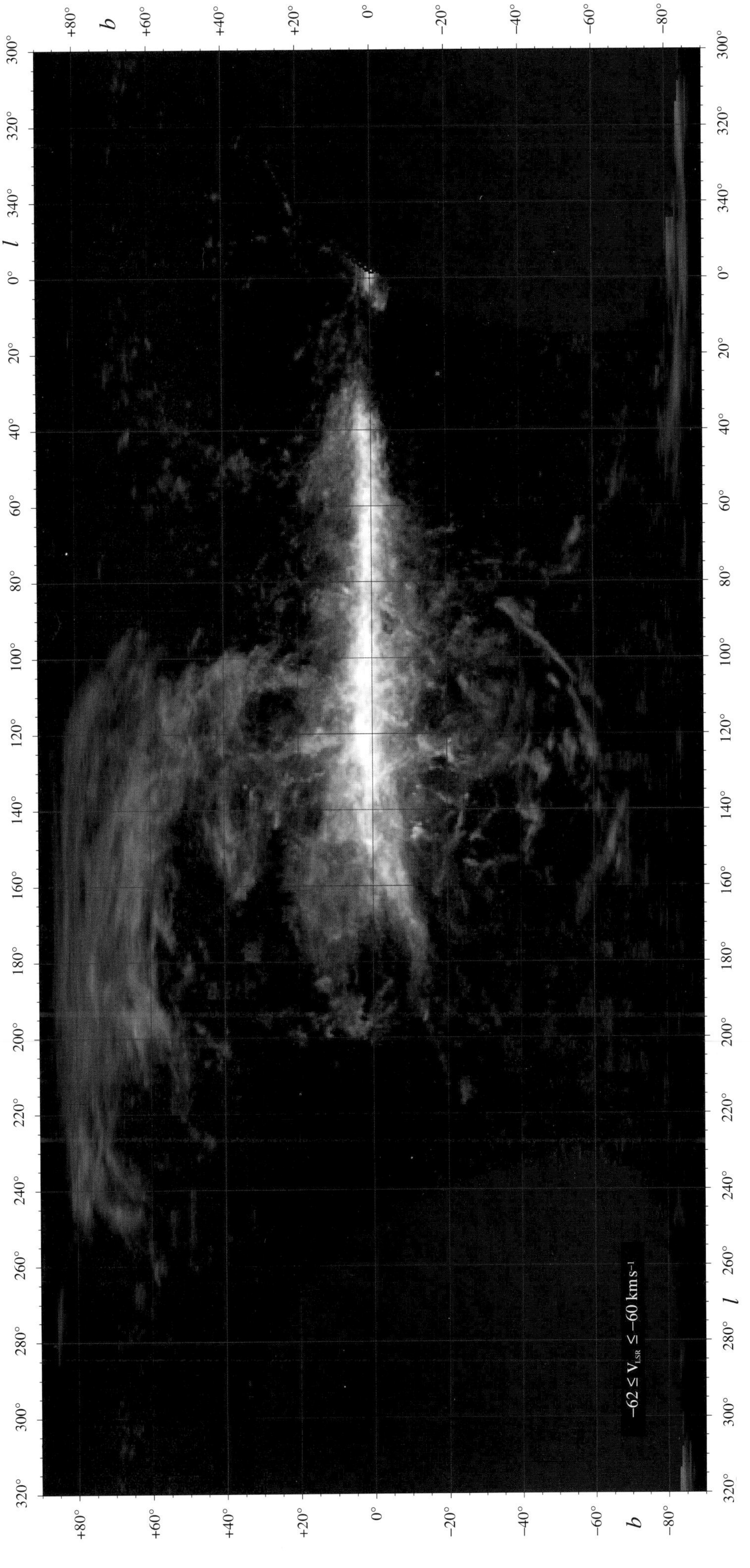
$-62 \le V_{LSR} \le -60$ km s^{-1}
l
b

$-60 \leq V_{LSR} \leq -58$ km s^{-1}

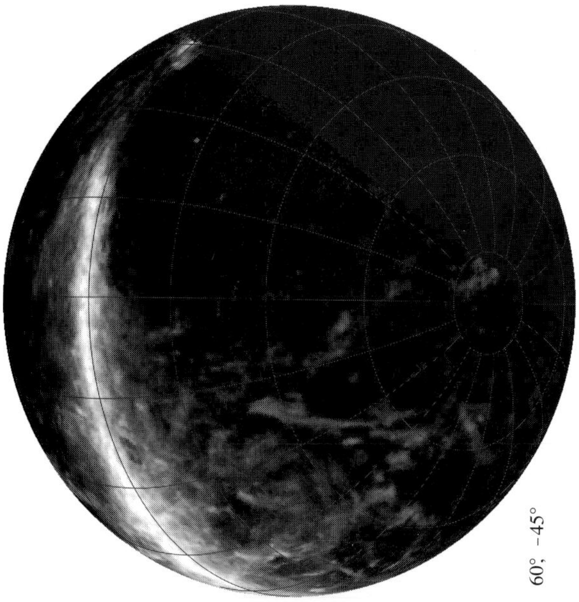

60°, −45°

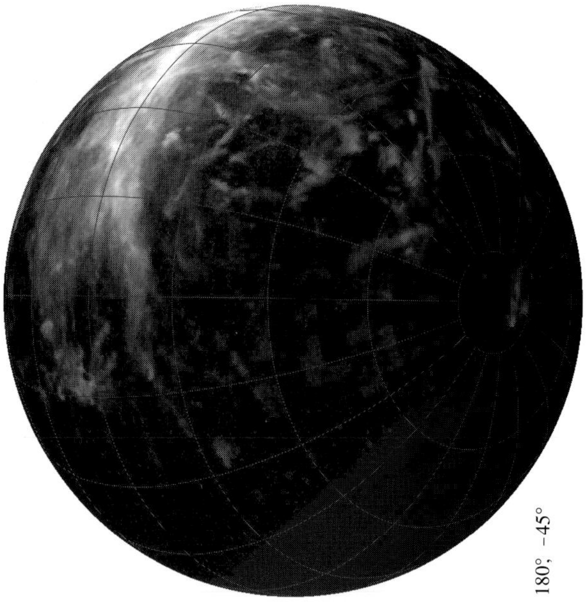

180°, −45°

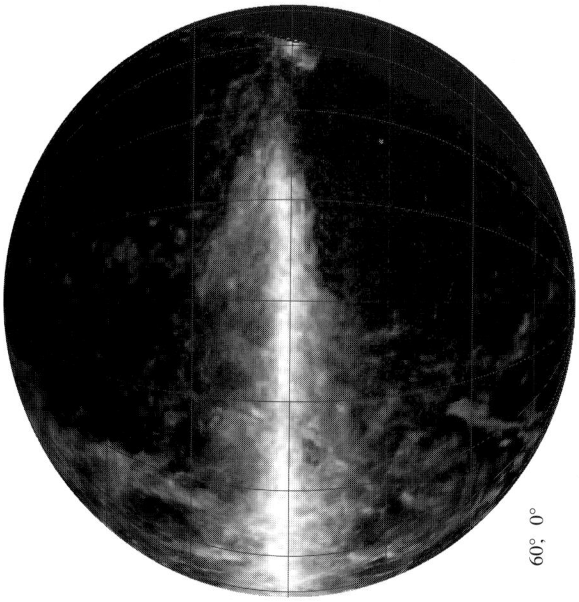

60°, 0°

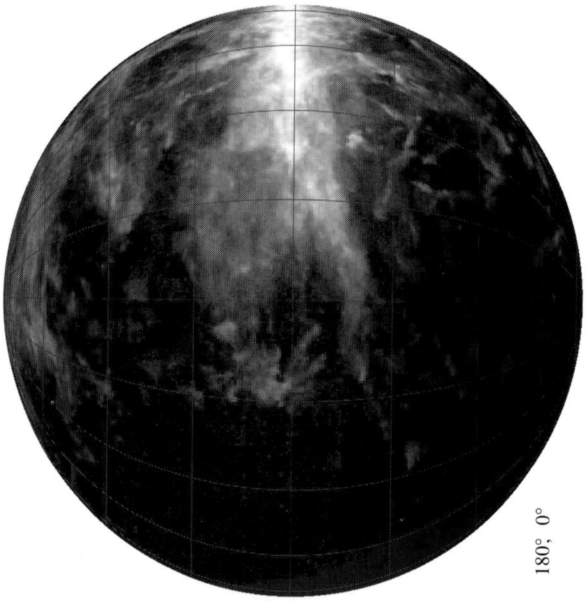

180°, 0°

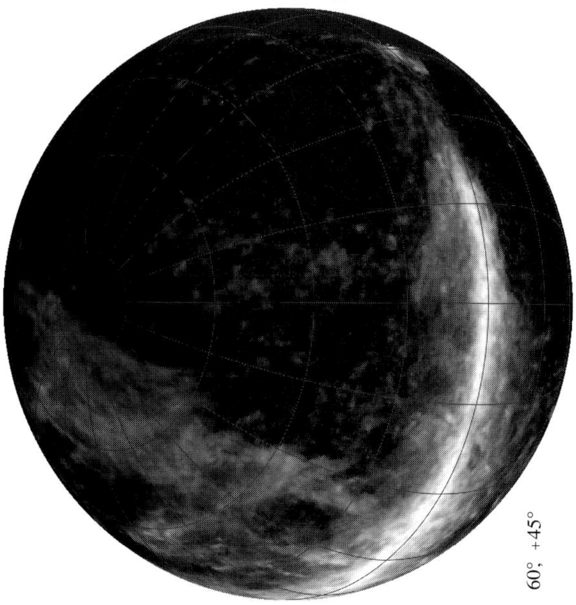

60°, +45°

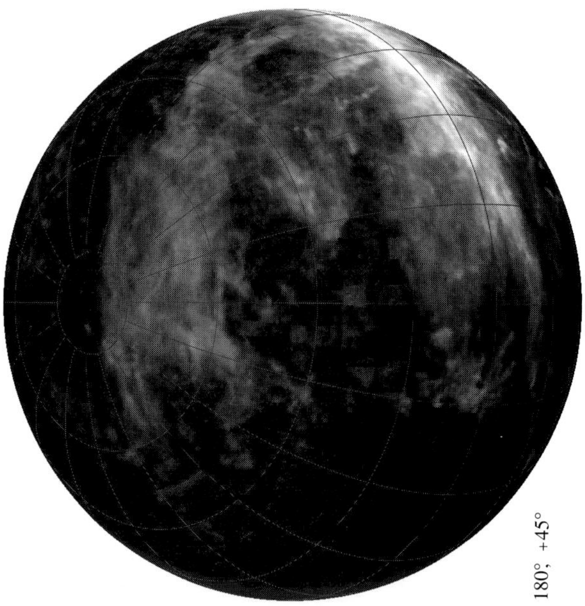

180°, +45°

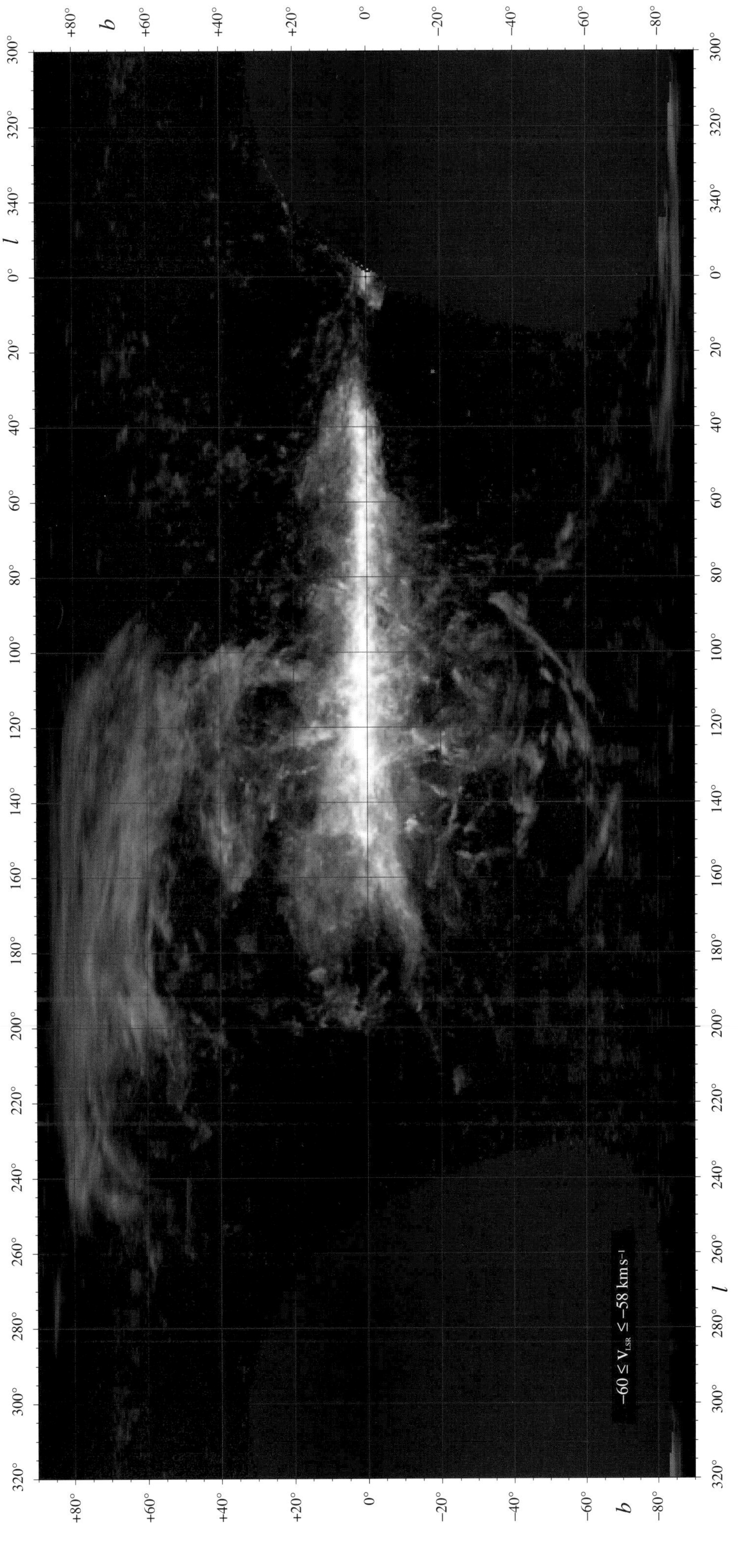
−60 ≤ V_LSR ≤ −58 km s−1

$-58 \le V_{LSR} \le -56$ km s^{-1}

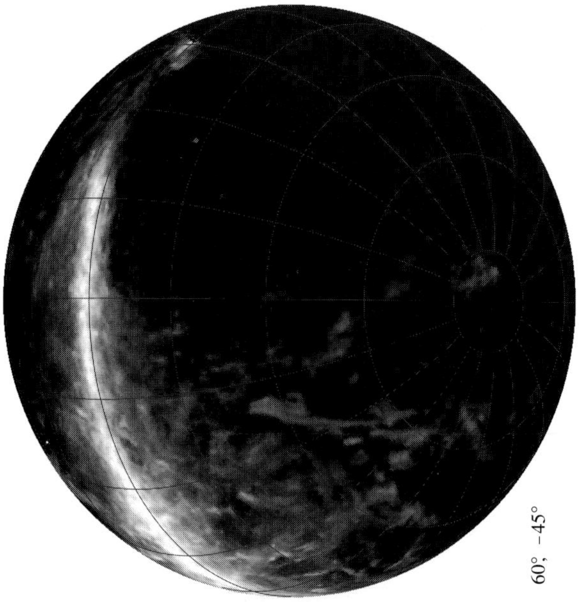

60°, −45°

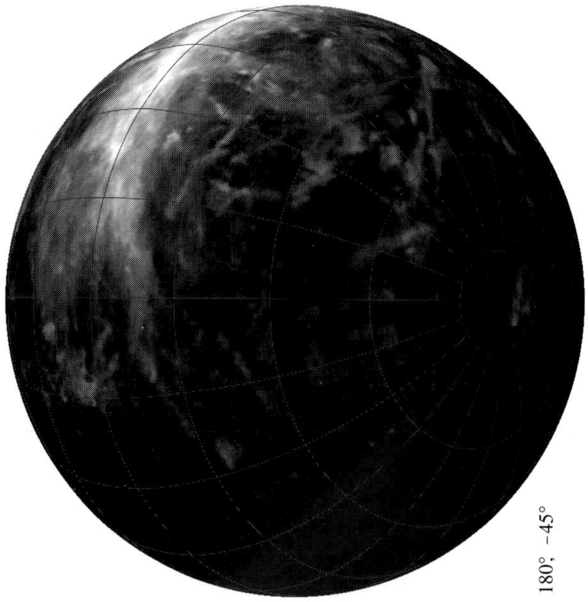

180°, −45°

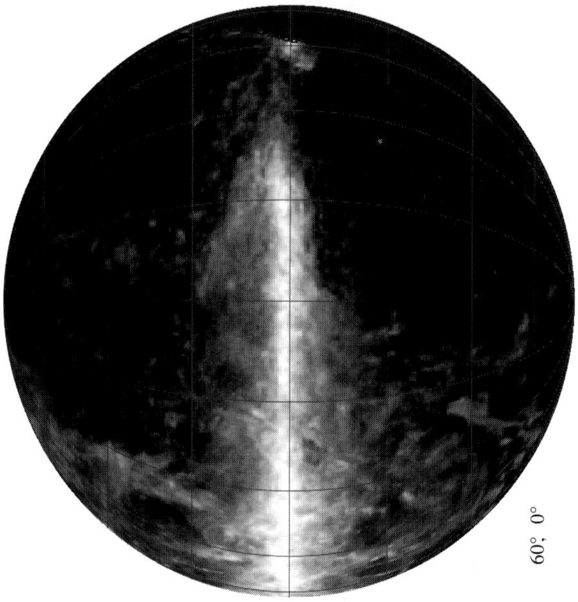

60°, 0°

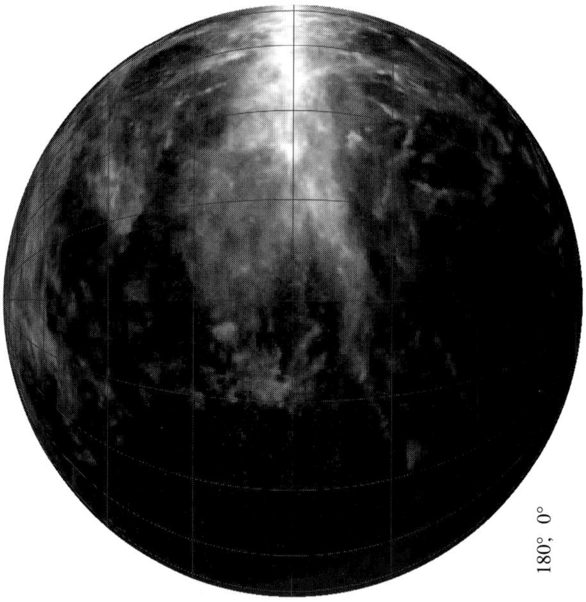

180°, 0°

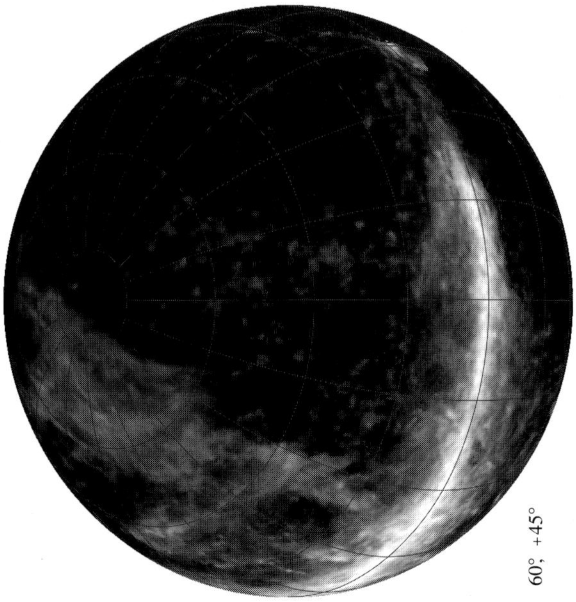

60°, +45°

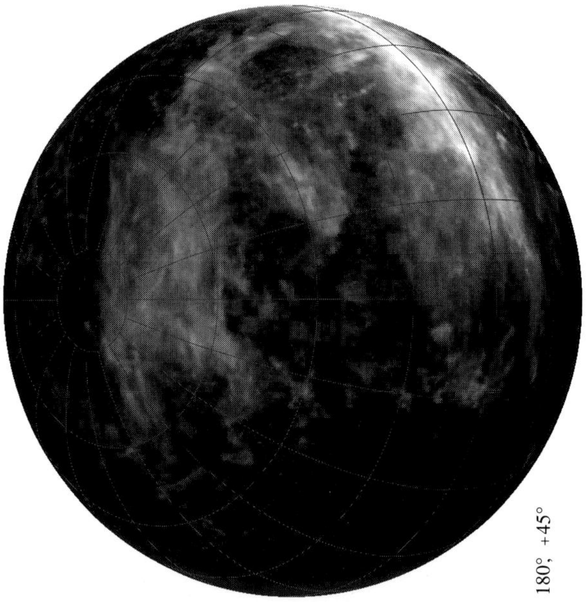

180°, +45°

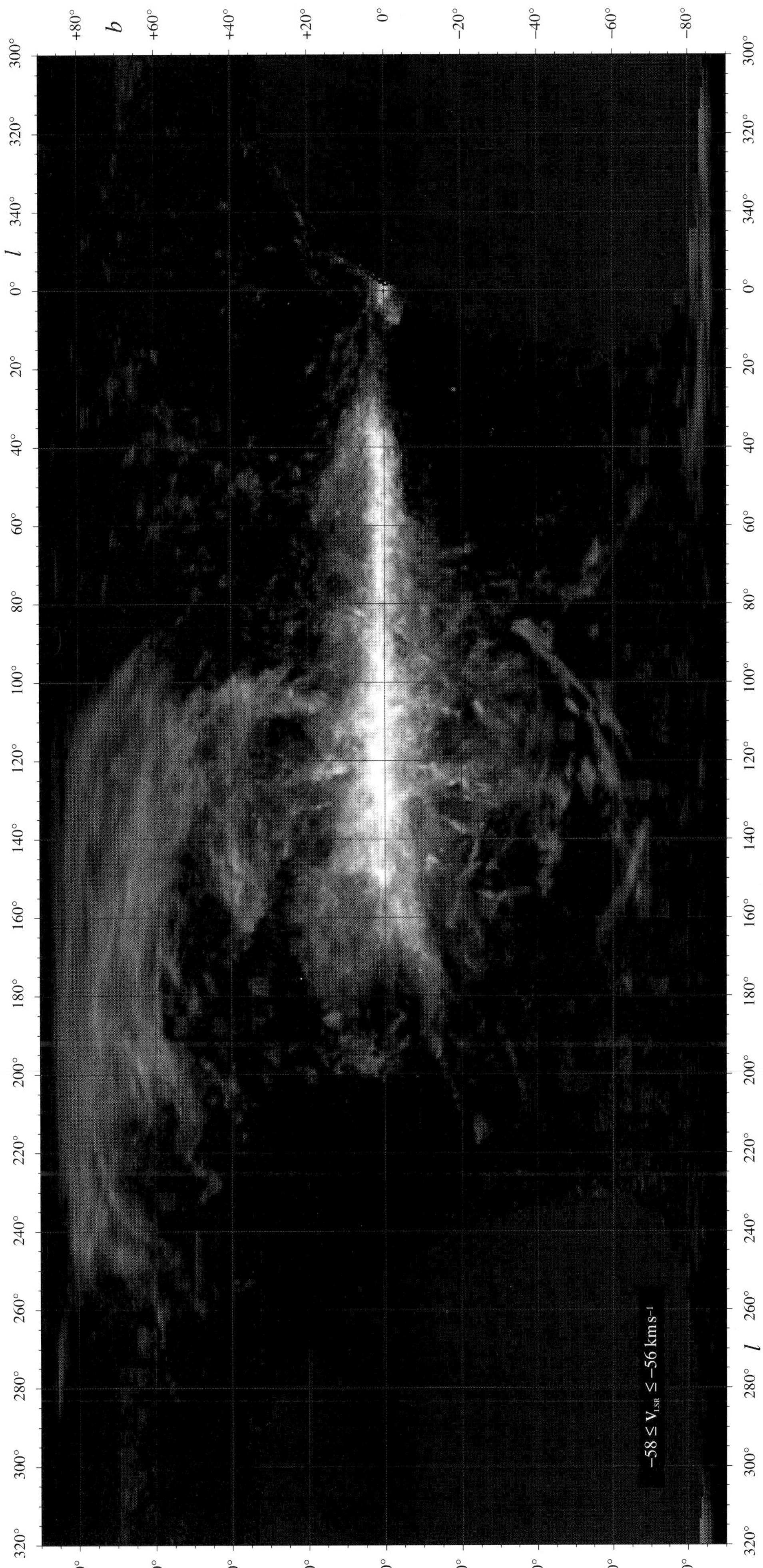
−58 ≤ V_LSR ≤ −56 km s⁻¹
l
b

$-56 \leq V_{LSR} \leq -54$ km s^{-1}

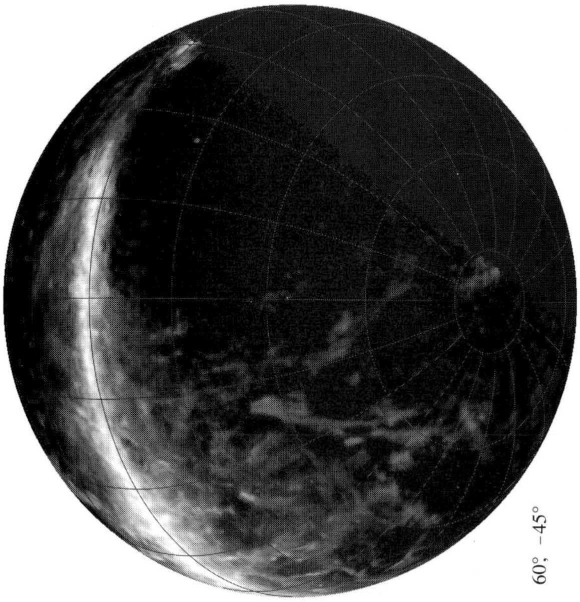

60°, −45°

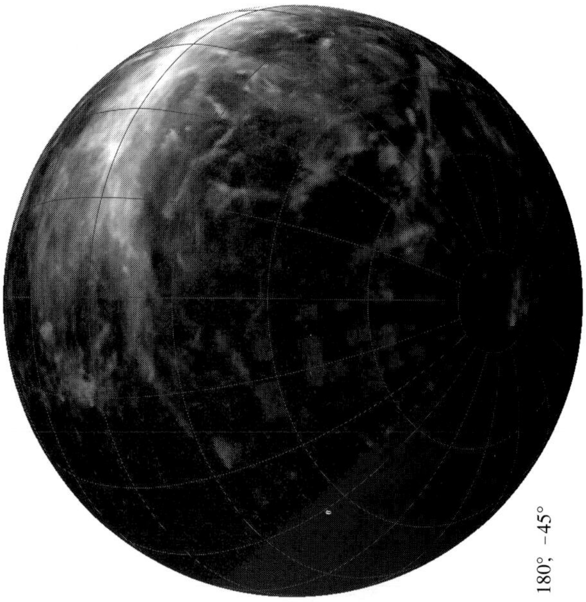

180°, −45°

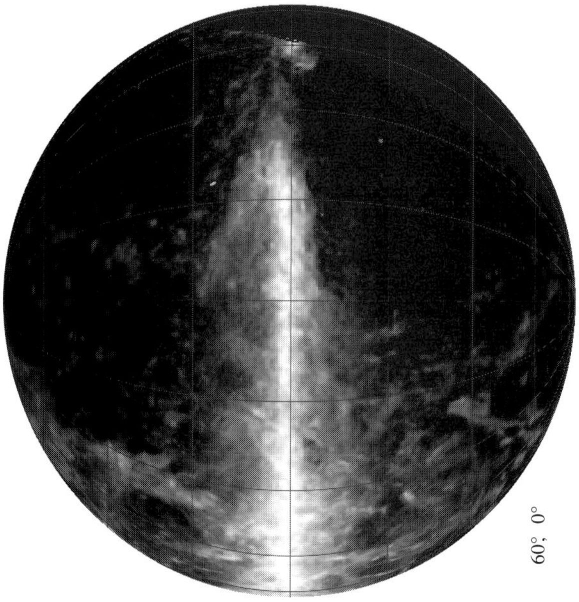

60°, 0°

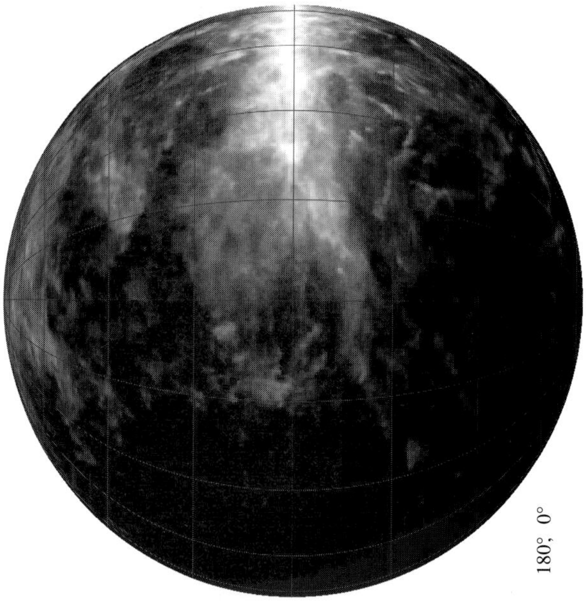

180°, 0°

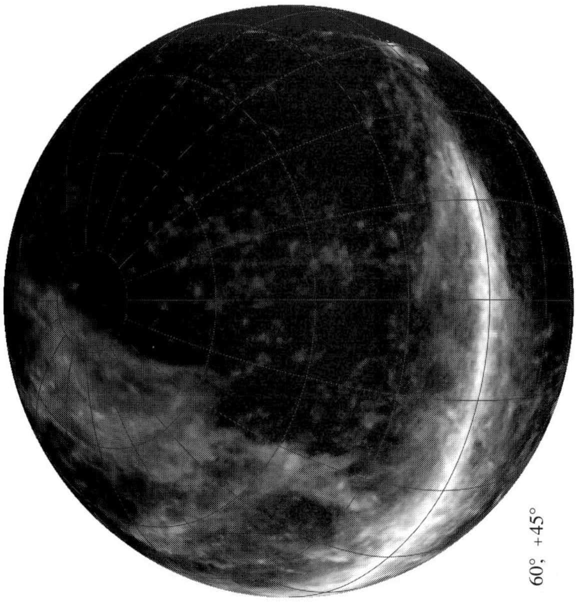

60°, +45°

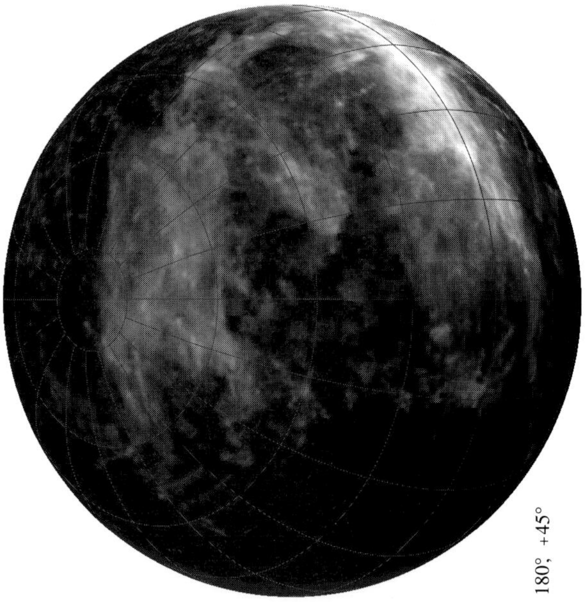

180°, +45°

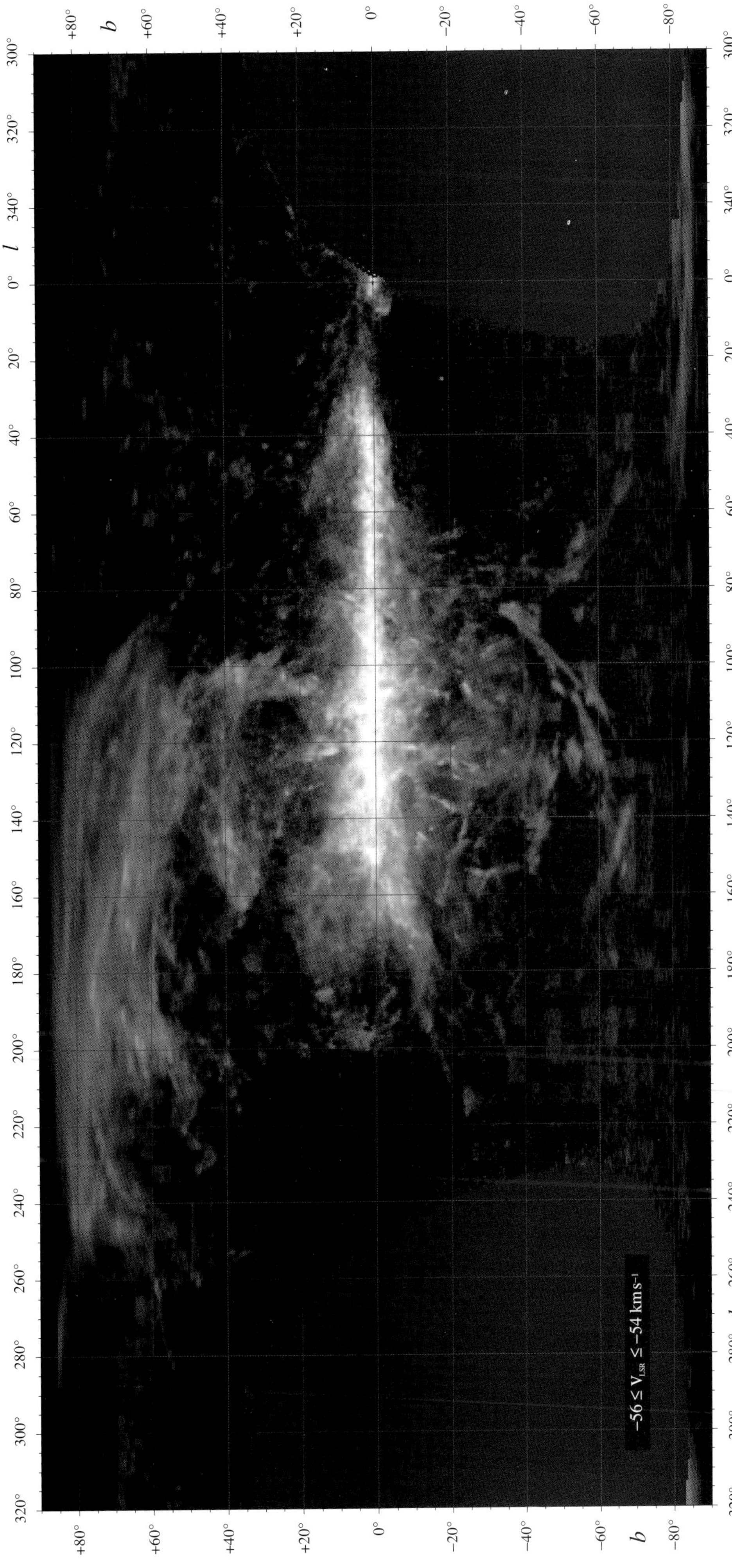
$-56 \leq V_{LSR} \leq -54$ kms^{-1}

$-54 \leq V_{LSR} \leq -52$ km s^{-1}

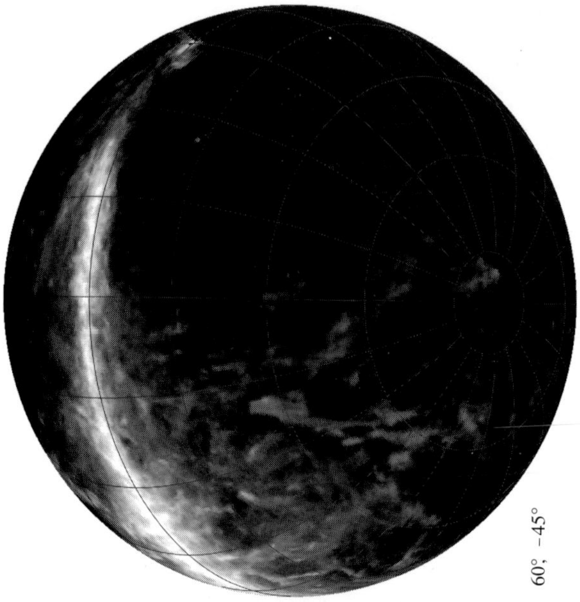

60°, −45°

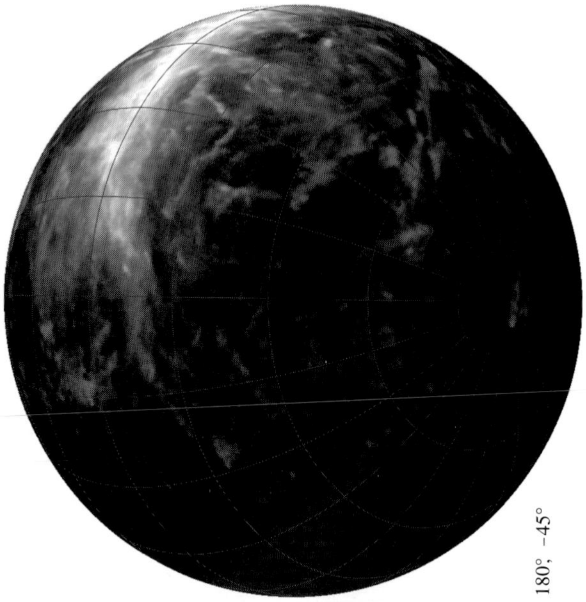

180°, −45°

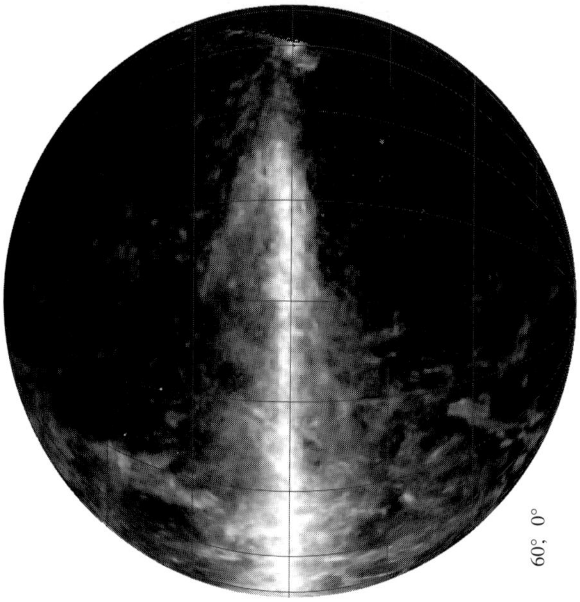

60°, 0°

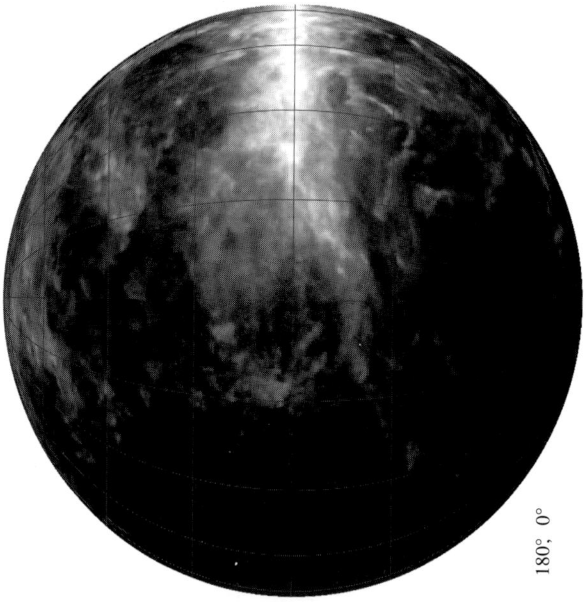

180°, 0°

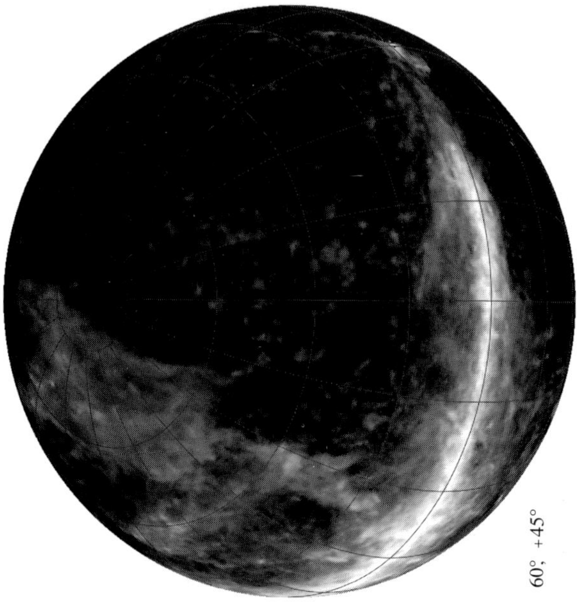

60°, +45°

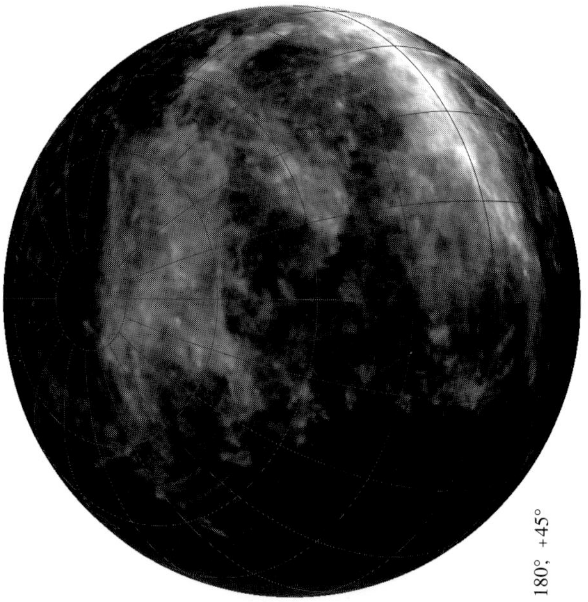

180°, +45°

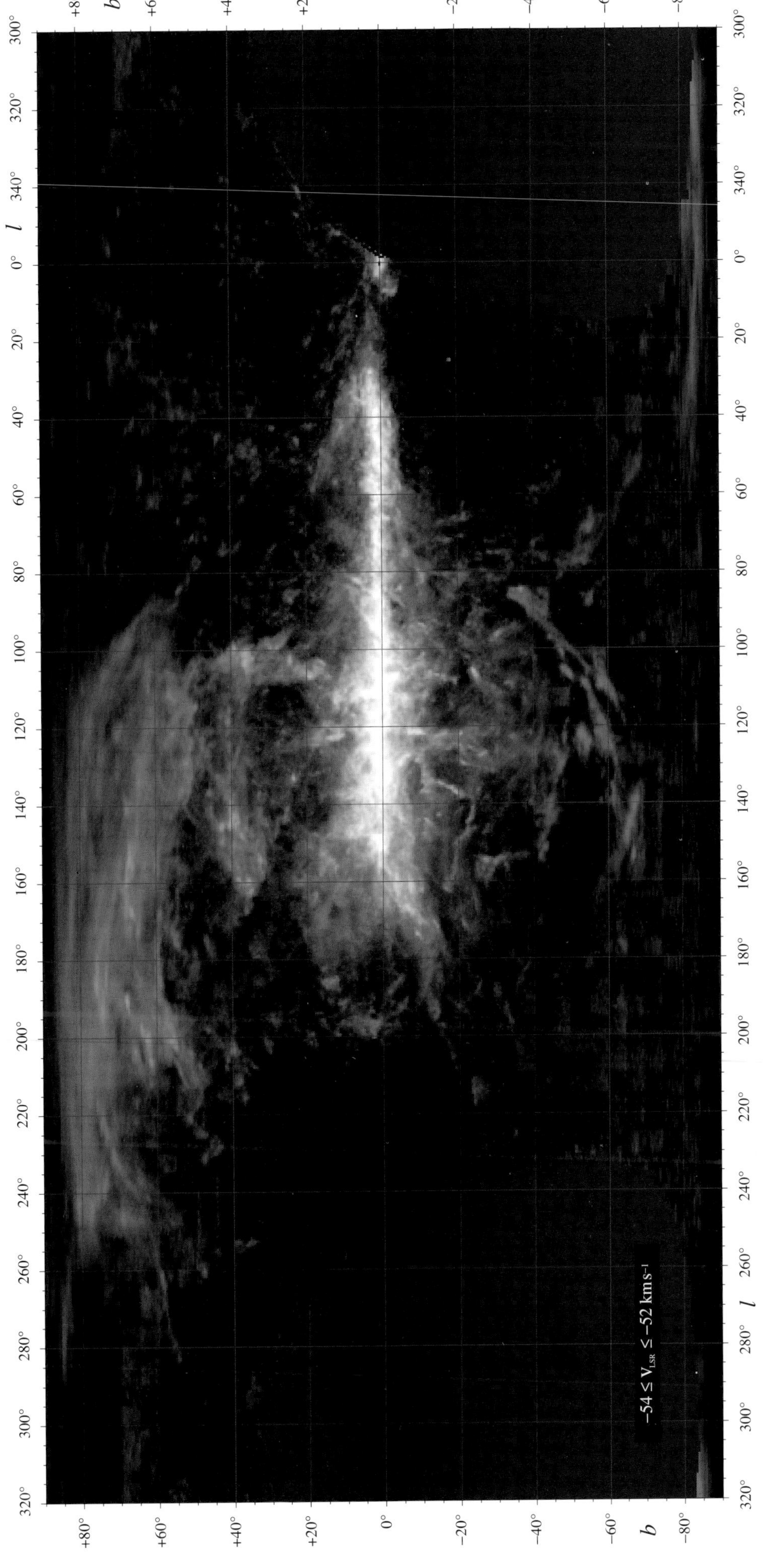

−54 ≤ V_LSR ≤ −52 kms−1
l
b
300° 320° 340° 0° 20° 40° 60° 80° 100° 120° 140° 160° 180° 200° 220° 240° 260° 280° 300° 320°
+80° +60° +40° +20° 0° −20° −40° −60° −80°

$-52 \leq V_{LSR} \leq -50$ km s^{-1}

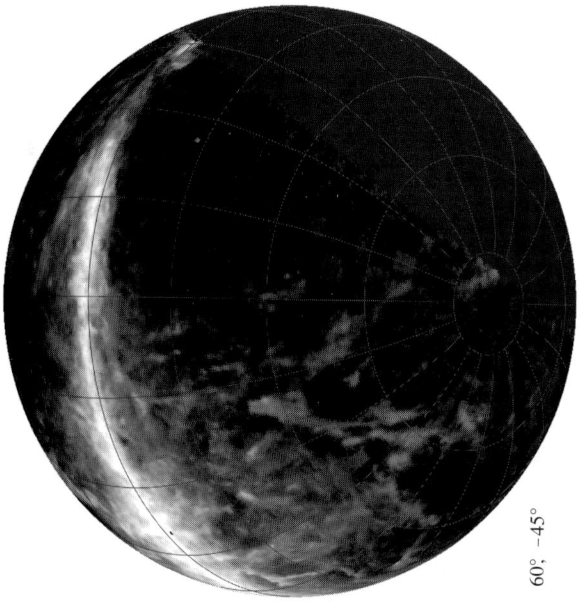

60°, −45°

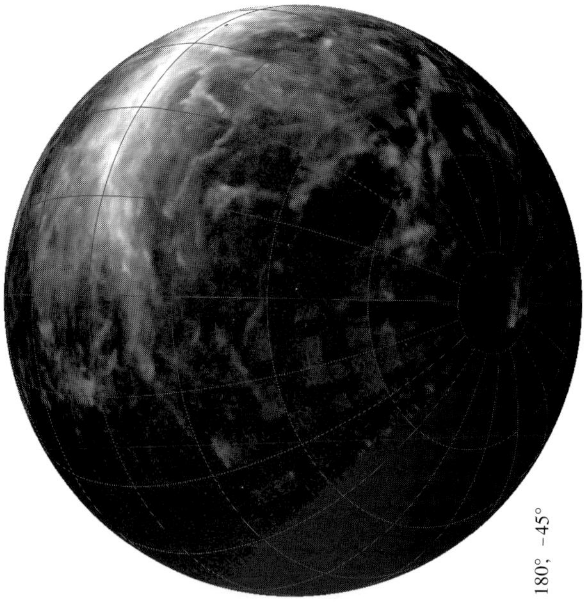

180°, −45°

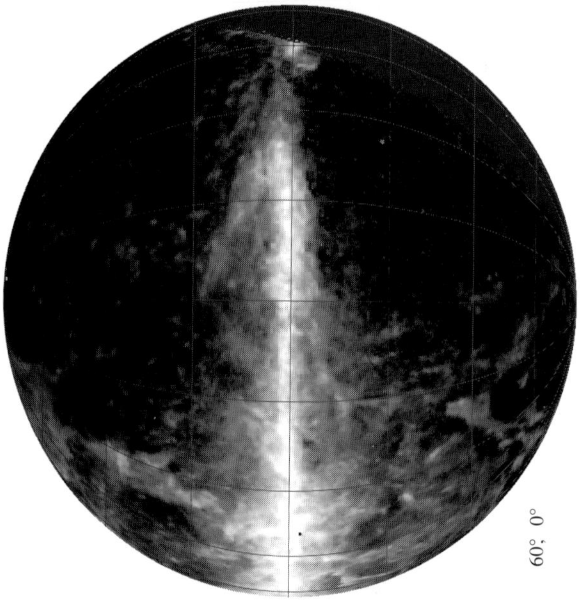

60°, 0°

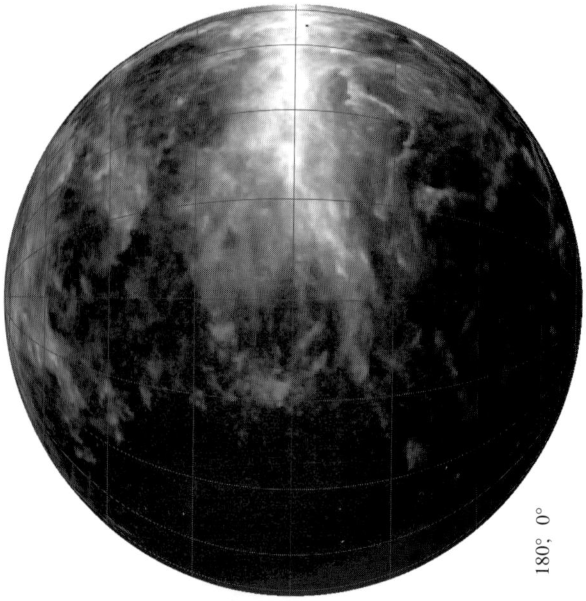

180°, 0°

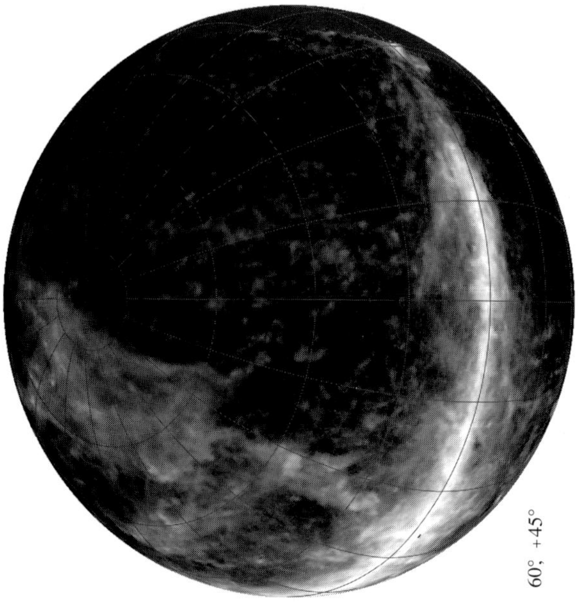

60°, +45°

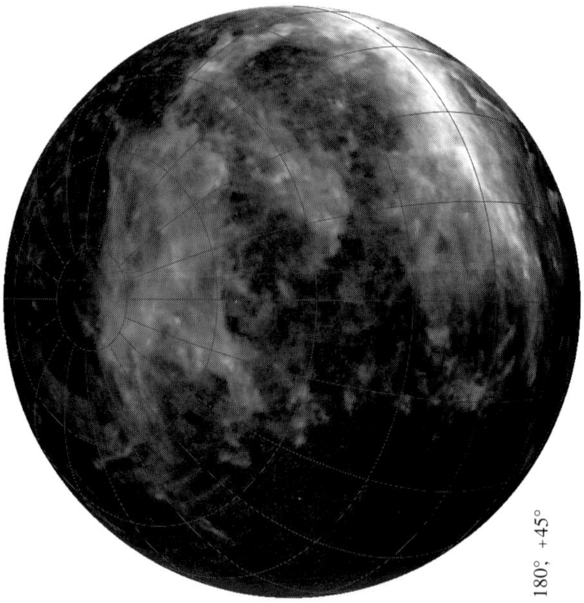

180°, +45°

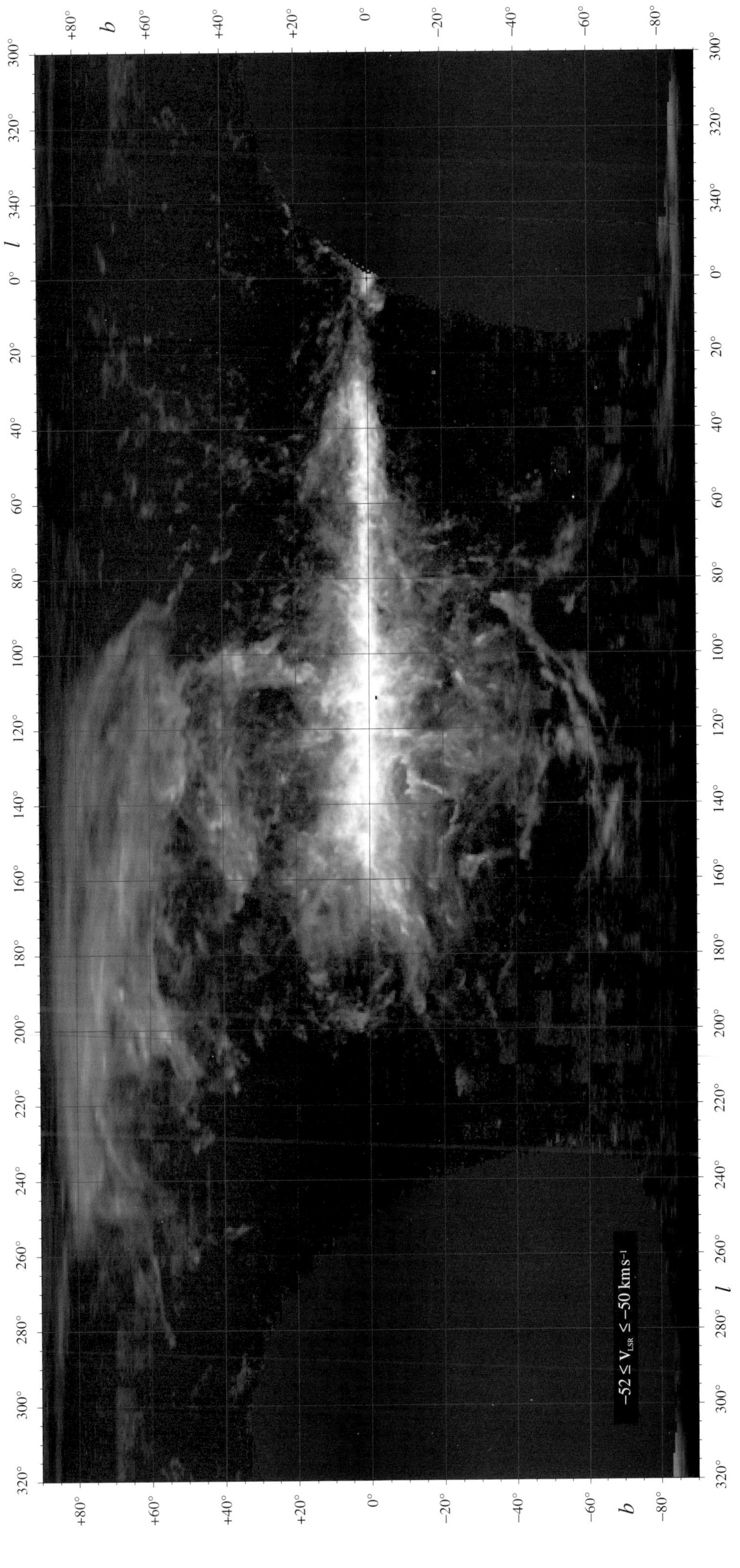
−52 ≤ V_{LSR} ≤ −50 km s^{-1}
l
b

$-50 \leq V_{LSR} \leq -48$ km s^{-1}

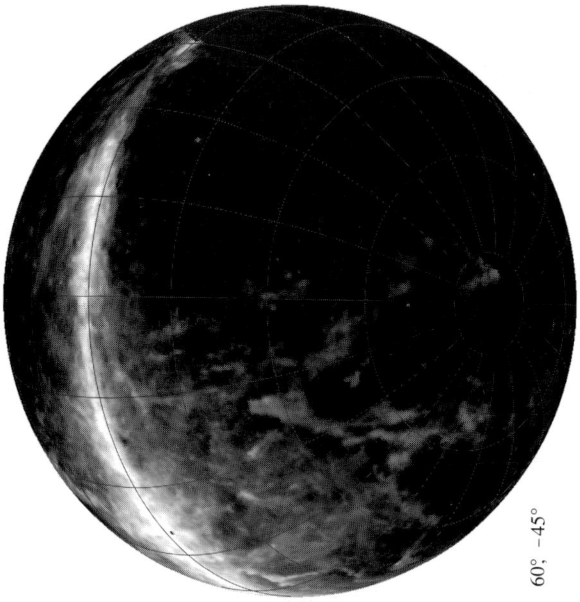

60°, −45°

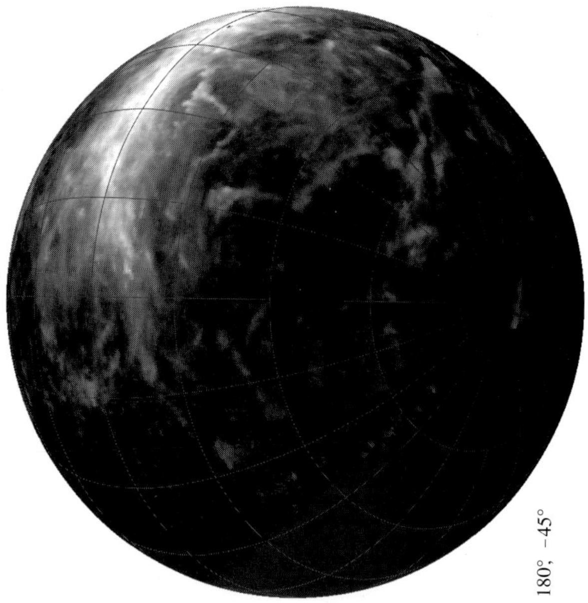

180°, −45°

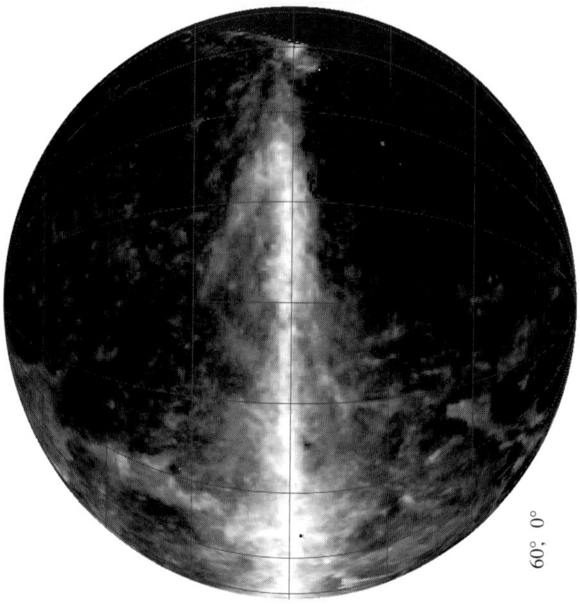

60°, 0°

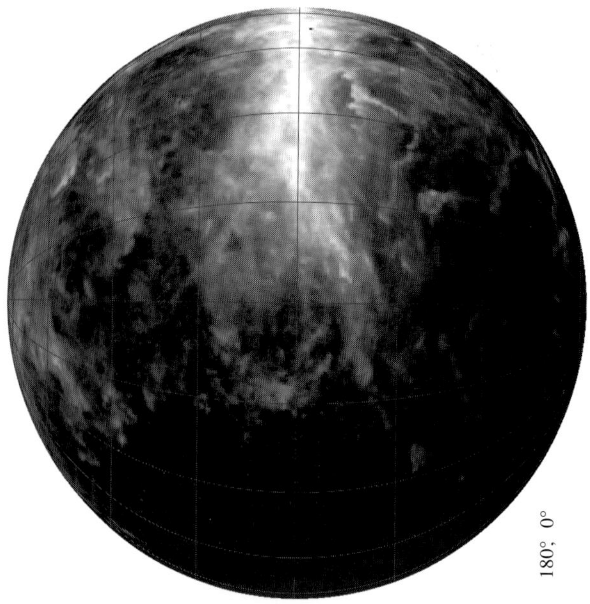

180°, 0°

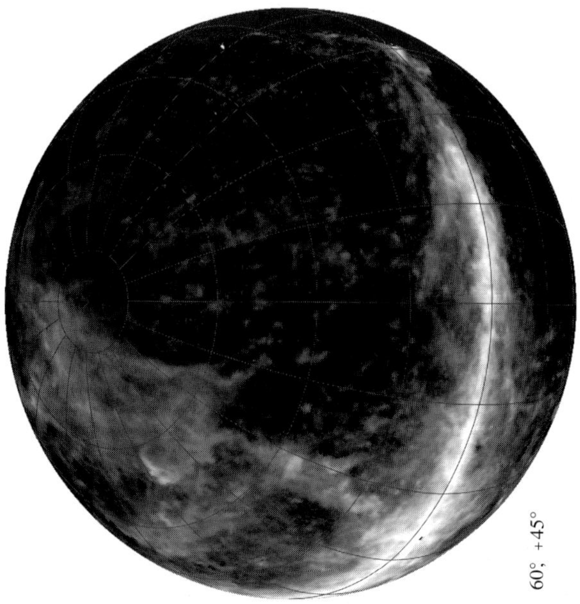

60°, +45°

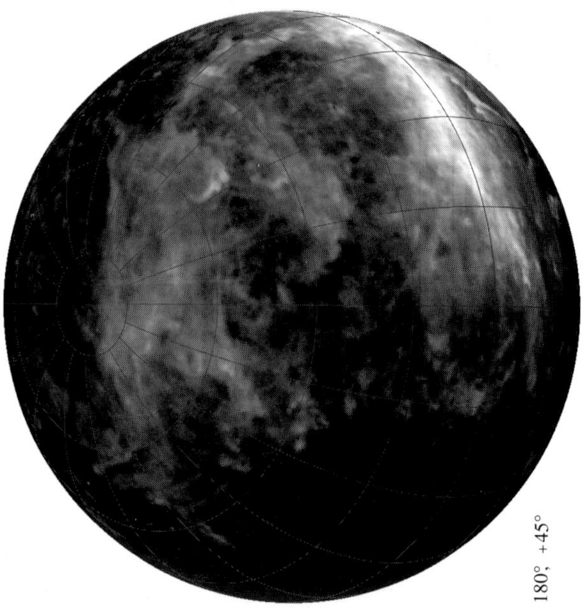

180°, +45°

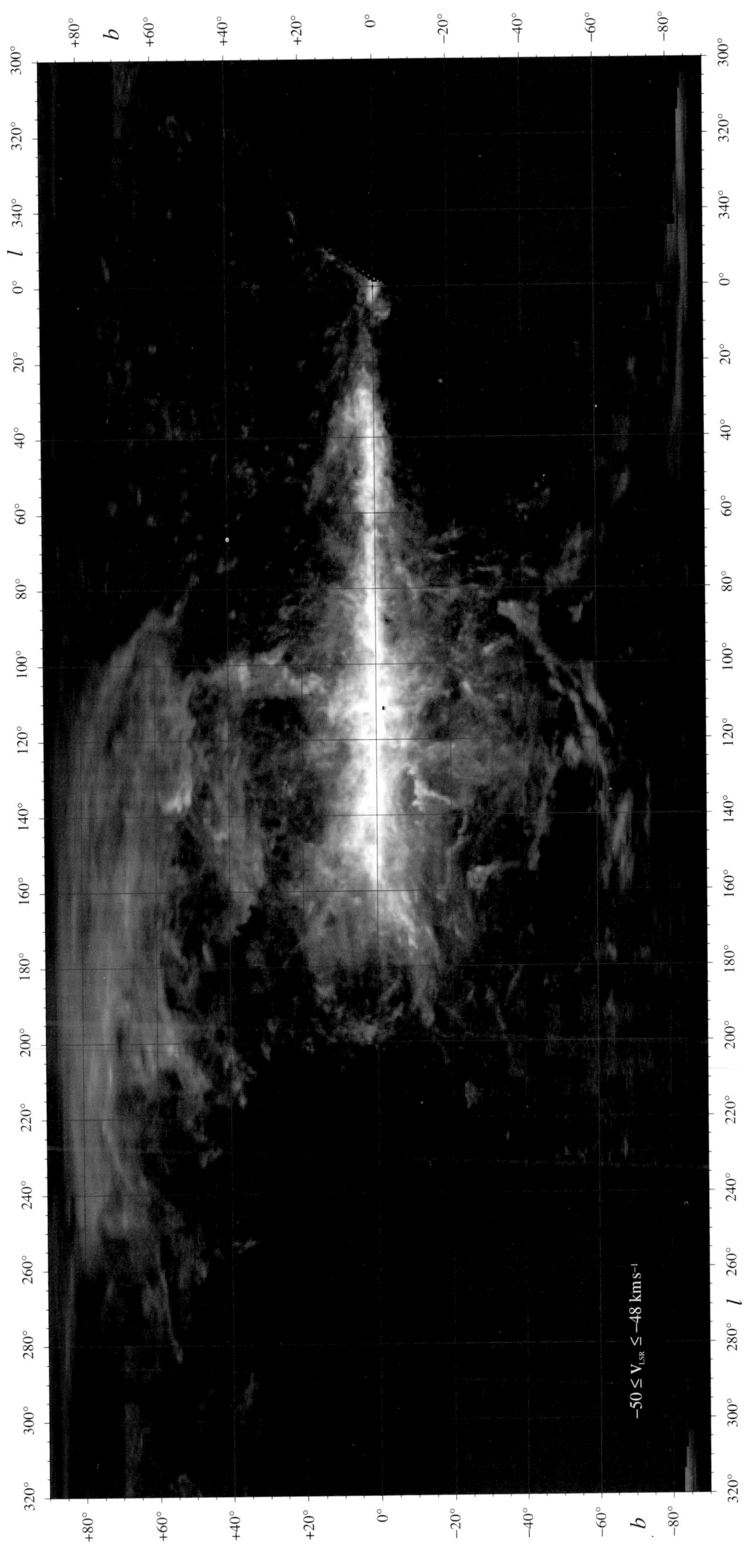
−50 ≤ V_LSR ≤ −48 kms−1

$-48 \leq V_{LSR} \leq -46$ km s^{-1}

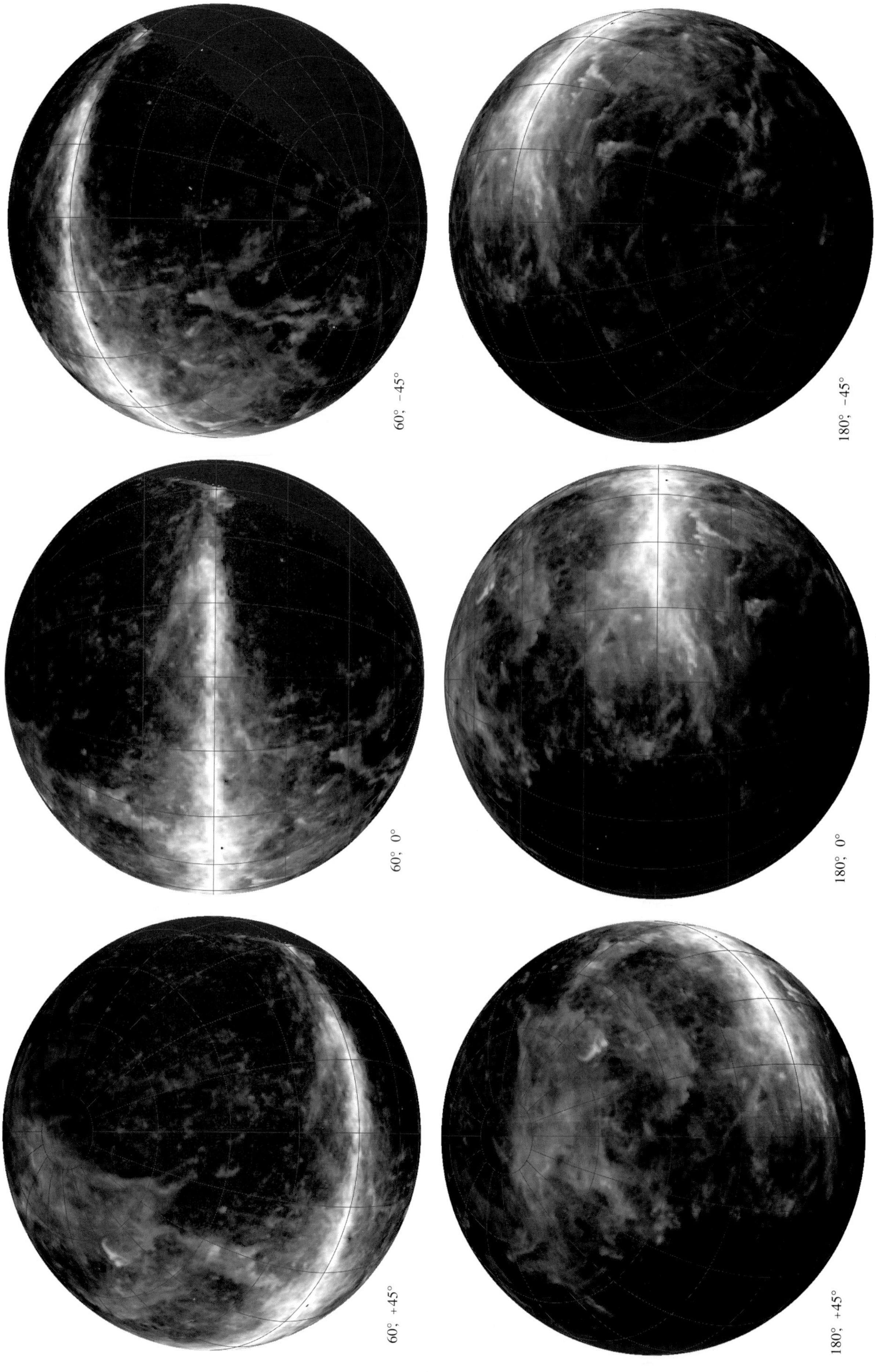

60°, −45°

180°, −45°

60°, 0°

180°, 0°

60°, +45°

180°, +45°

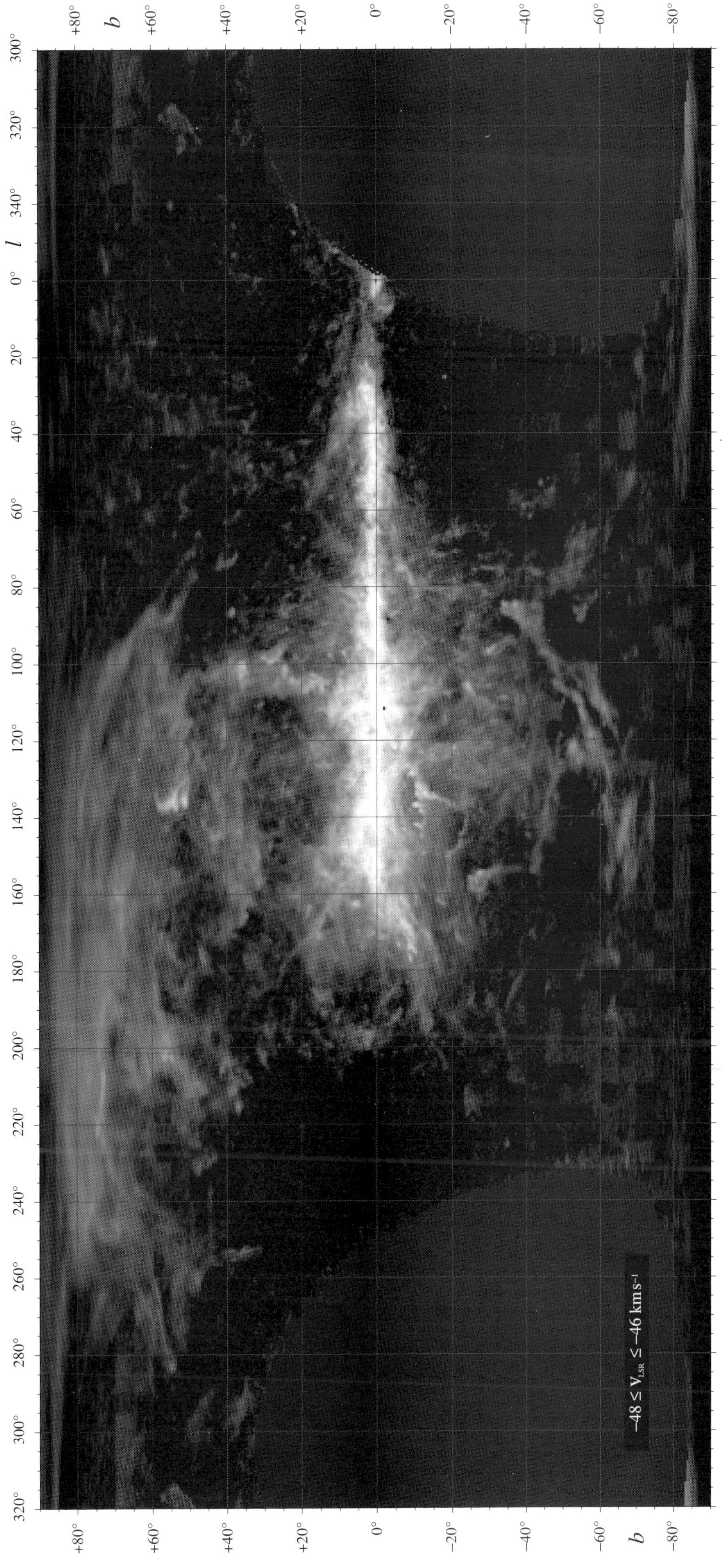

l
300° 320° 340° 0° 20° 40° 60° 80° 100° 120° 140° 160° 180° 200° 220° 240° 260° 280° 300° 320°
b
+80° +60° +40° +20° 0° −20° −40° −60° −80°
$-48 \le V_{LSR} \le -46$ km s^{-1}

$-46 \le V_{LSR} \le -44$ km s^{-1}

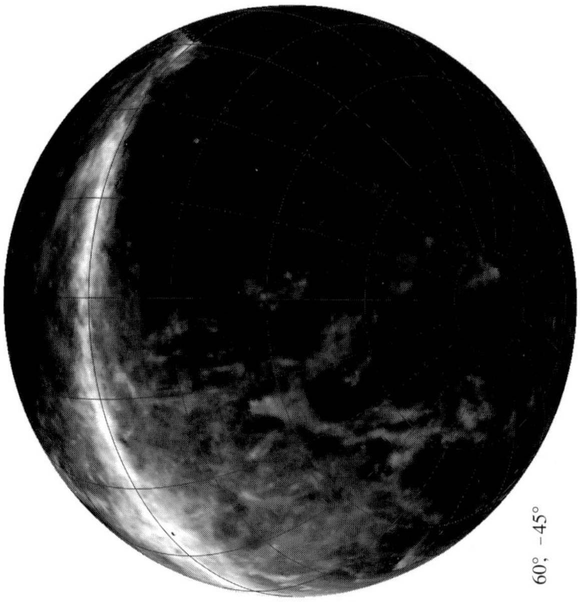

60°, −45°

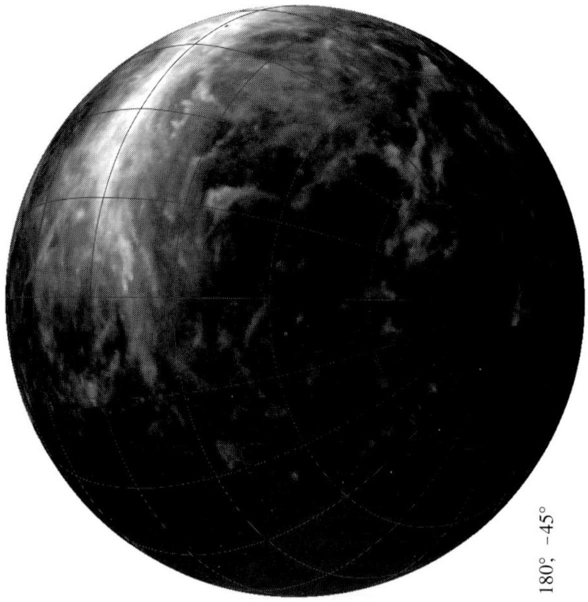

180°, −45°

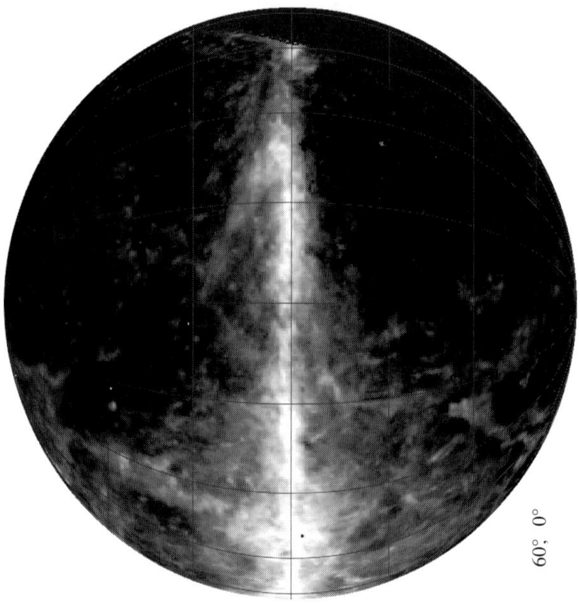

60°, 0°

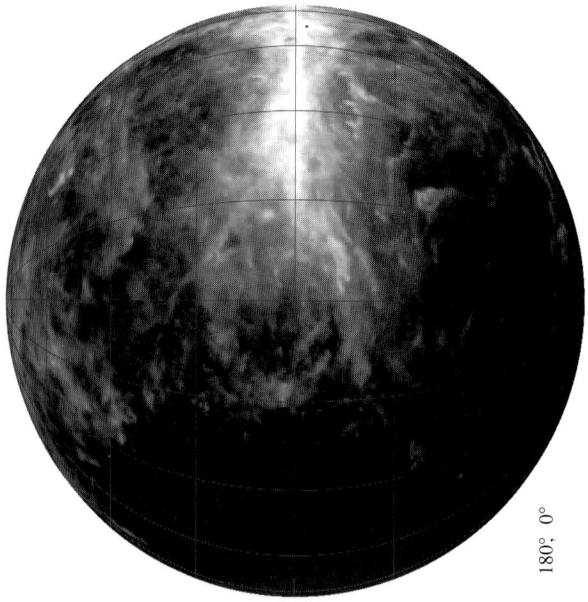

180°, 0°

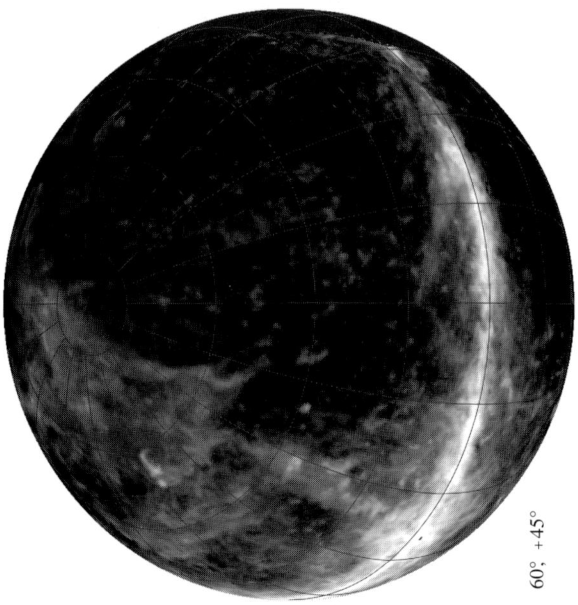

60°, +45°

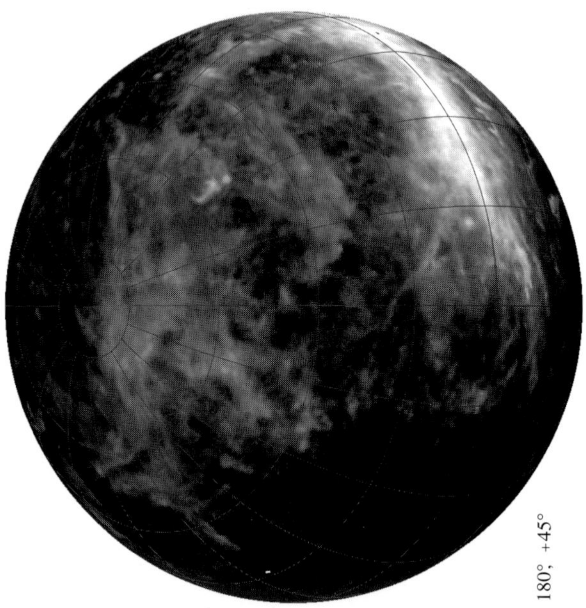

180°, +45°

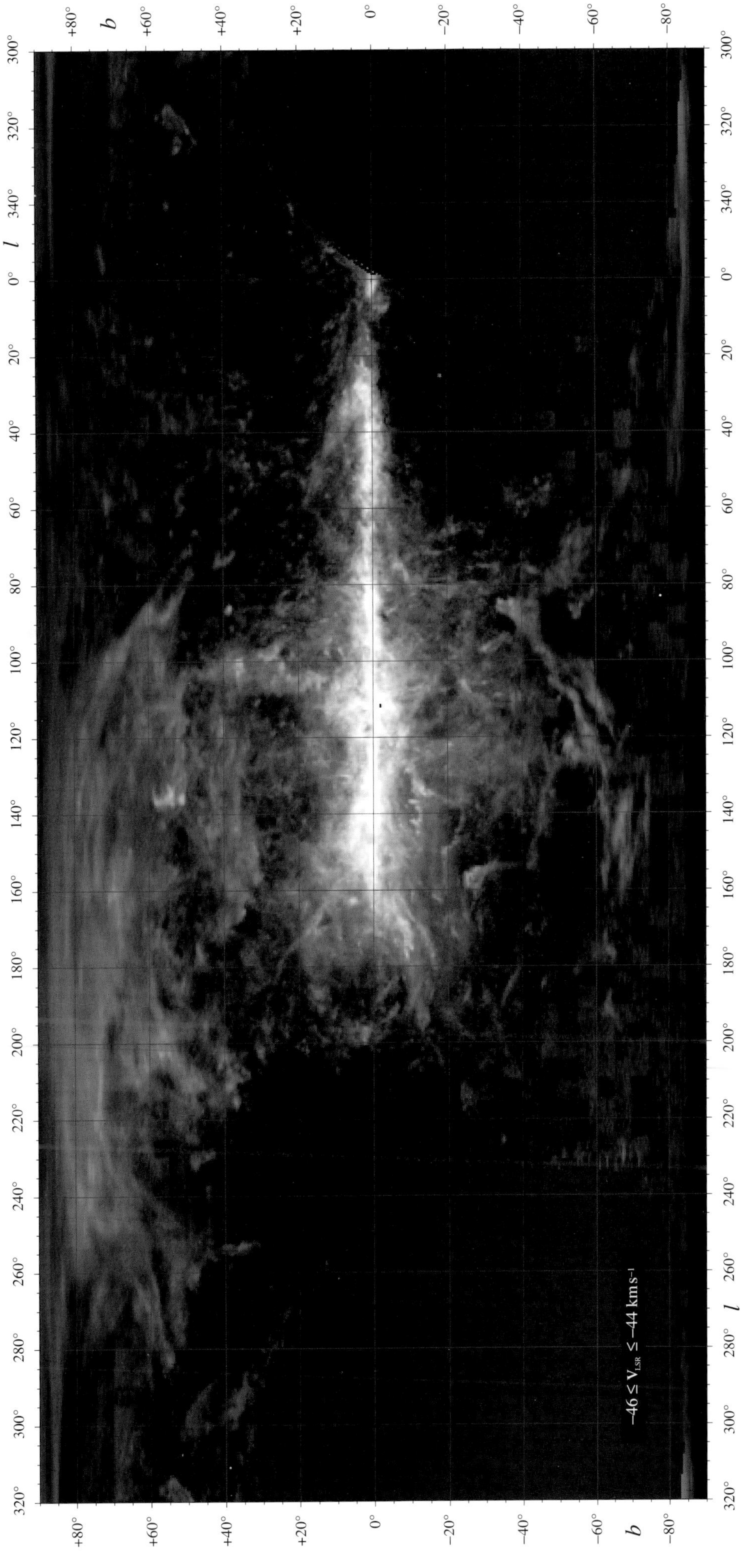

l
b
300° 320° 340° 0° 20° 40° 60° 80° 100° 120° 140° 160° 180° 200° 220° 240° 260° 280° 300° 320°
+80° +60° +40° +20° 0° −20° −40° −60° −80°
−46 ≤ V_LSR ≤ −44 kms⁻¹

$-44 \leq V_{LSR} \leq -42$ km s^{-1}

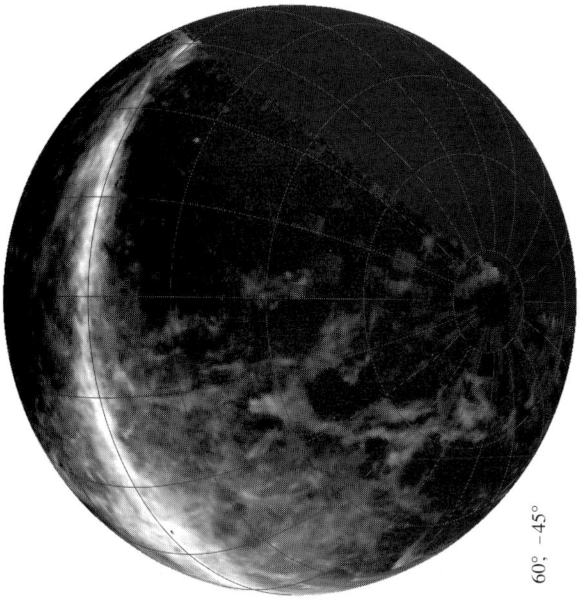

60°, −45°

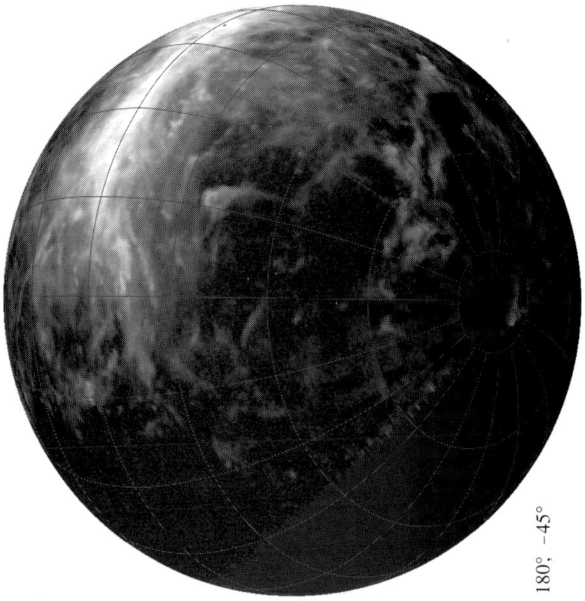

180°, −45°

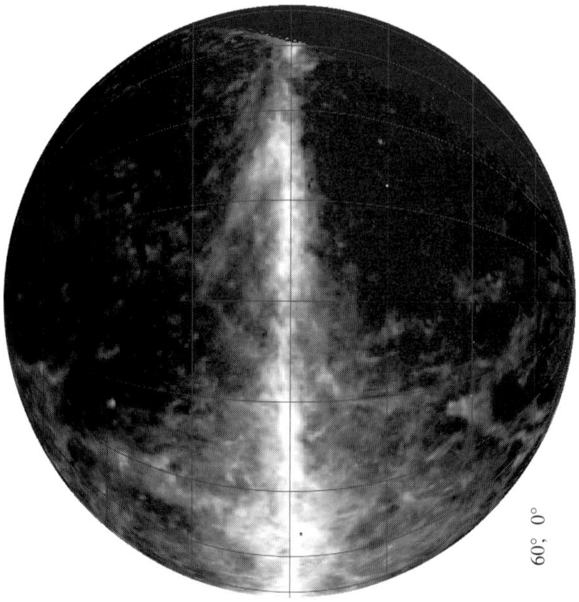

60°, 0°

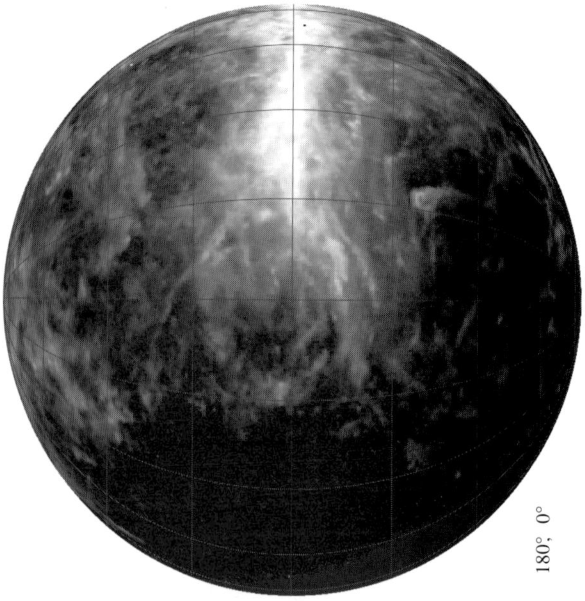

180°, 0°

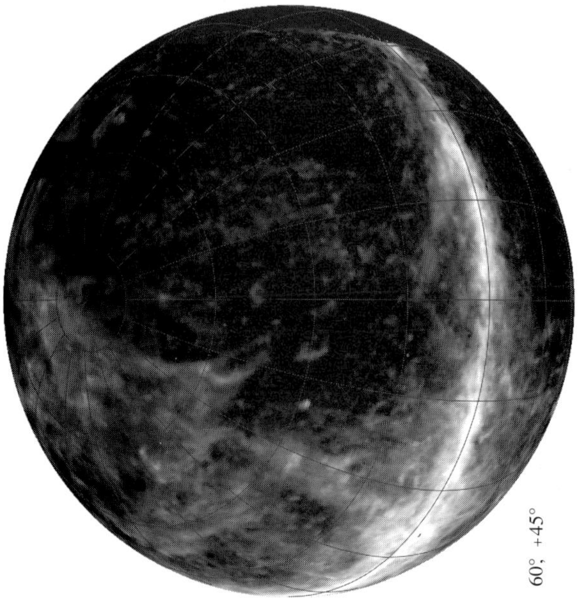

60°, +45°

180°, +45°

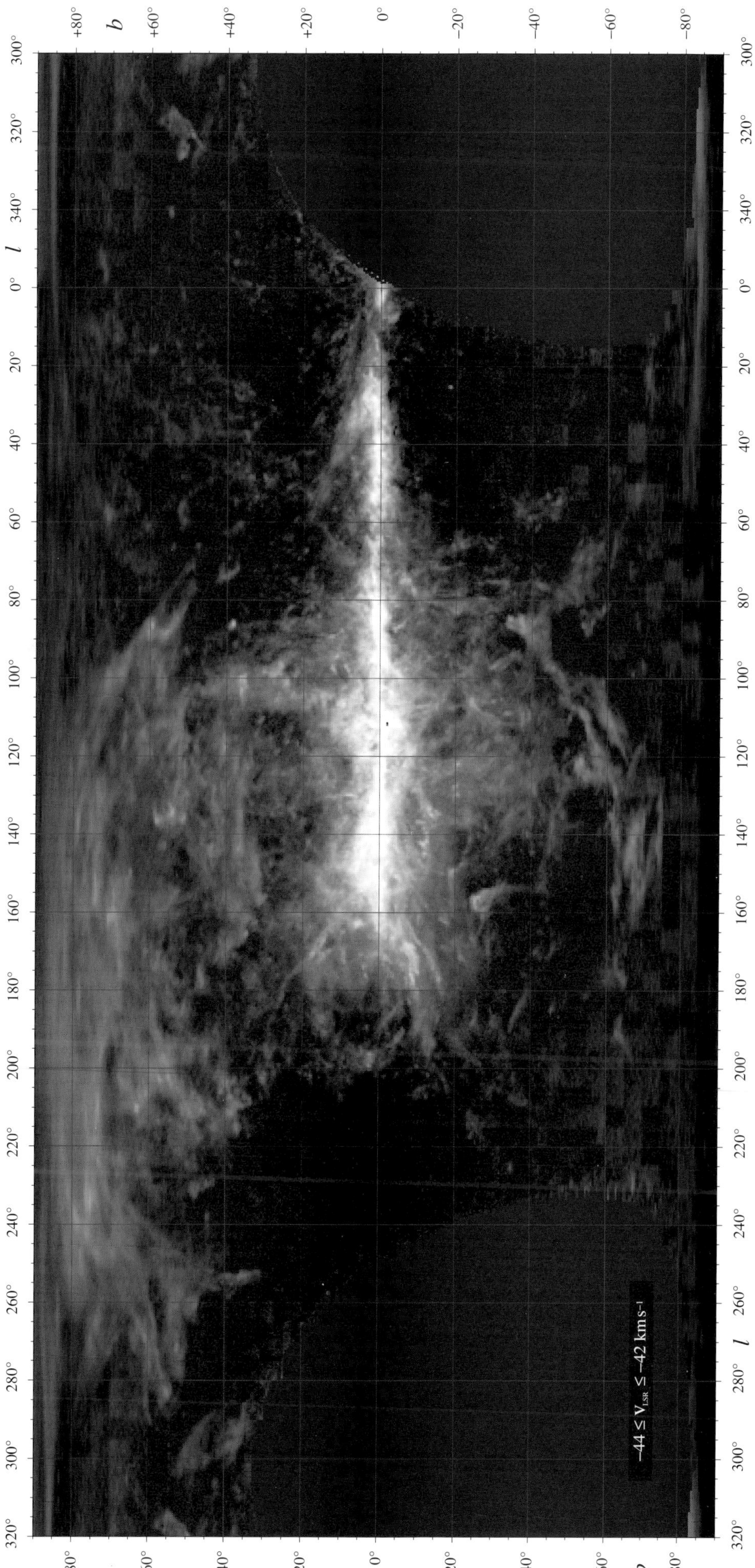

−44 ≤ V_LSR ≤ −42 kms−1
l
b

$-42 \leq V_{LSR} \leq -40$ km s^{-1}

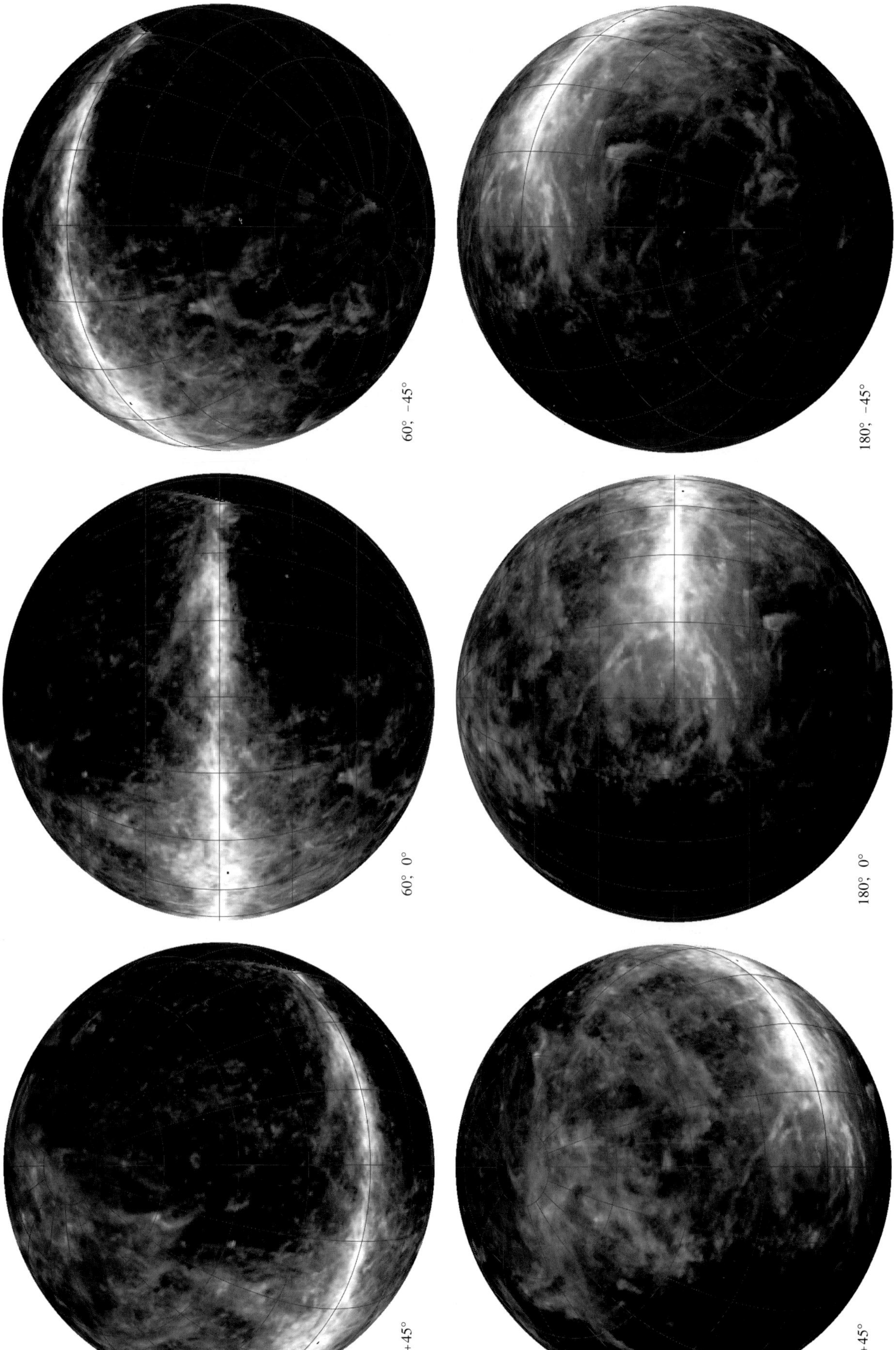

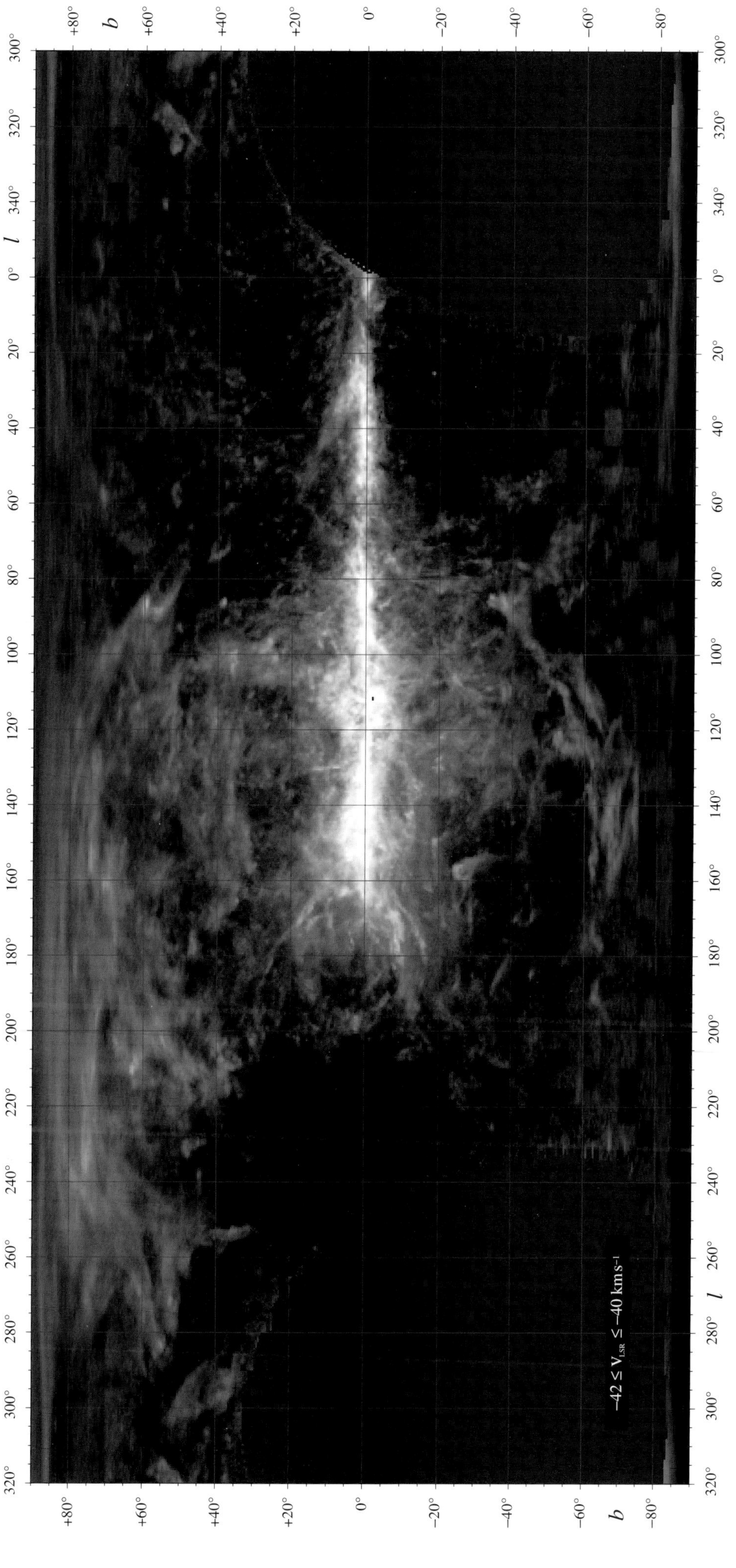
l
b
$-42 \leq V_{LSR} \leq -40$ km s^{-1}

$-40 \leq V_{LSR} \leq -38$ km s^{-1}

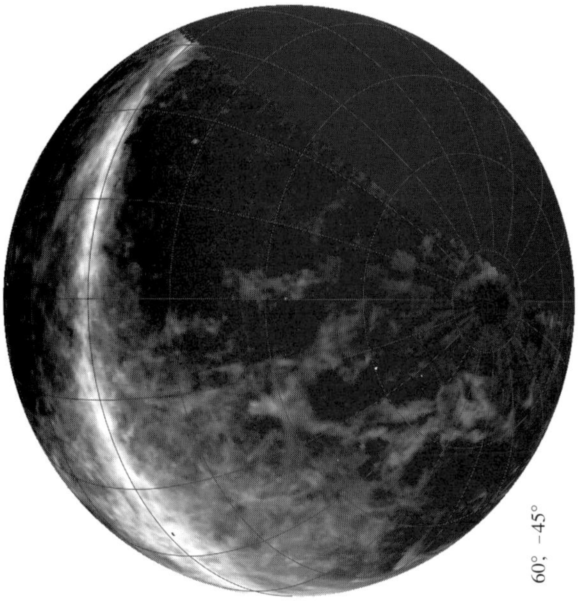

60°, −45°

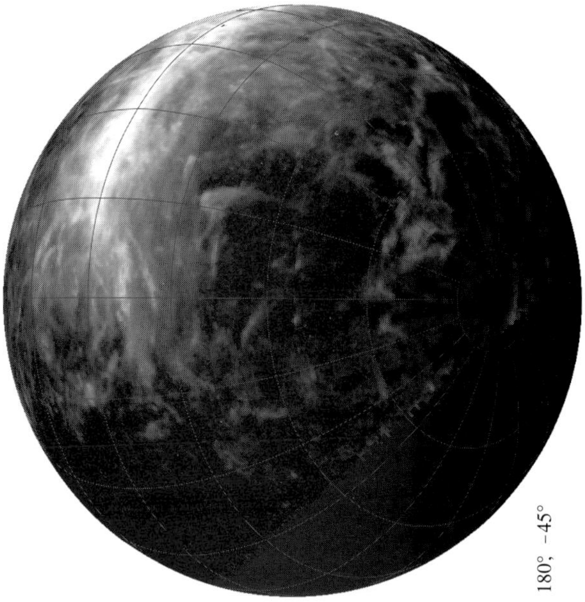

180°, −45°

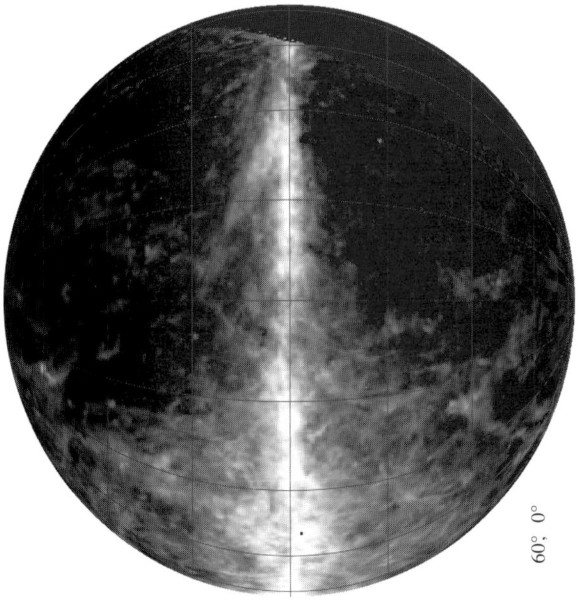

60°, 0°

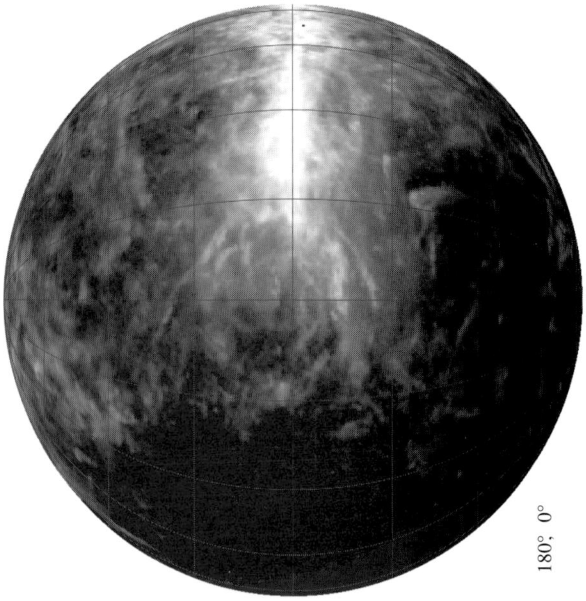

180°, 0°

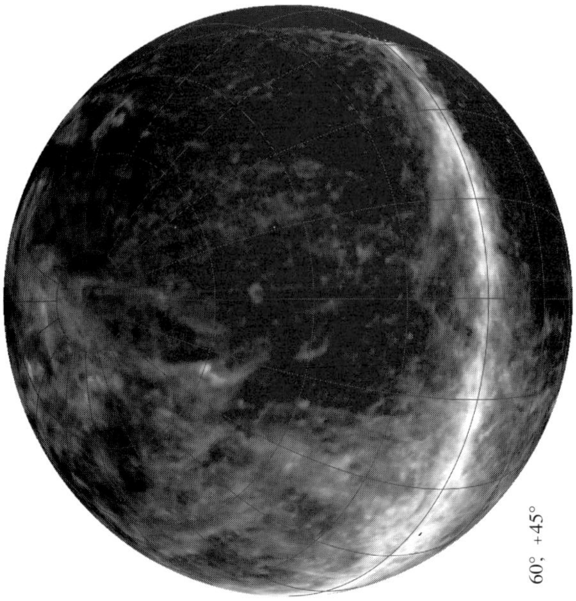

60°, +45°

180°, +45°

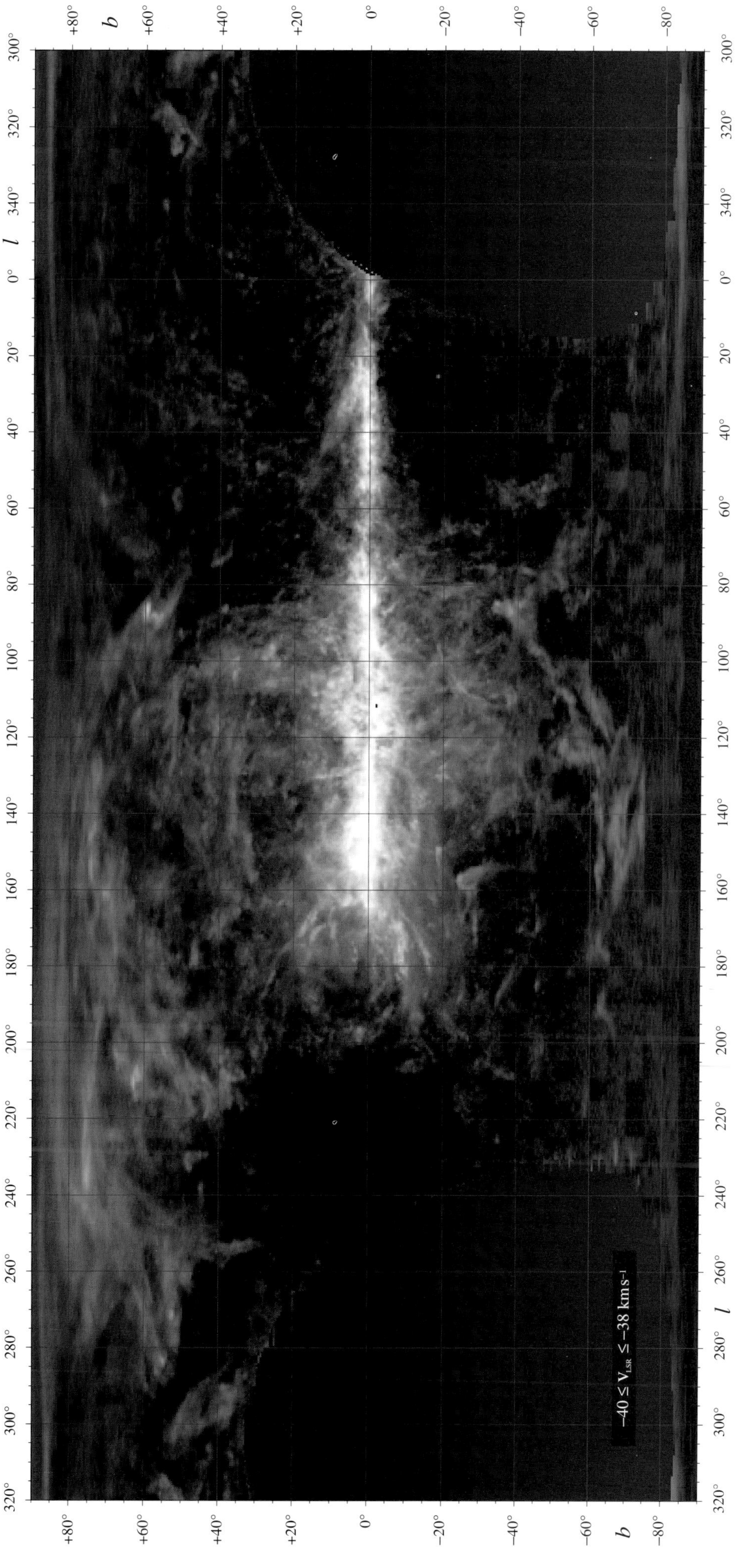
−40 ≤ V_LSR ≤ −38 kms−1
l
b

$-38 \leq V_{LSR} \leq -36$ km s^{-1}

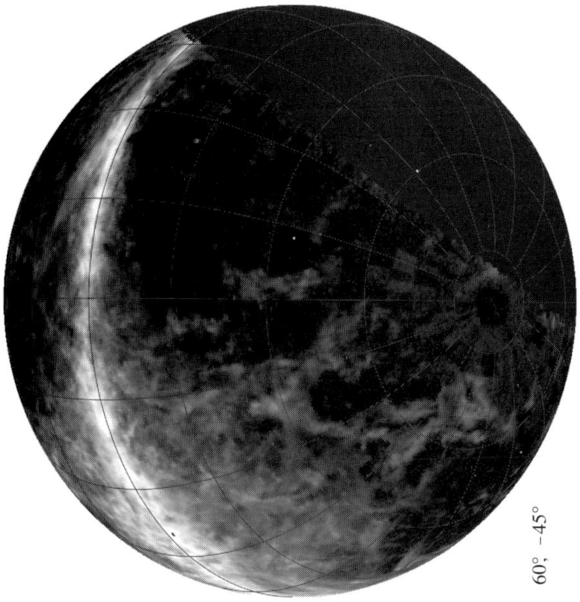

60°, −45°

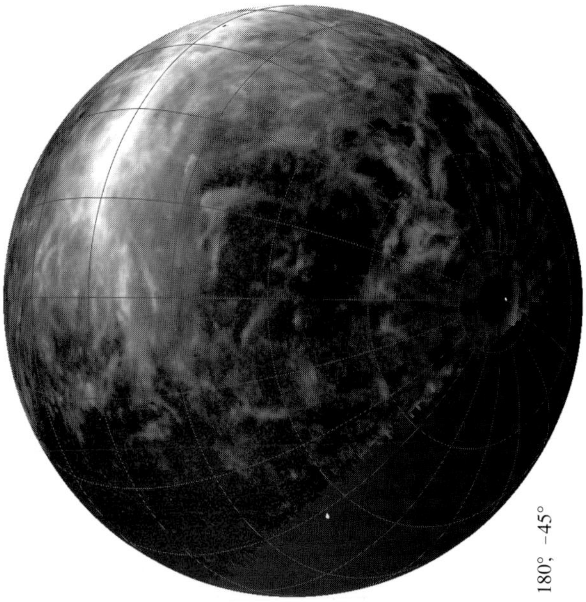

180°, −45°

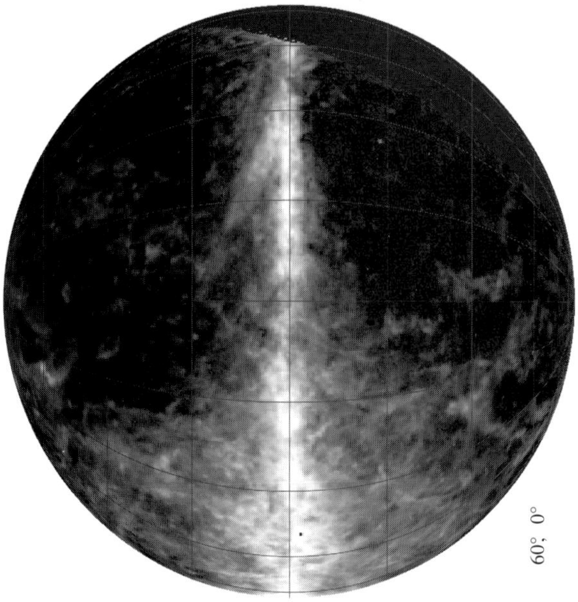

60°, 0°

180°, 0°

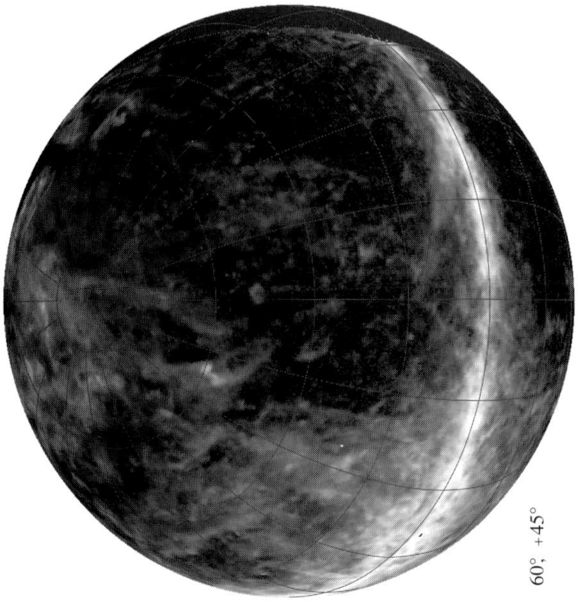

60°, +45°

180°, +45°

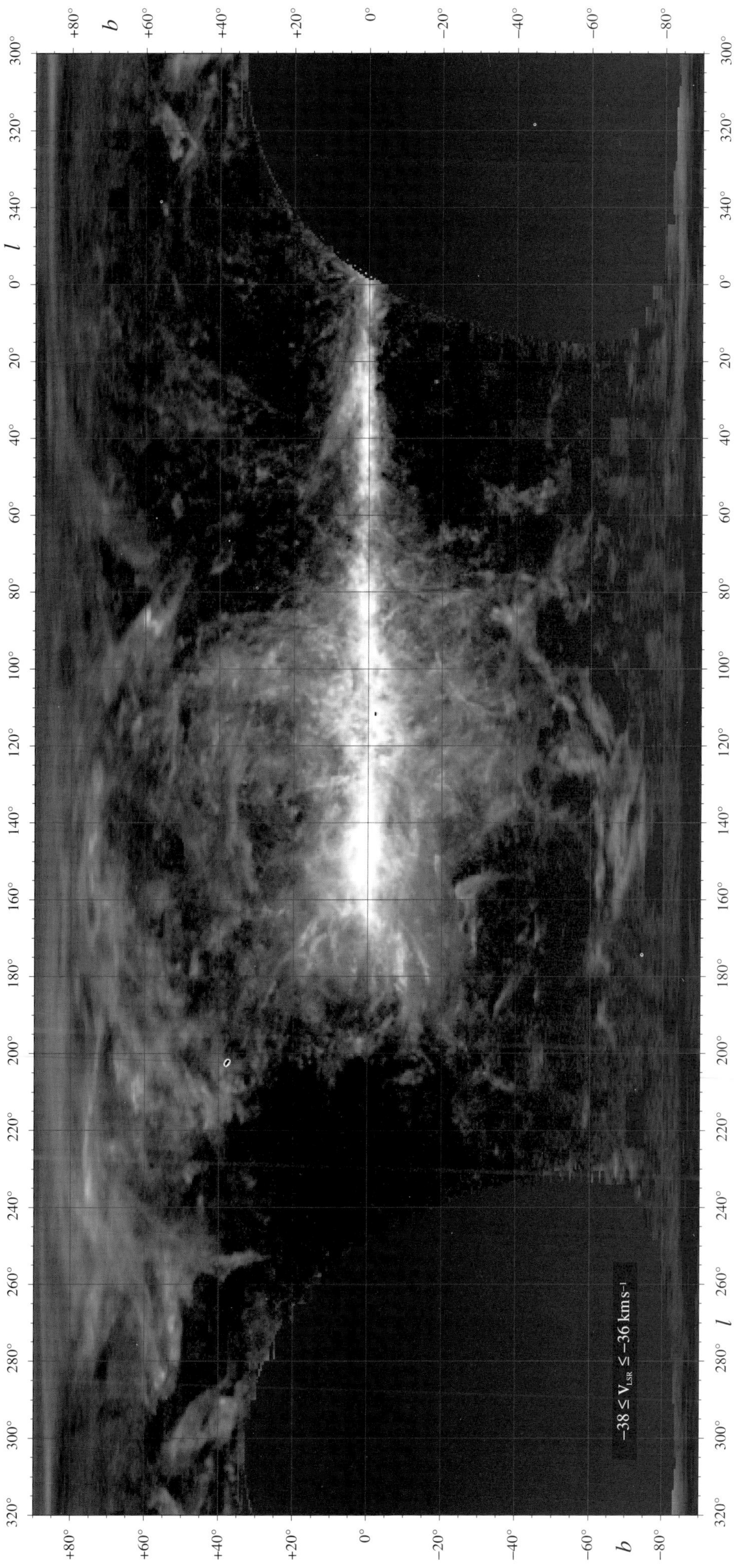
−38 ≤ V_LSR ≤ −36 km s⁻¹

$-36 \leq V_{LSR} \leq -34$ km s^{-1}

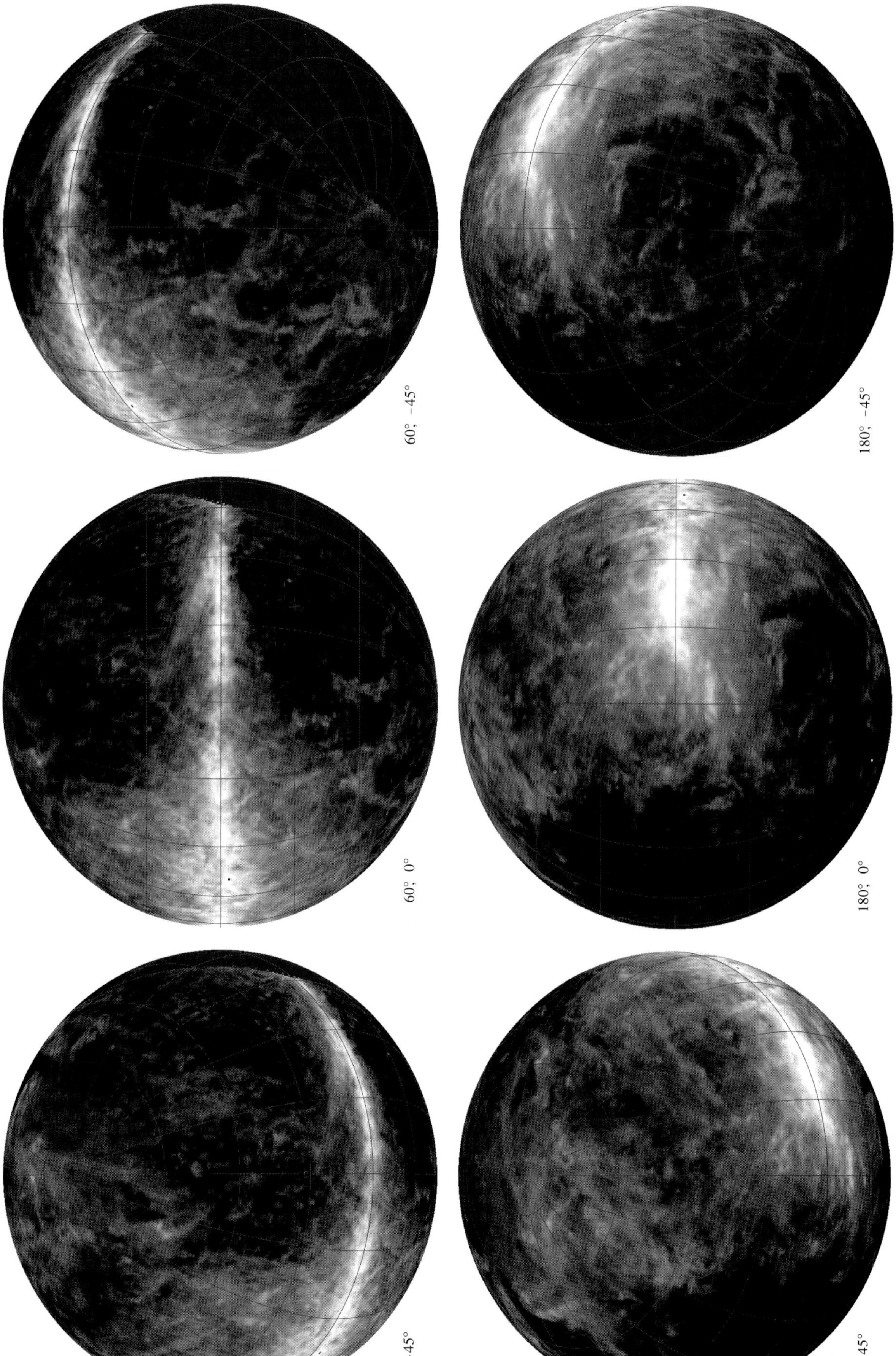

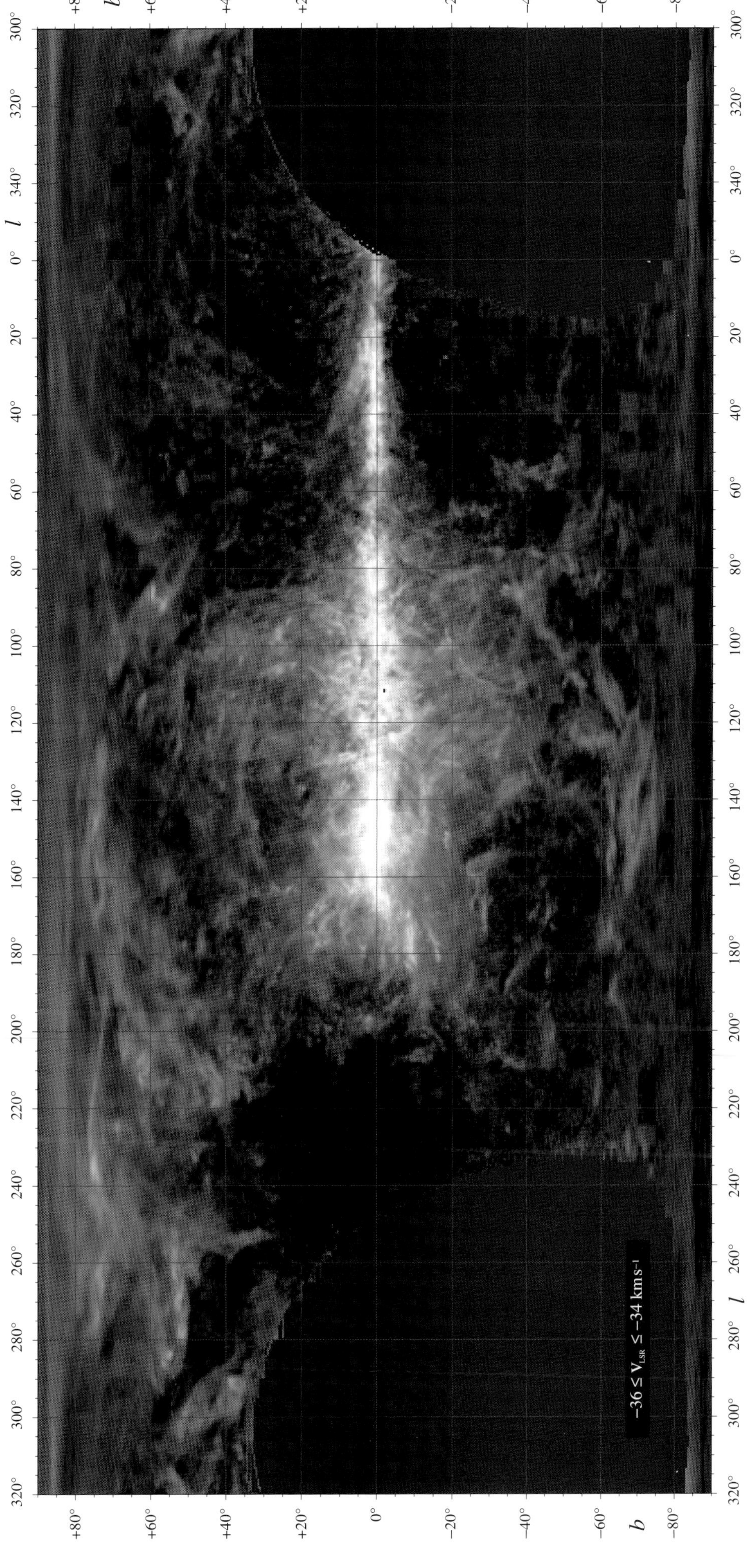

−36 ≤ V_LSR ≤ −34 kms⁻¹

$-34 \leq V_{LSR} \leq -32$ km s^{-1}

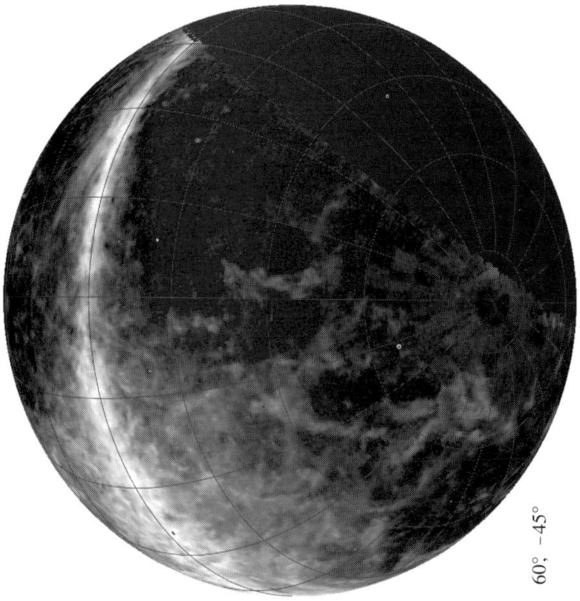

60°, −45°

180°, −45°

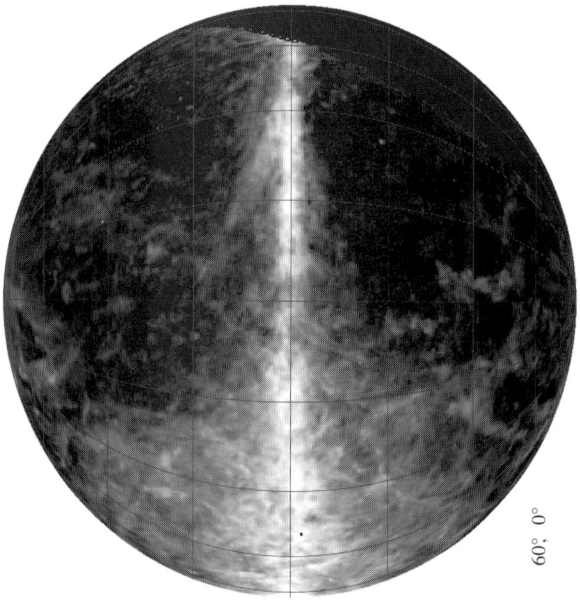

60°, 0°

180°, 0°

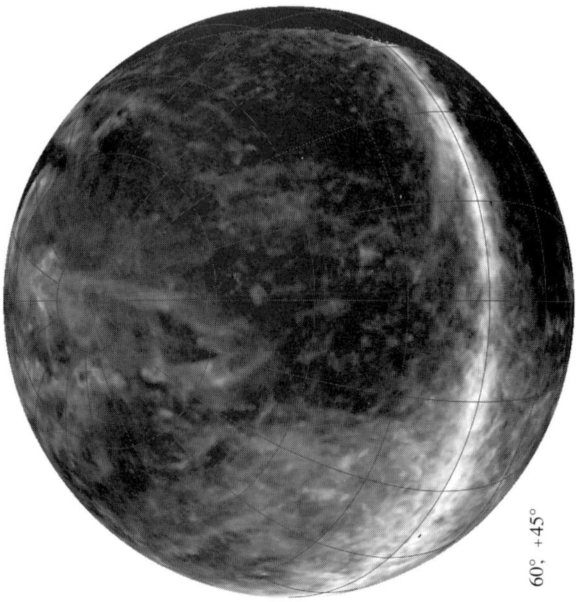

60°, +45°

180°, +45°

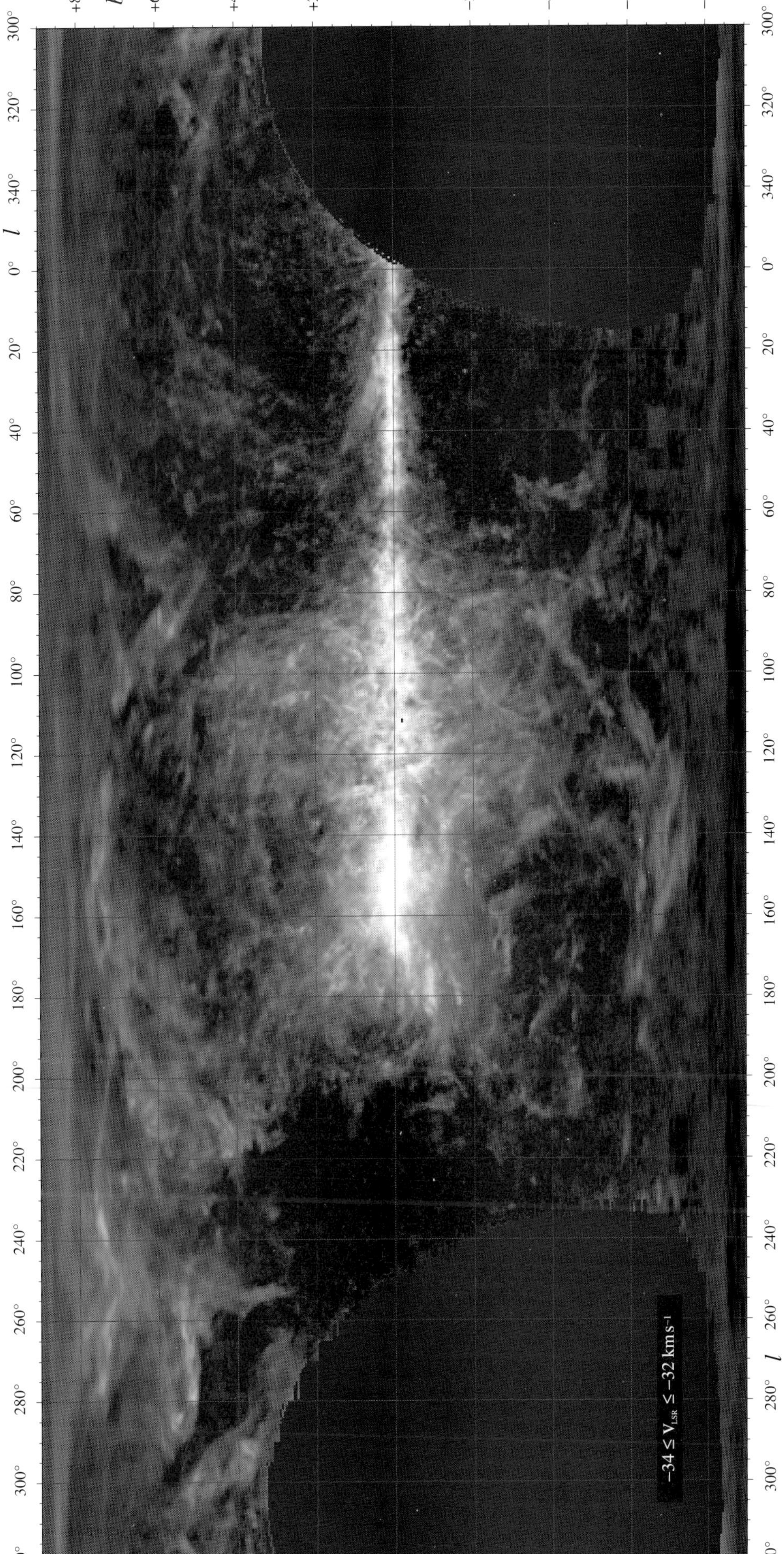
$-34 \le V_{LSR} \le -32$ km s^{-1}
l
b
300° 320° 340° 0° 20° 40° 60° 80° 100° 120° 140° 160° 180° 200° 220° 240° 260° 280° 300° 320°
+80° +60° +40° +20° 0° −20° −40° −60° −80°

$-32 \leq V_{LSR} \leq -30$ km s^{-1}

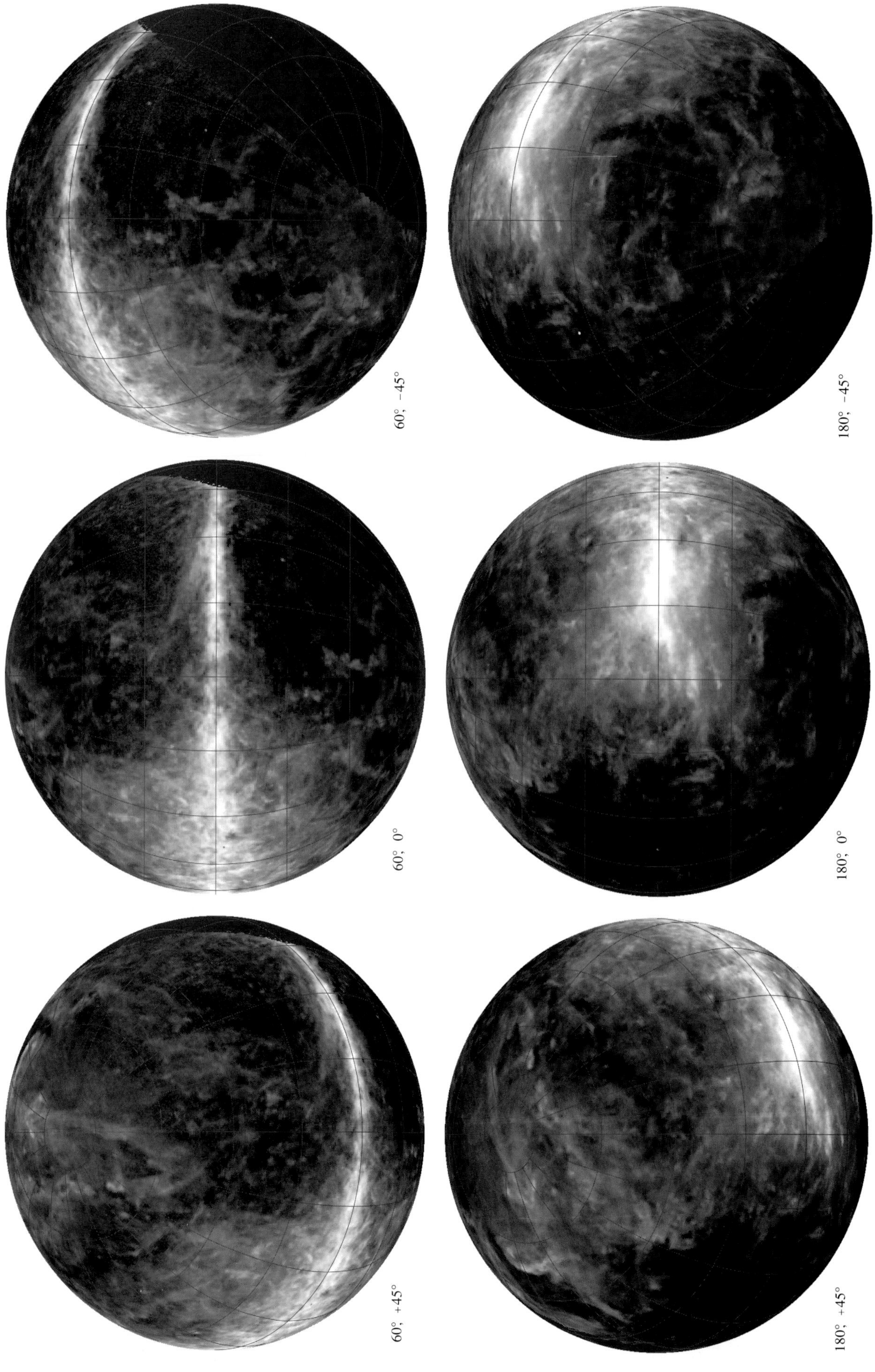

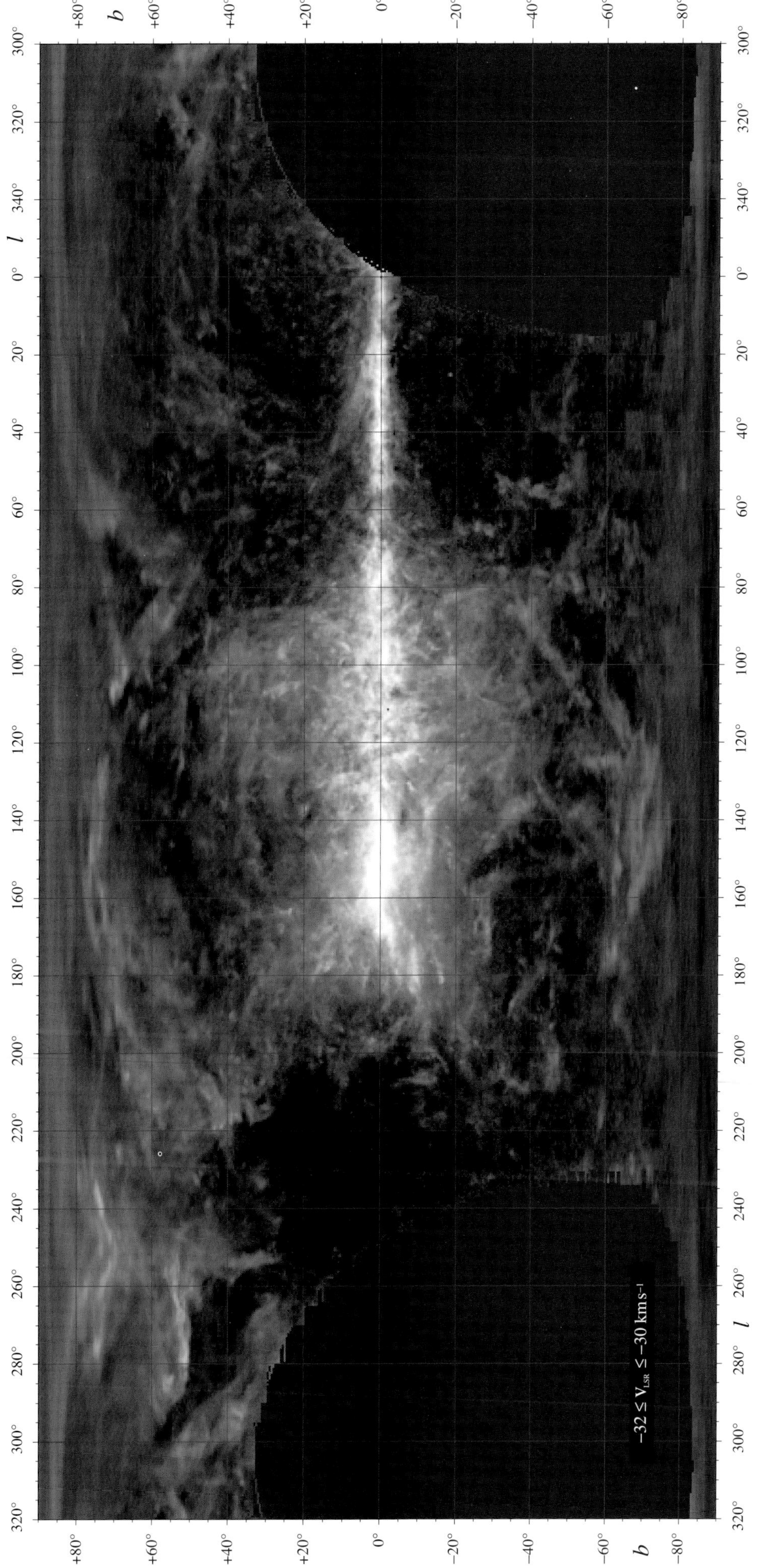
−32 ≤ V_{LSR} ≤ −30 km s⁻¹
l
b
300° 320° 340° 0° 20° 40° 60° 80° 100° 120° 140° 160° 180° 200° 220° 240° 260° 280° 300° 320°
+80° +60° +40° +20° 0° −20° −40° −60° −80°

$-30 \le V_{LSR} \le -28$ km s^{-1}

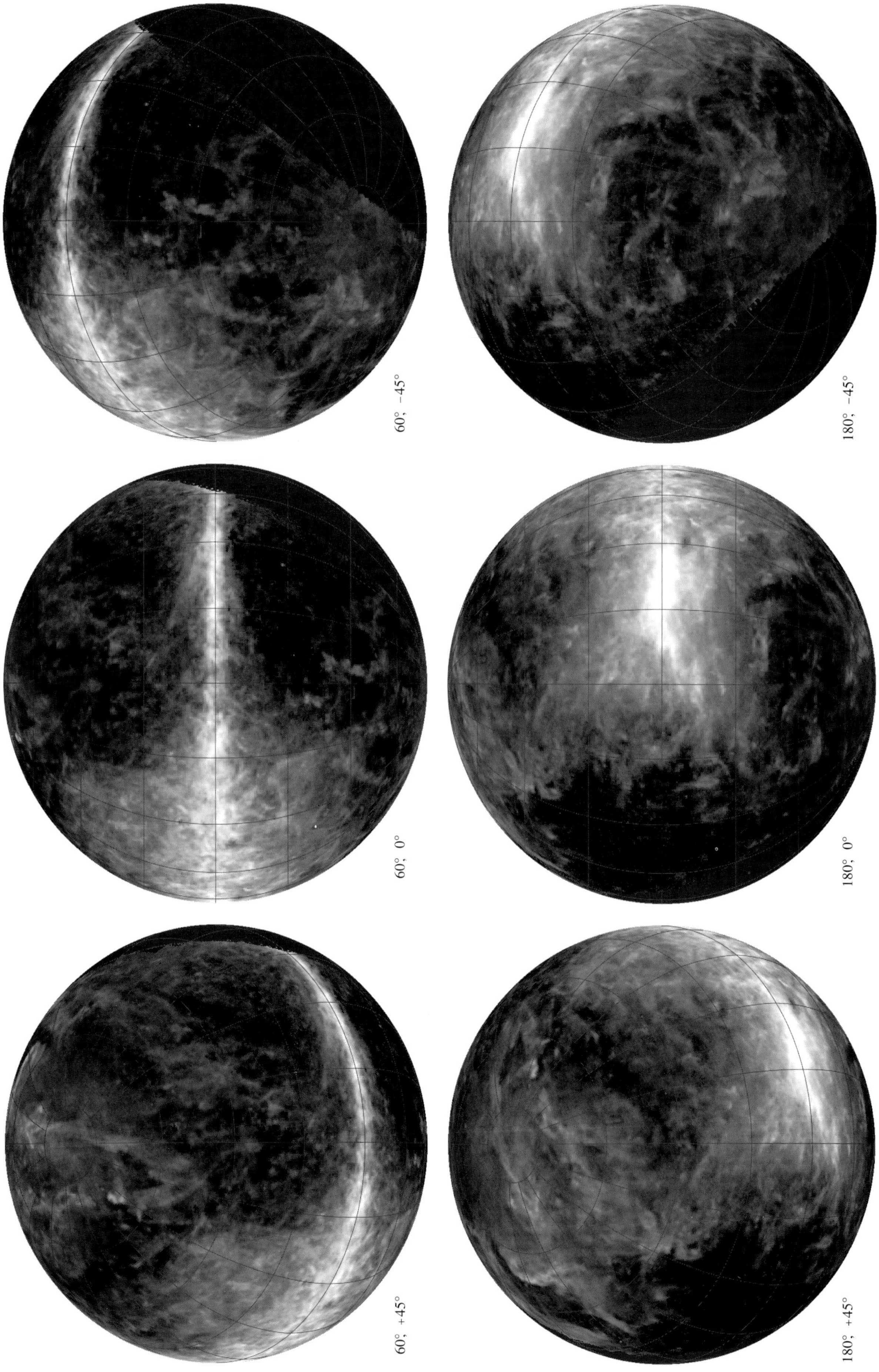

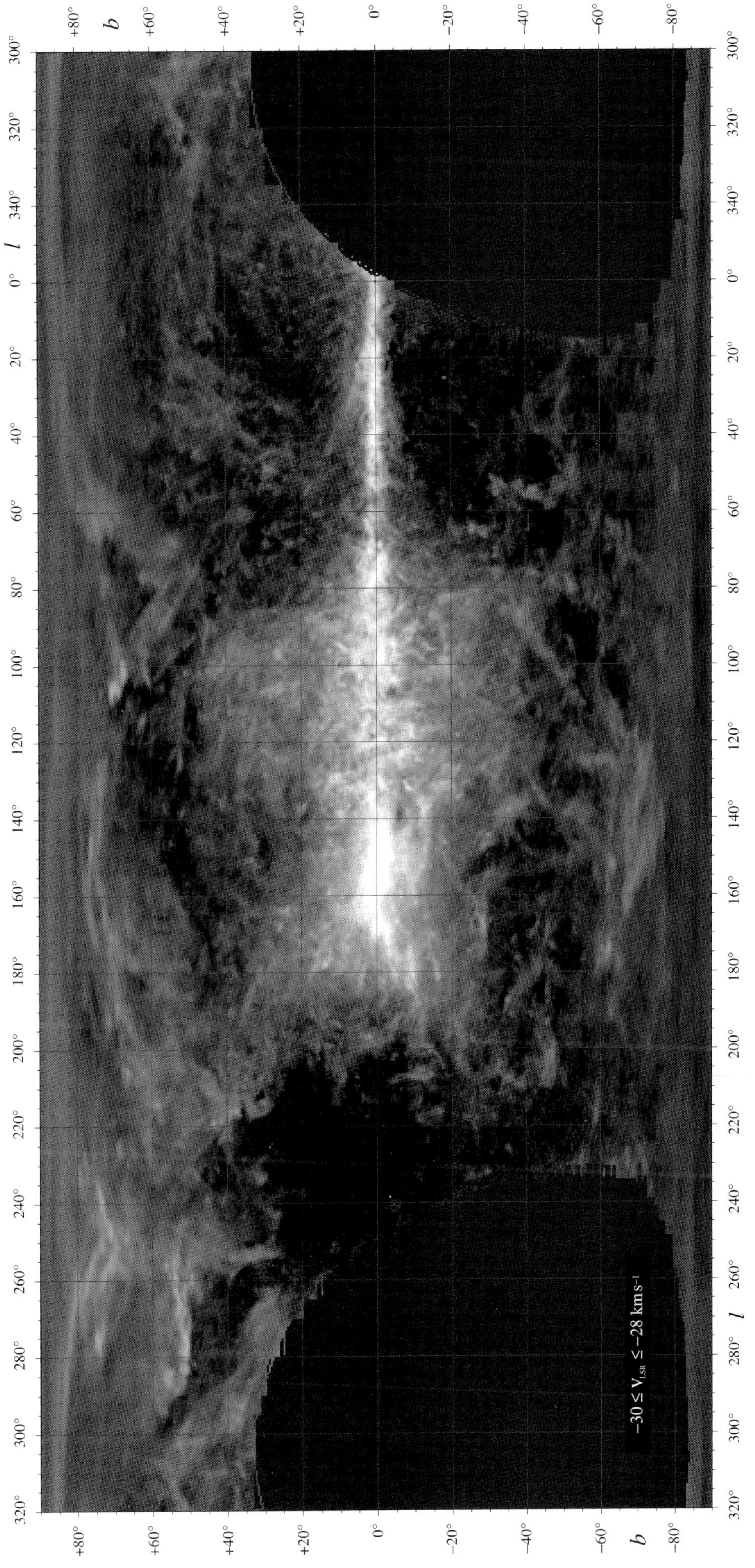
$-30 \le V_{LSR} \le -28$ km s^{-1}
l
b

$-28 \leq V_{LSR} \leq -26$ km s^{-1}

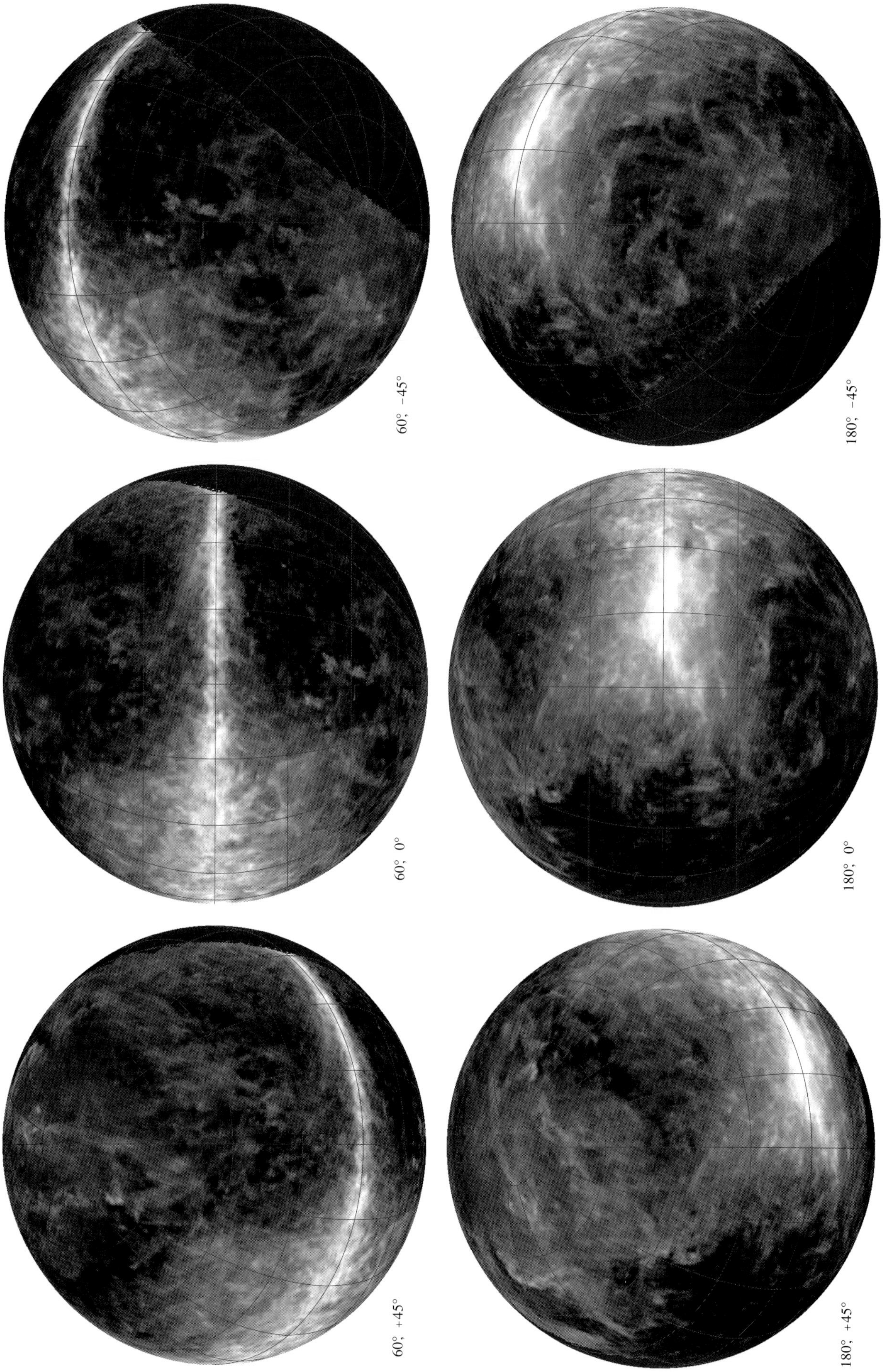

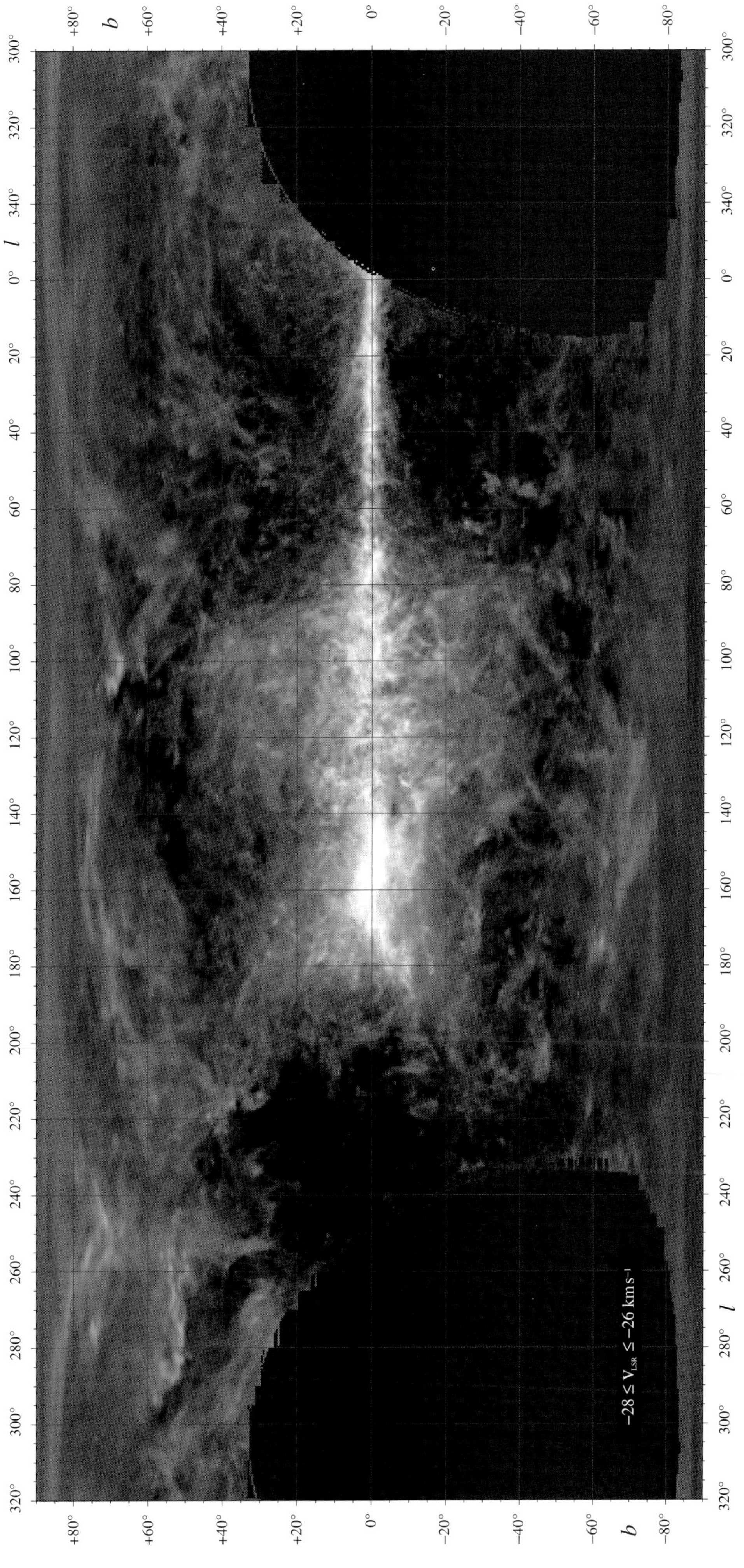
−28 ≤ V_LSR ≤ −26 km s⁻¹
l
b

$-26 \leq V_{LSR} \leq -24$ km s^{-1}

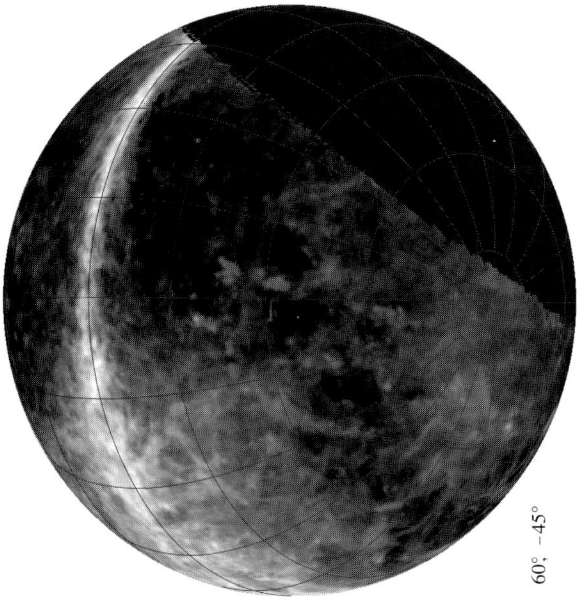

60°, −45°

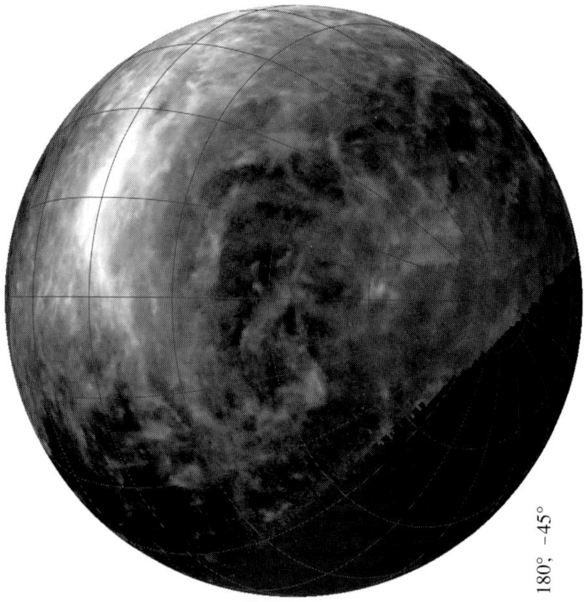

180°, −45°

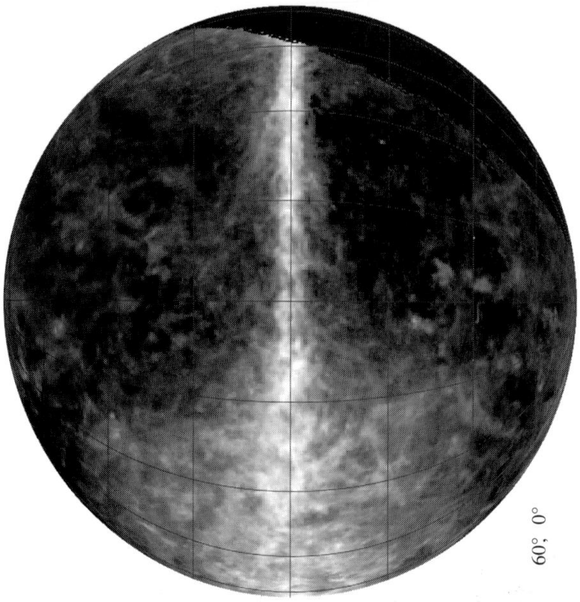

60°, 0°

180°, 0°

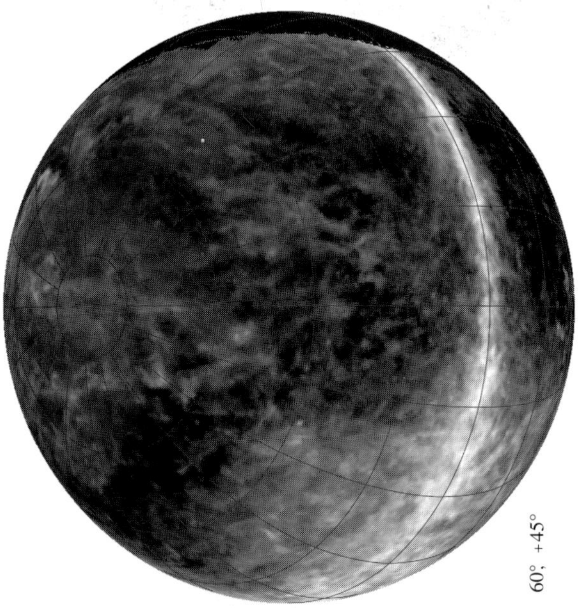

60°, +45°

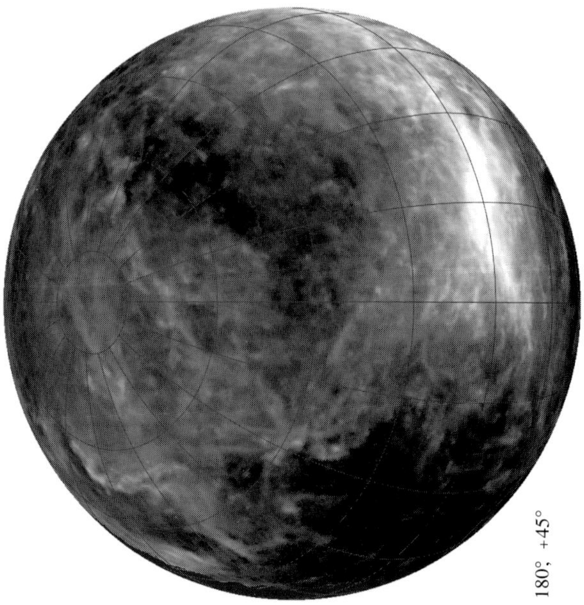

180°, +45°

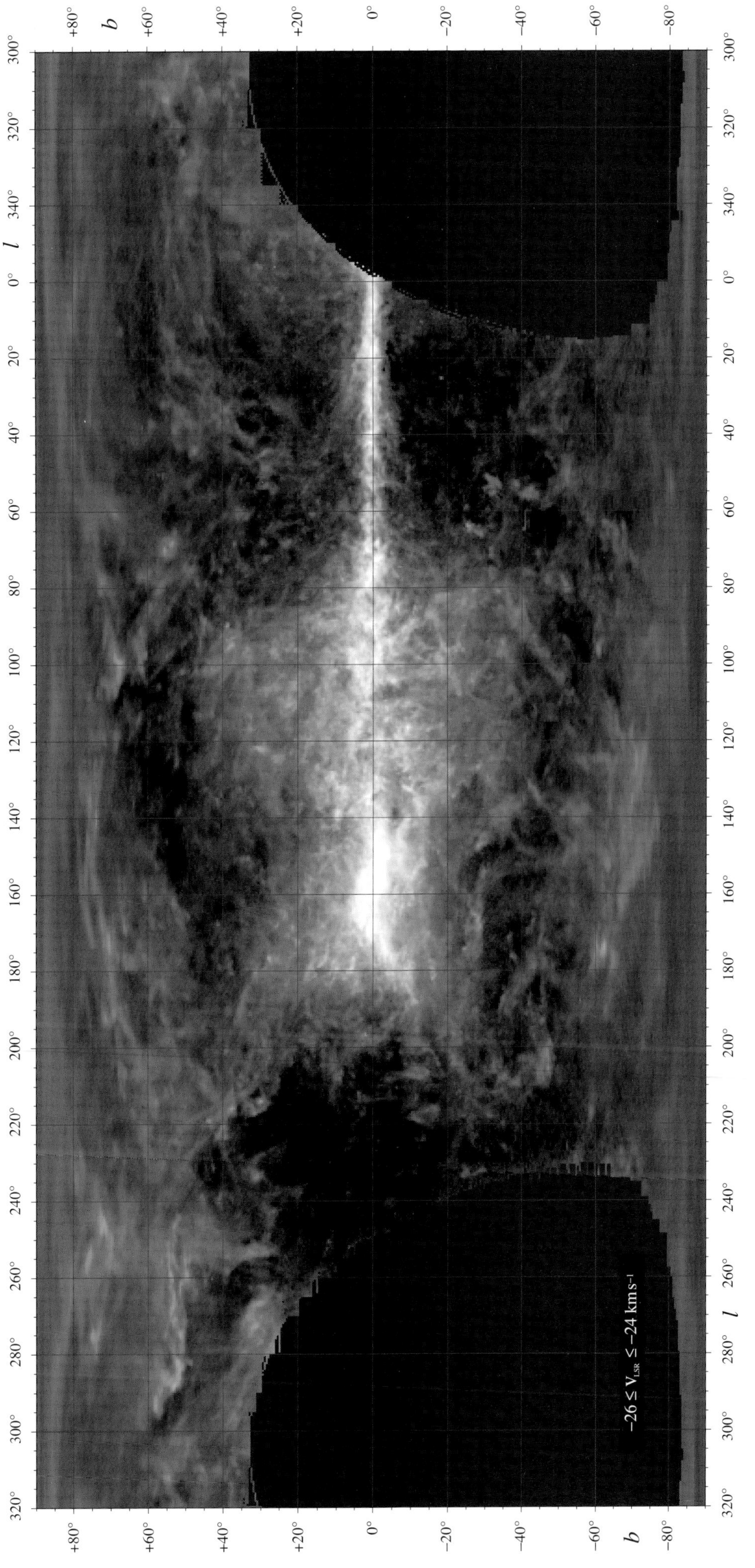

−26 ≤ V_LSR ≤ −24 kms−1
l
b
300° 320° 340° 0° 20° 40° 60° 80° 100° 120° 140° 160° 180° 200° 220° 240° 260° 280° 300° 320°
+80° +60° +40° +20° 0° −20° −40° −60° −80°

$-24 \le V_{LSR} \le -22\ \mathrm{km\,s^{-1}}$

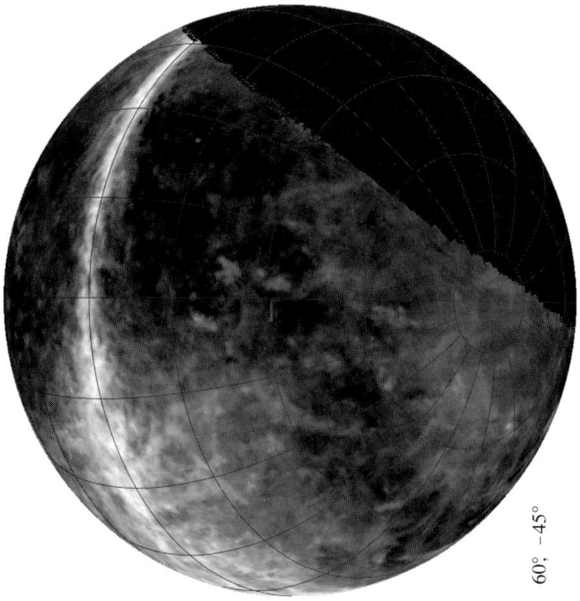

60°; −45°

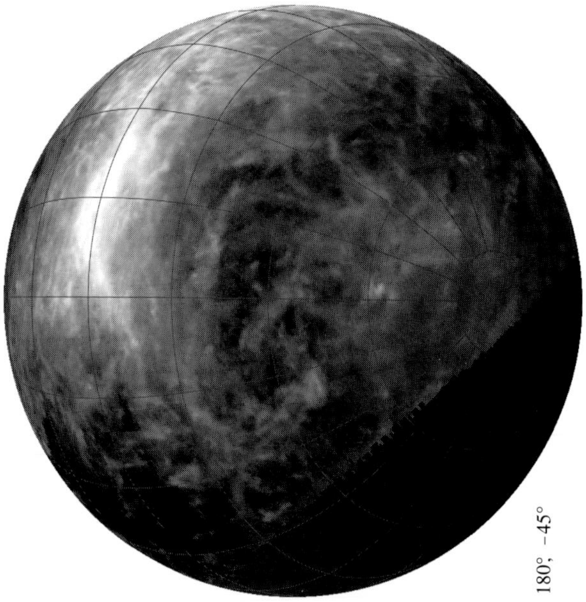

180°; −45°

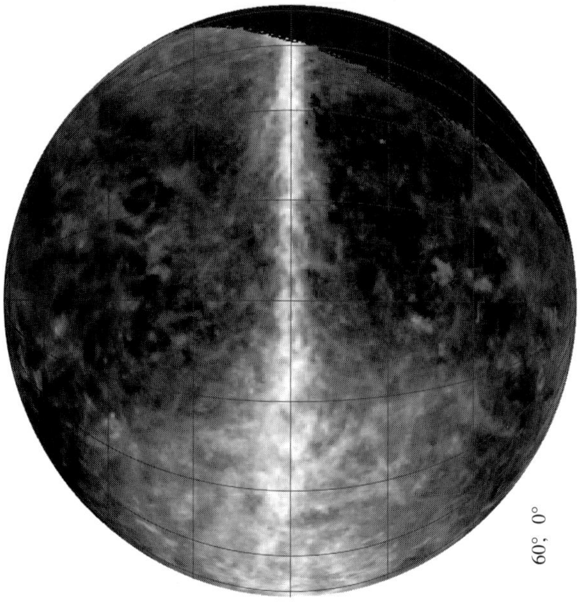

60°; 0°

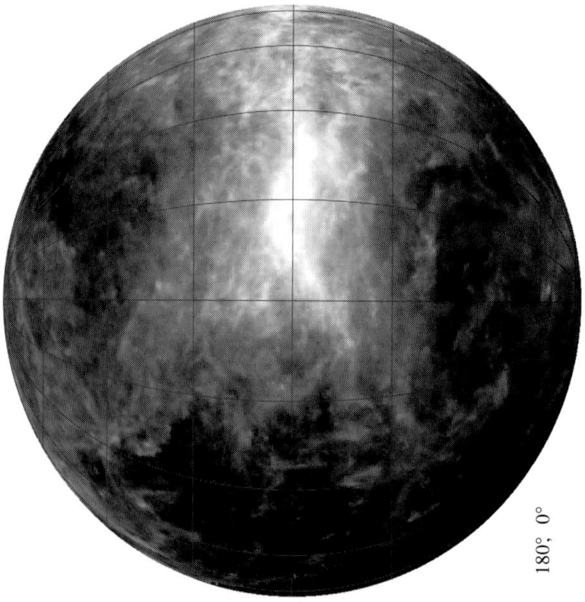

180°; 0°

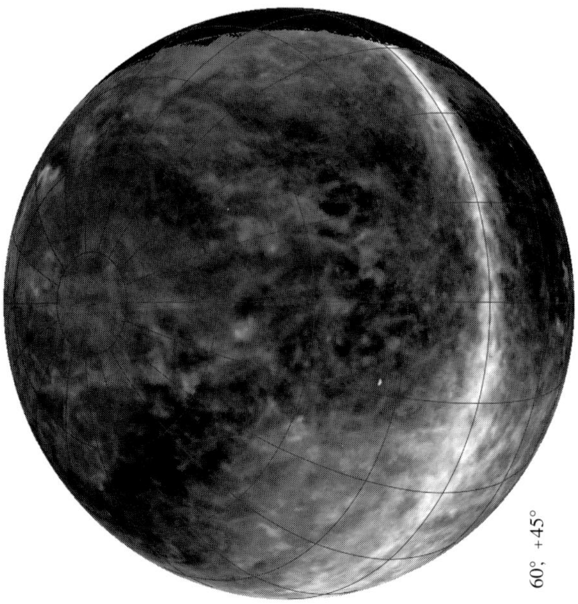

60°; +45°

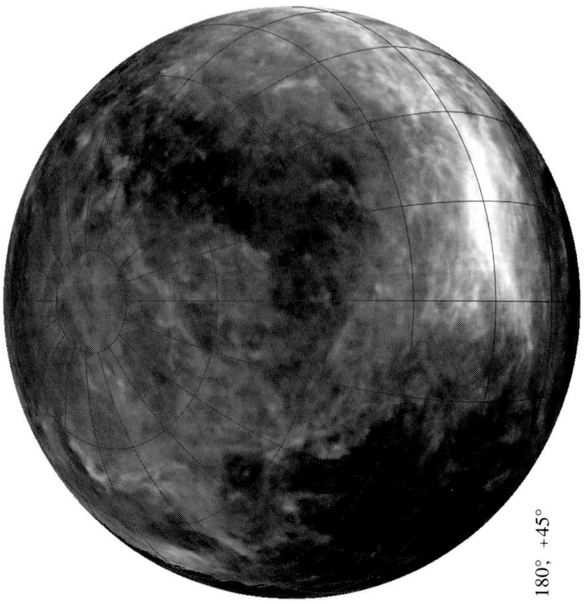

180°; +45°

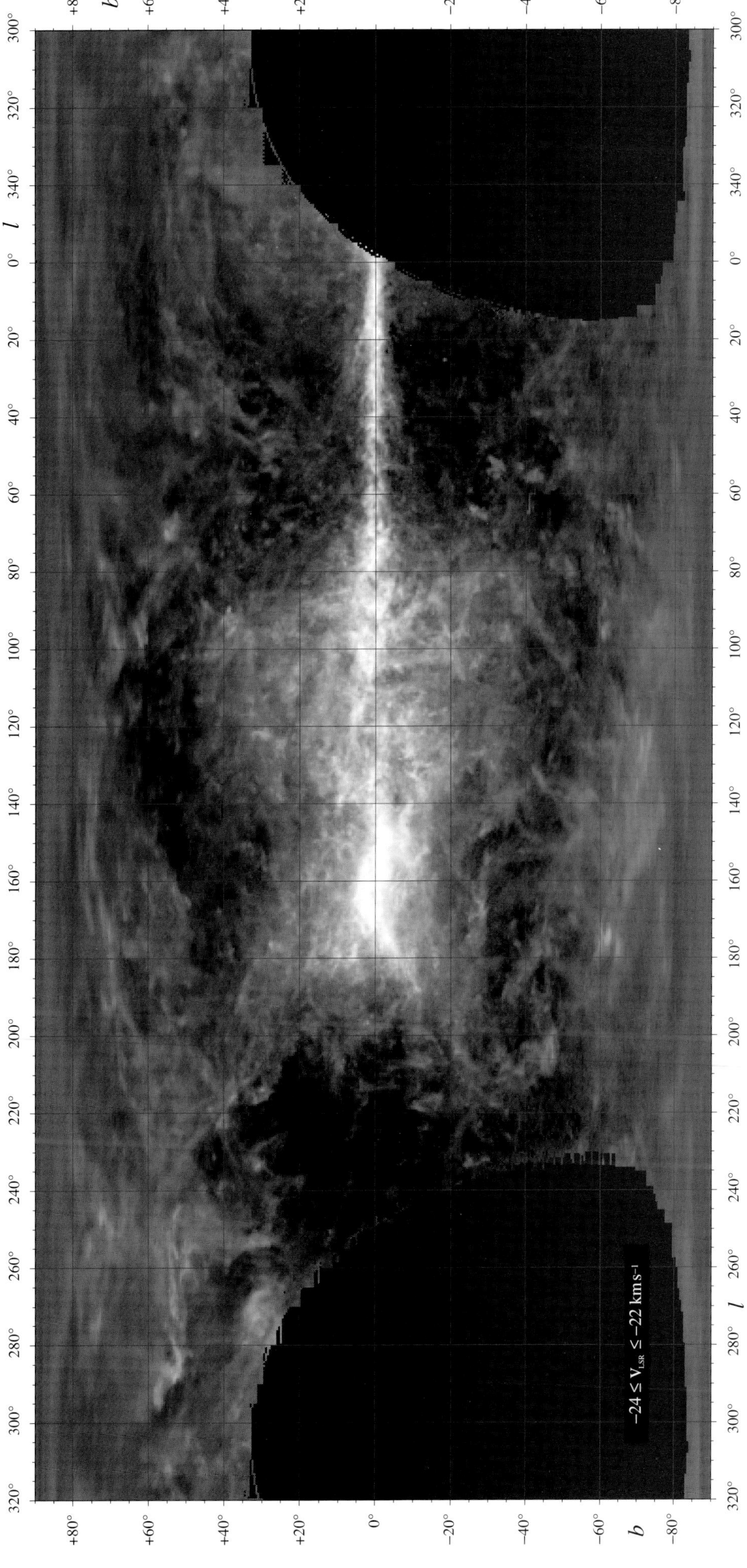
−24 ≤ V_{LSR} ≤ −22 km s⁻¹
l
b
300° 320° 340° 0° 20° 40° 60° 80° 100° 120° 140° 160° 180° 200° 220° 240° 260° 280° 300° 320°
+80° +60° +40° +20° 0° −20° −40° −60° −80°

$-22 \leq V_{LSR} \leq -20$ km s^{-1}

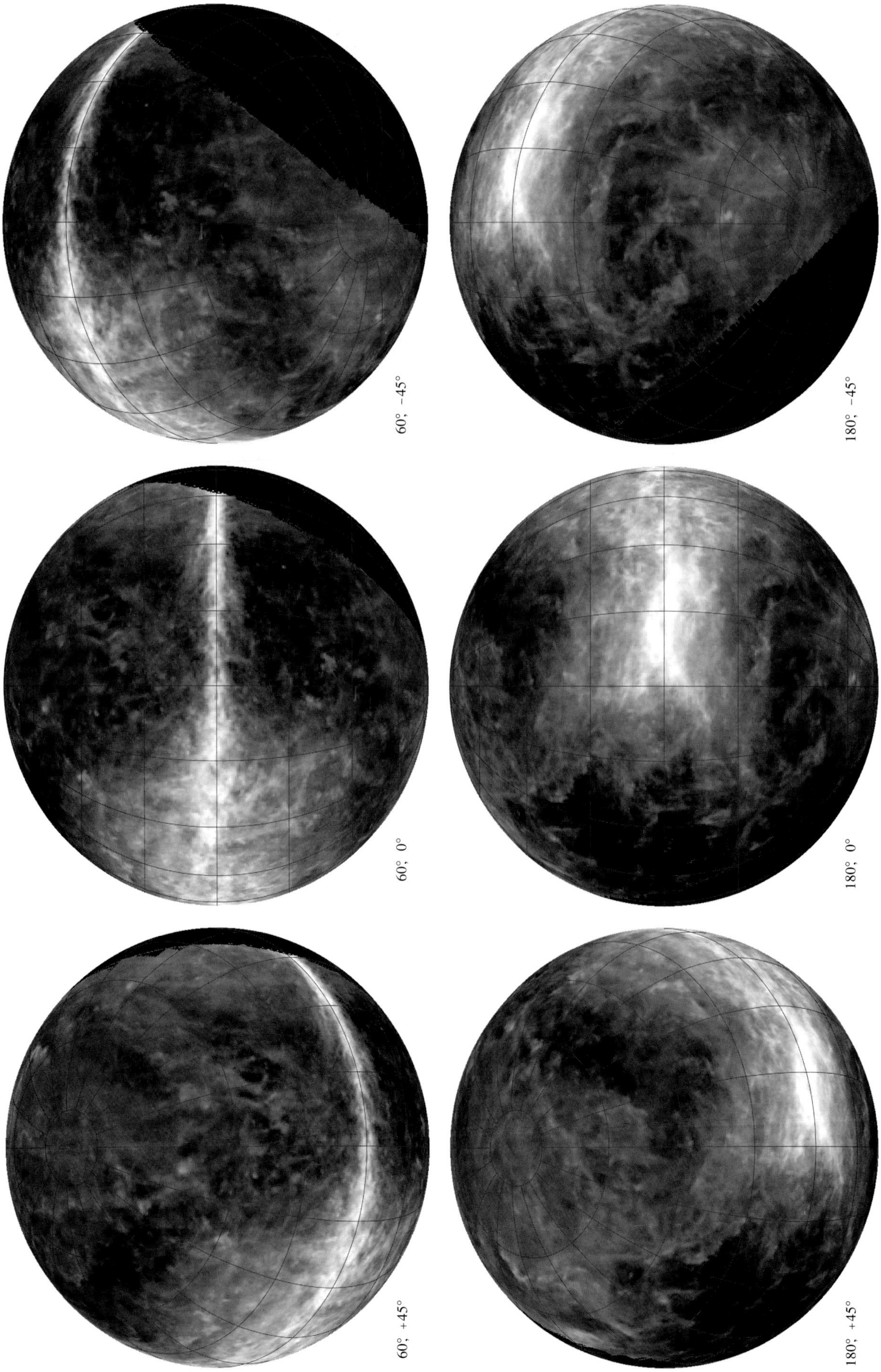

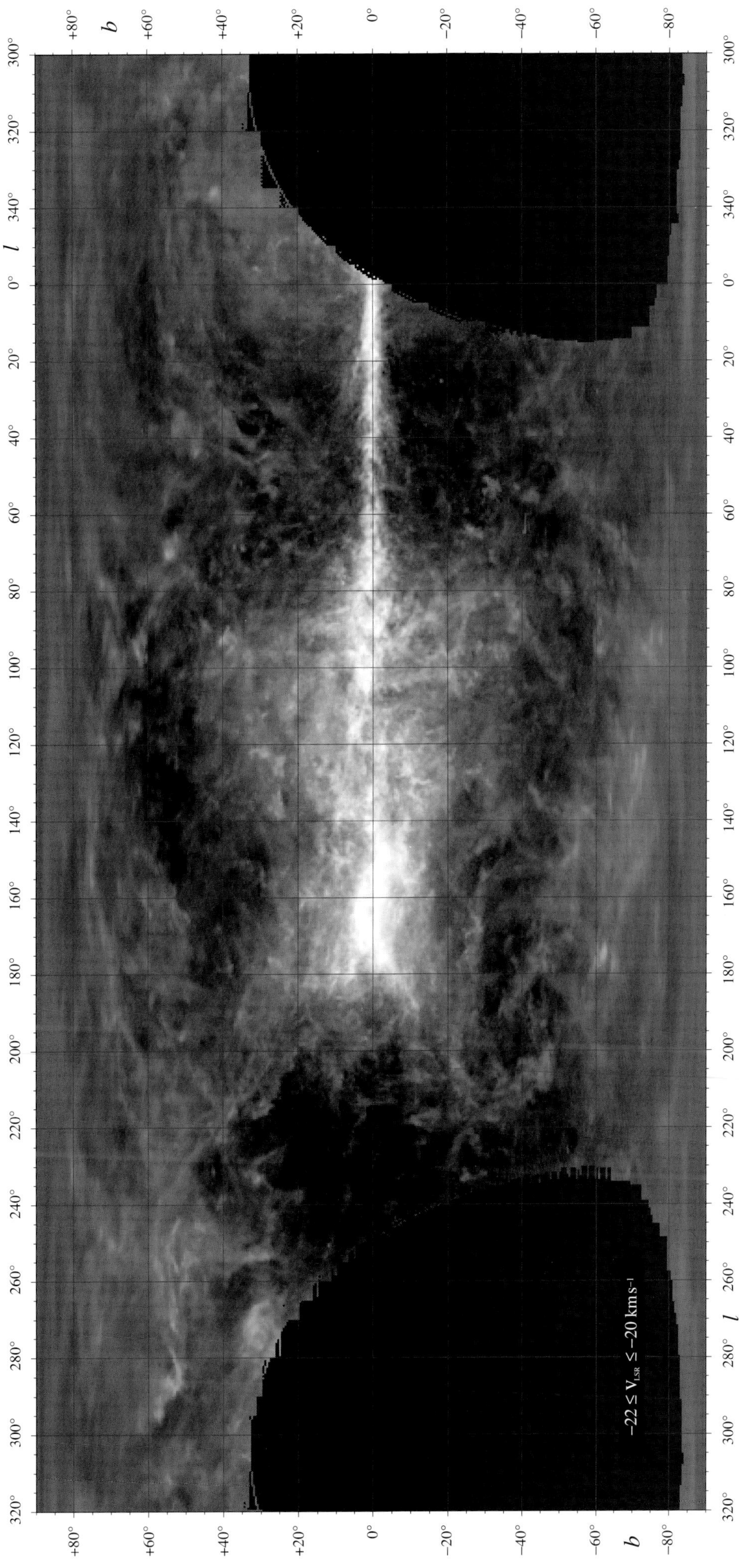
$-22 \le V_{LSR} \le -20$ km s^{-1}
l
b

$-20 \leq V_{LSR} \leq -18$ km s^{-1}

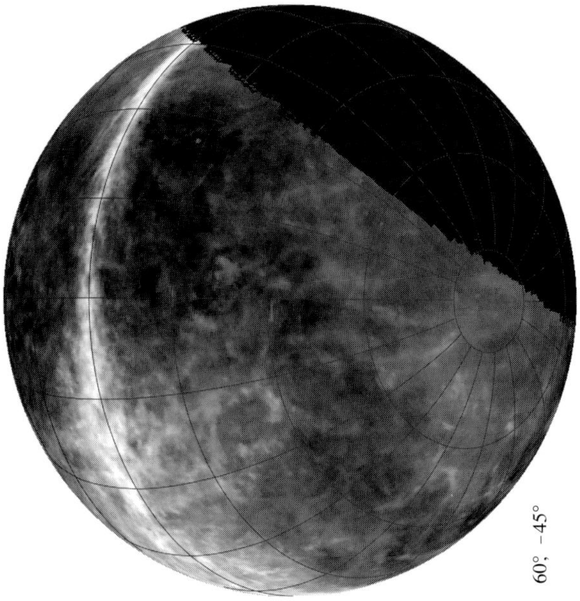

60°, −45°

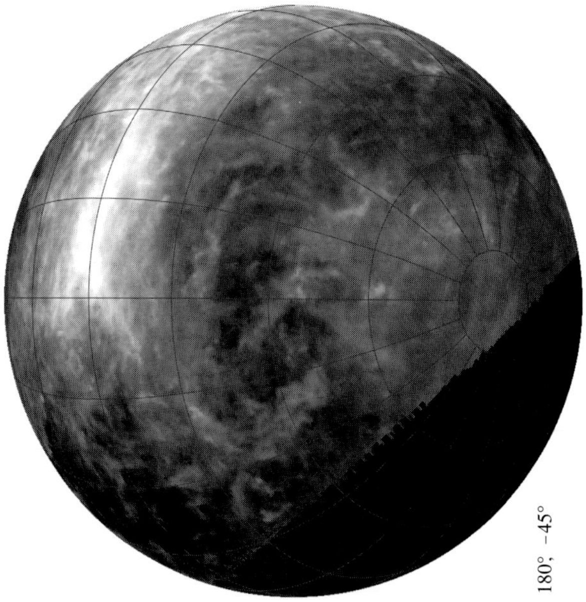

180°, −45°

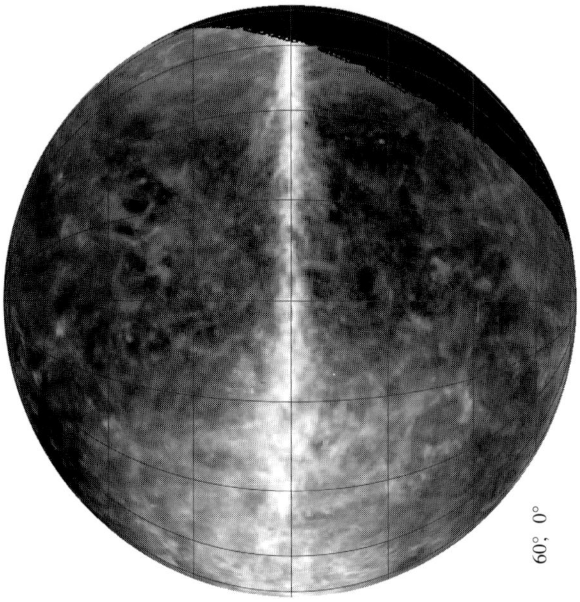

60°, 0°

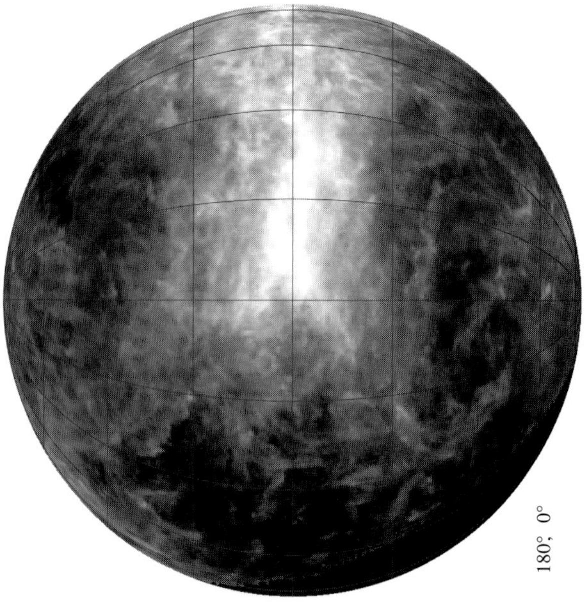

180°, 0°

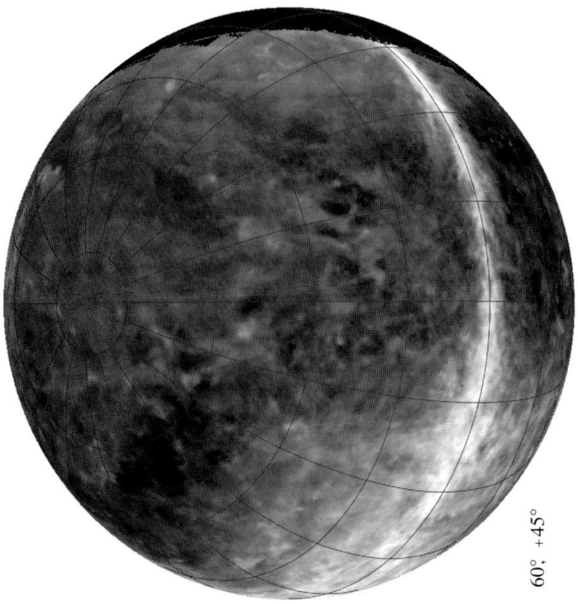

60°, +45°

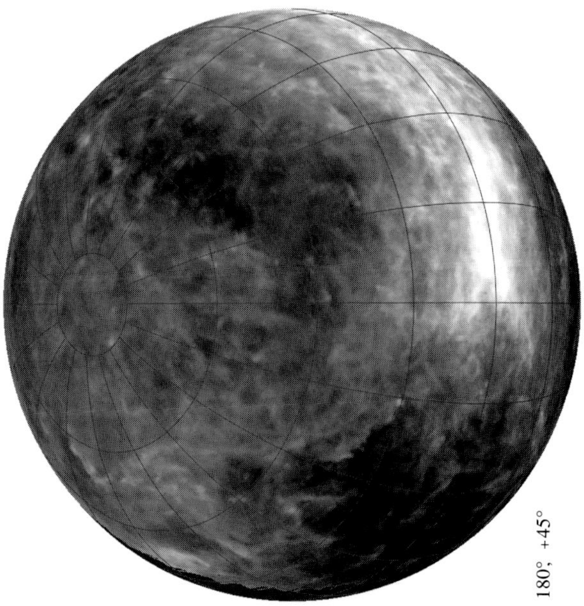

180°, +45°

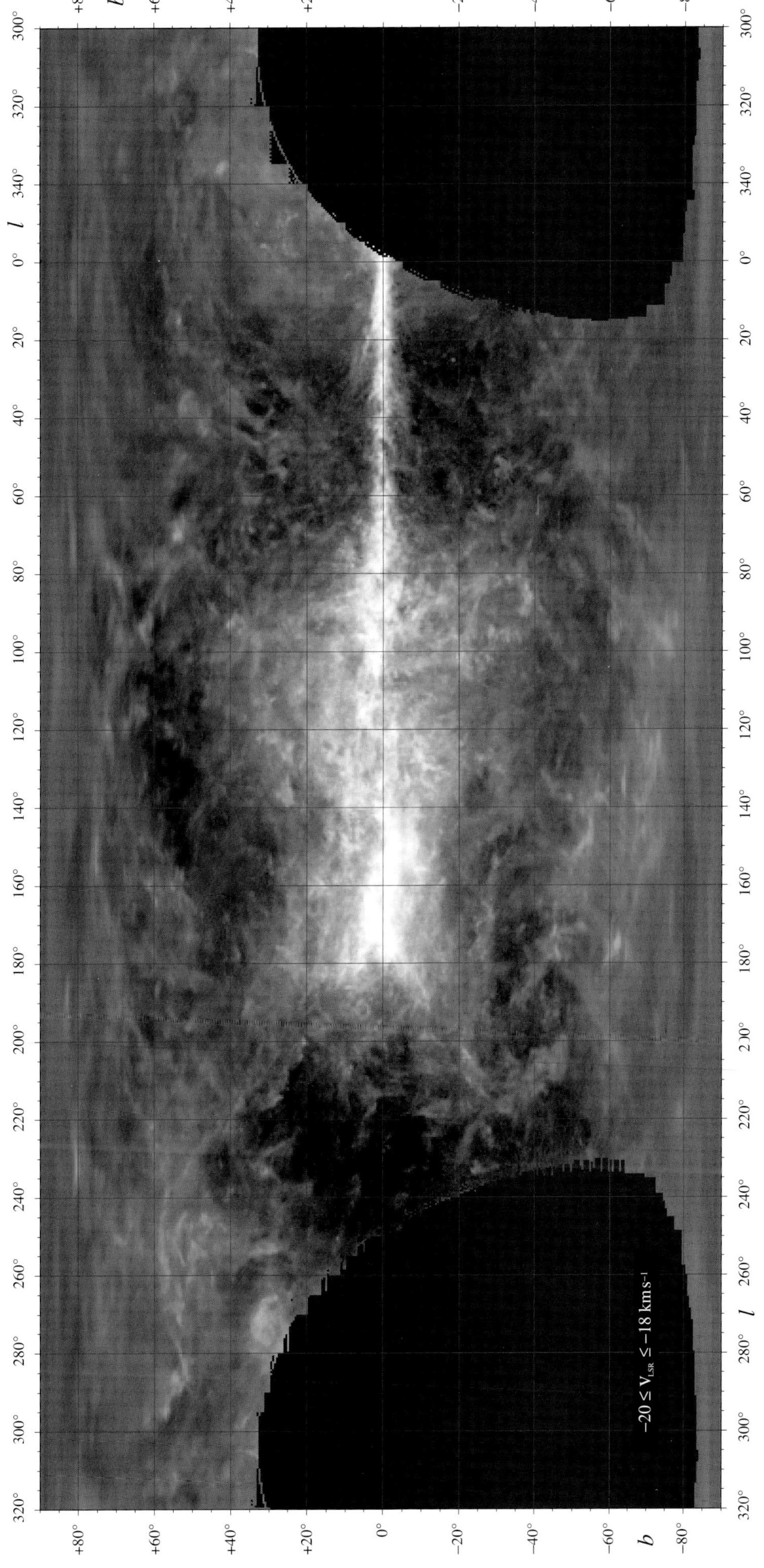
$-20 \le V_{LSR} \le -18$ kms^{-1}

$-18 \le V_{LSR} \le -16$ km s^{-1}

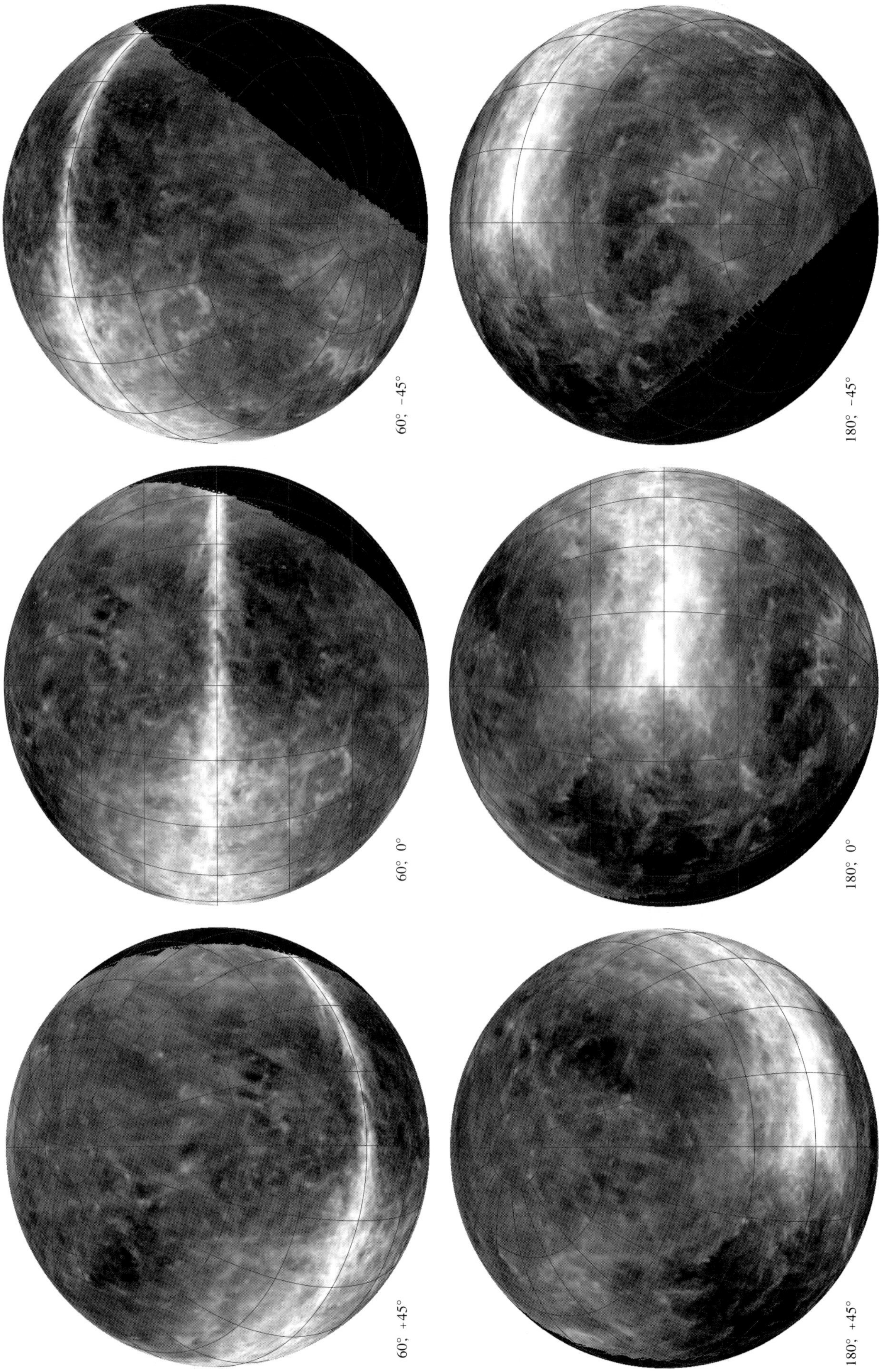

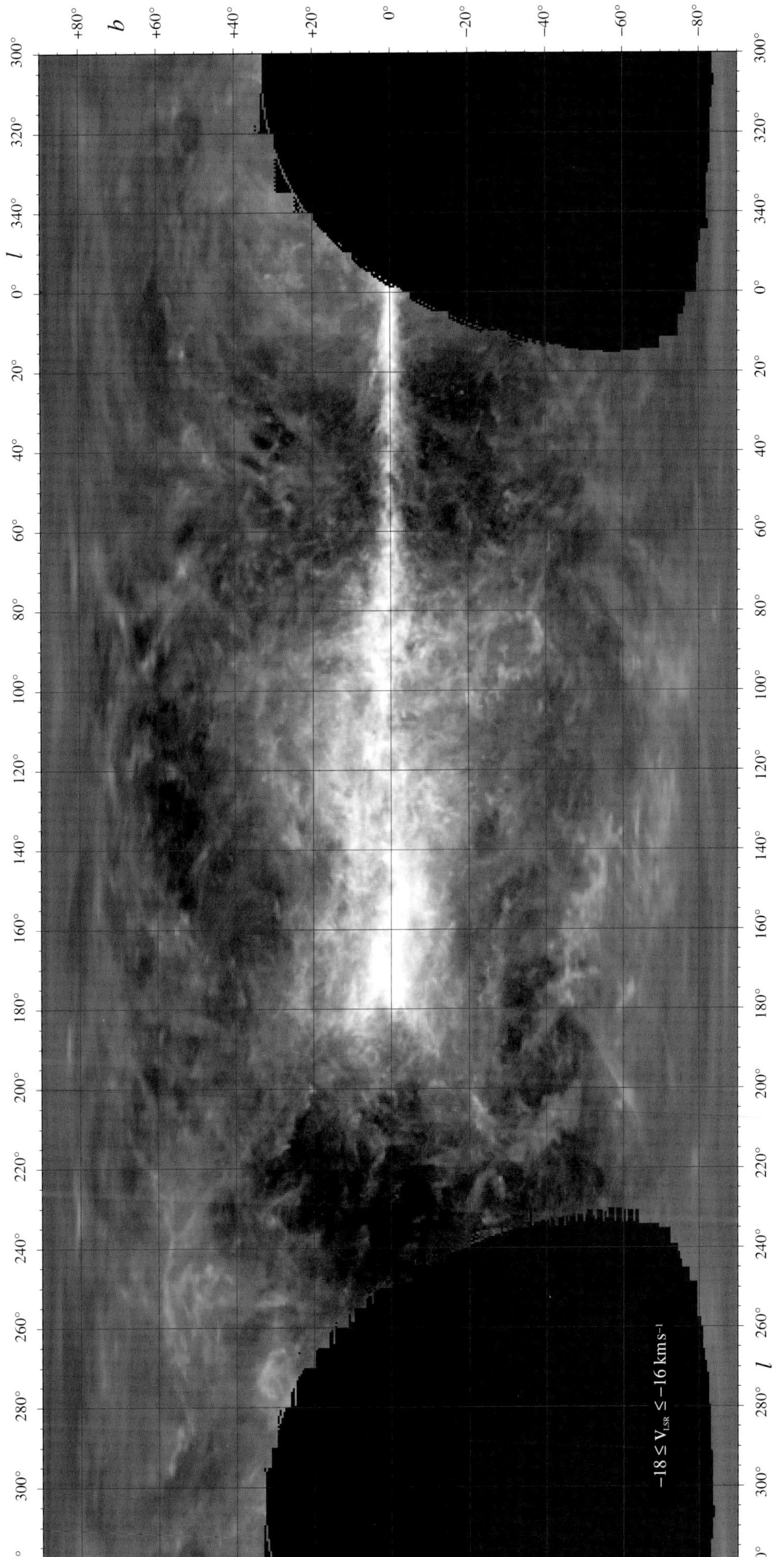
−18 ≤ V_LSR ≤ −16 km s−1

$-16 \leq V_{LSR} \leq -14$ km s^{-1}

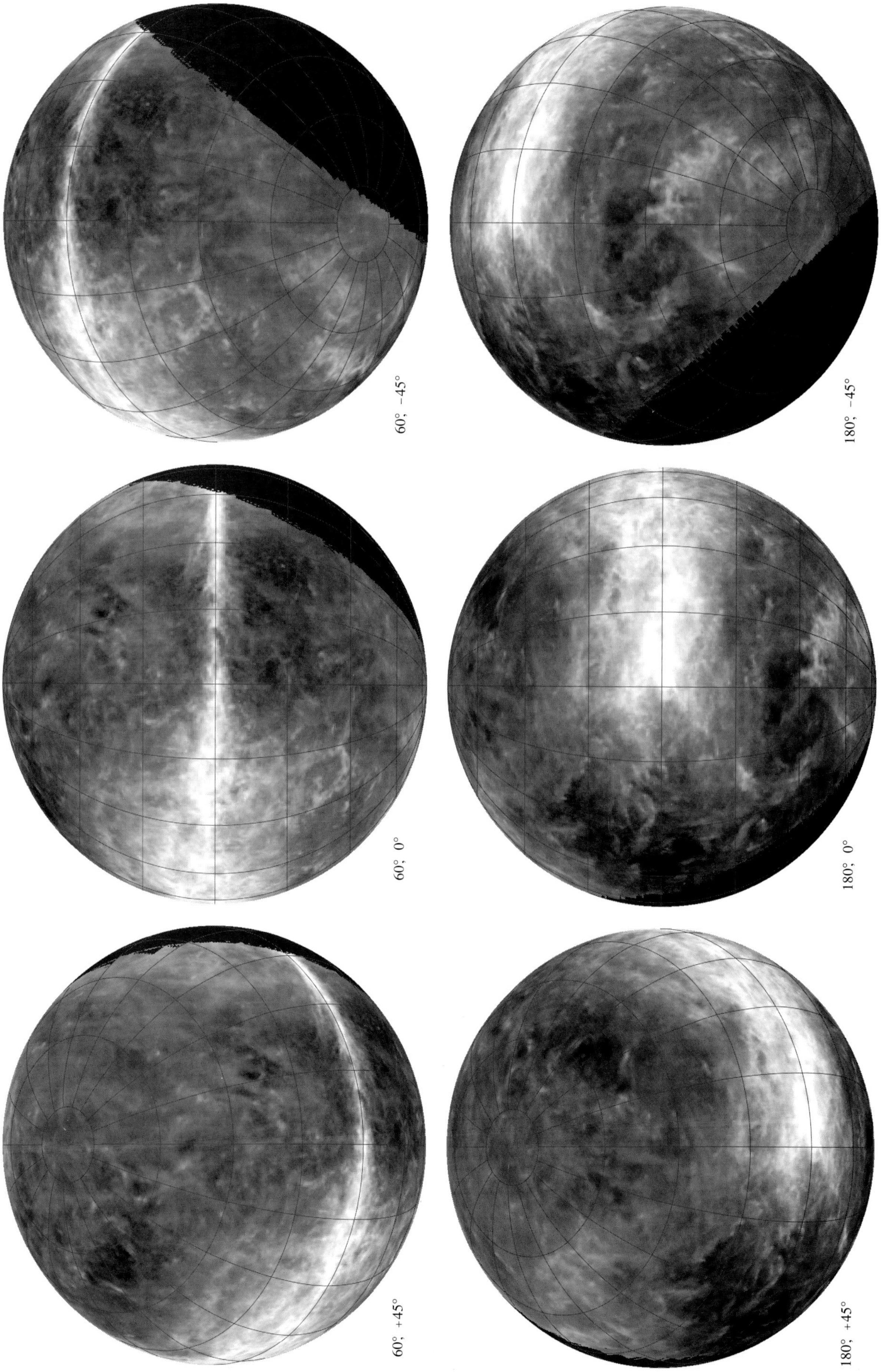

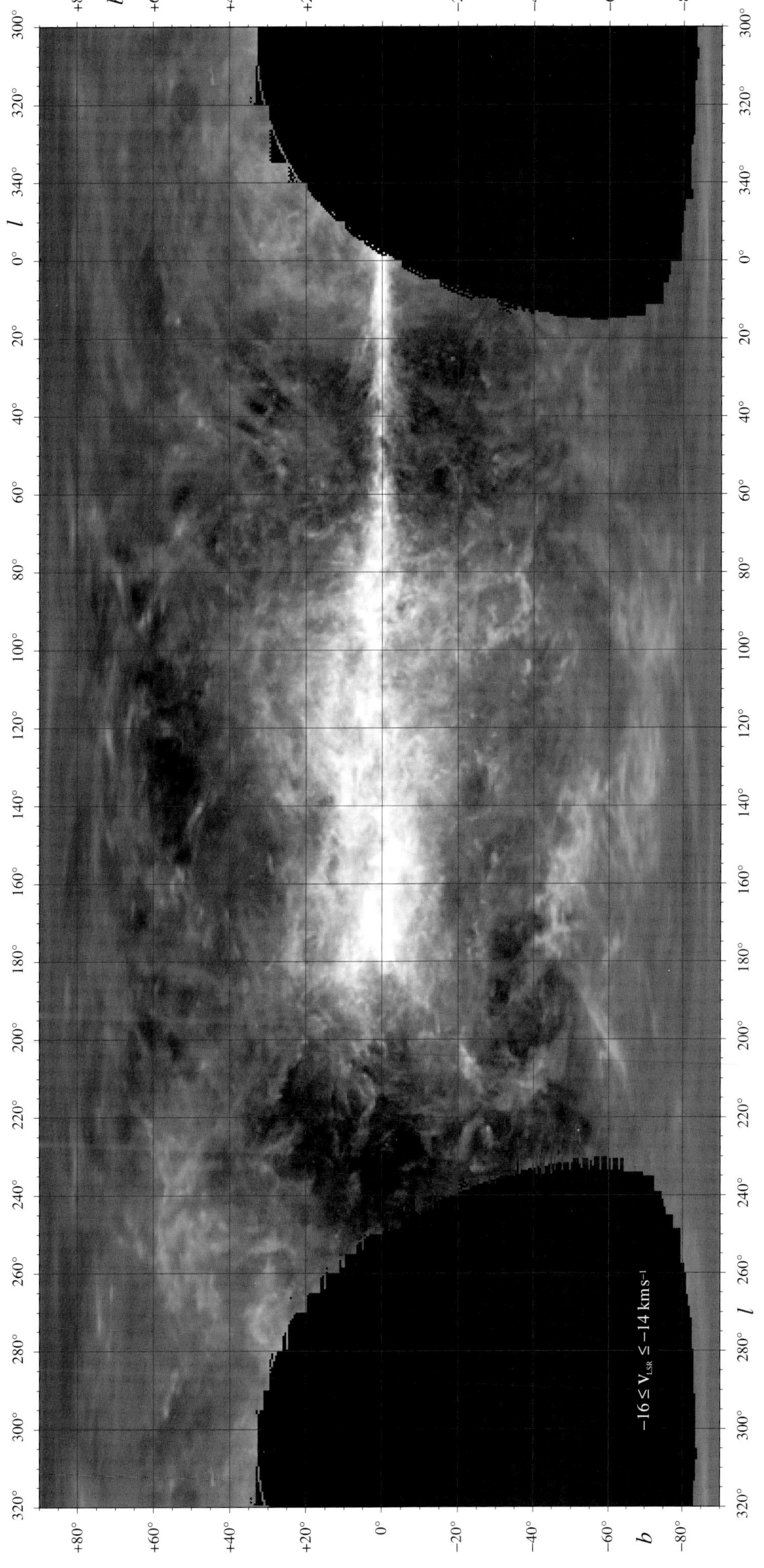

$-16 \le V_{LSR} \le -14$ km s^{-1}
l
b

$-14 \leq V_{LSR} \leq -12$ km s^{-1}

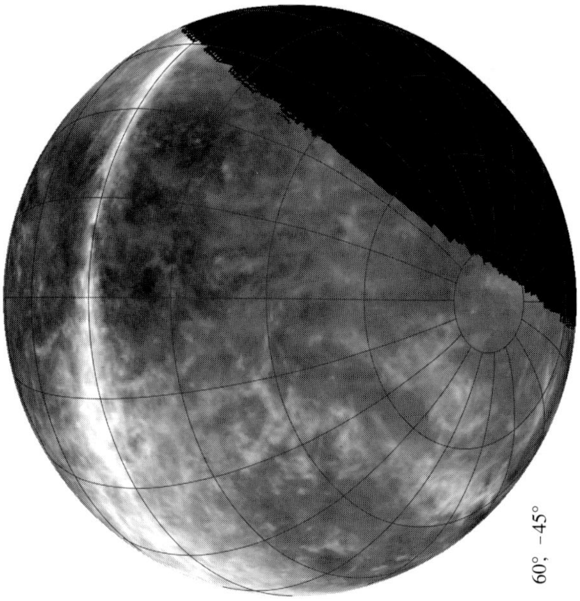

60°; −45°

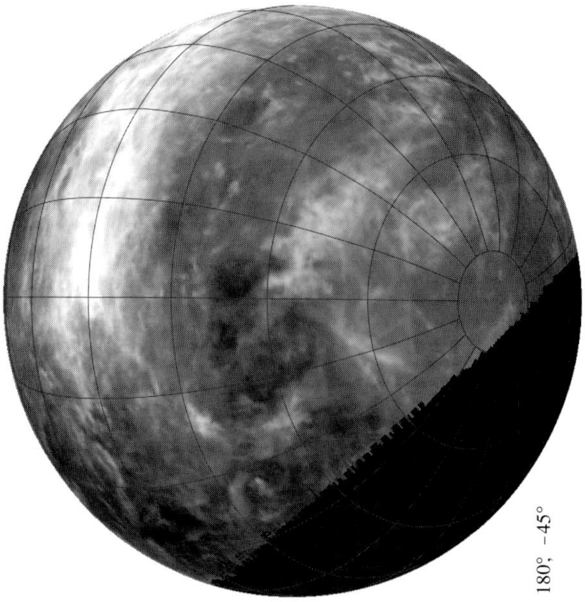

180°; −45°

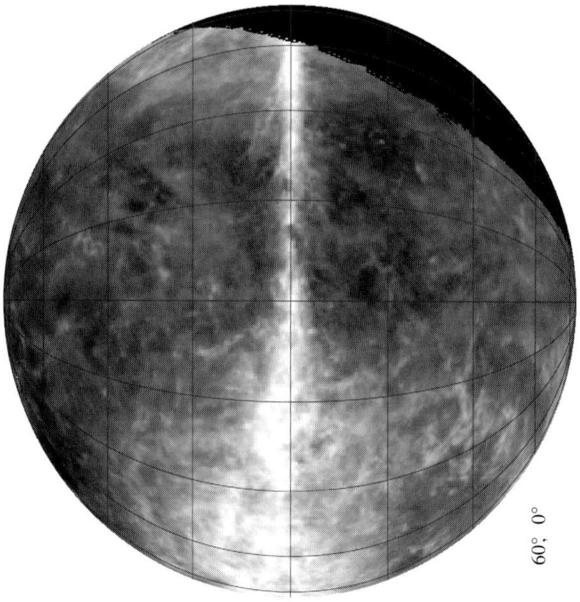

60°; 0°

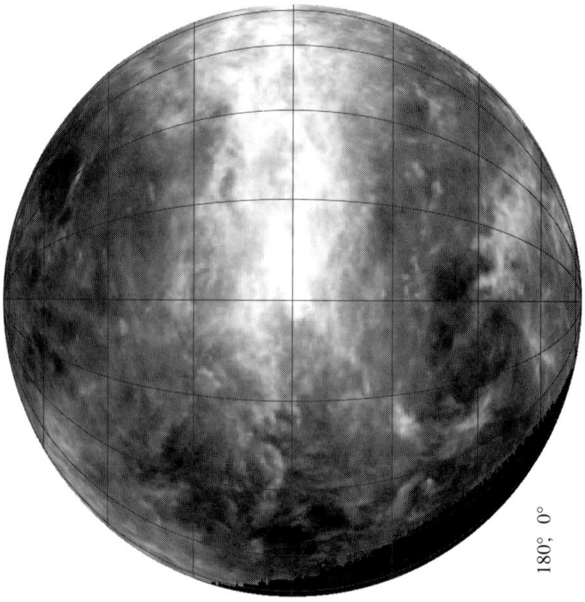

180°; 0°

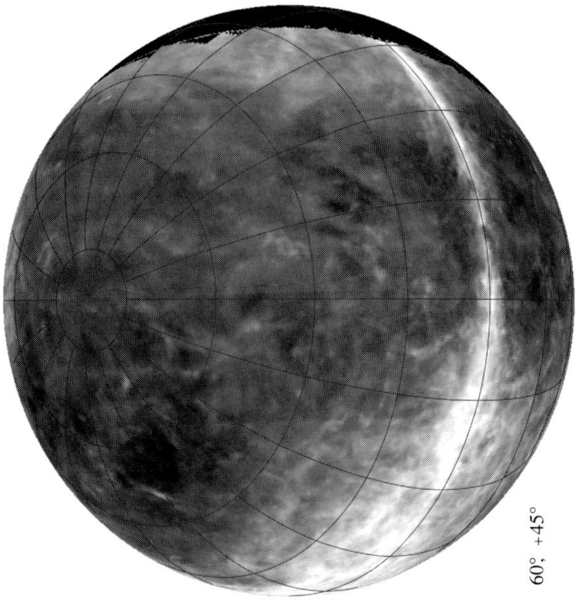

60°; +45°

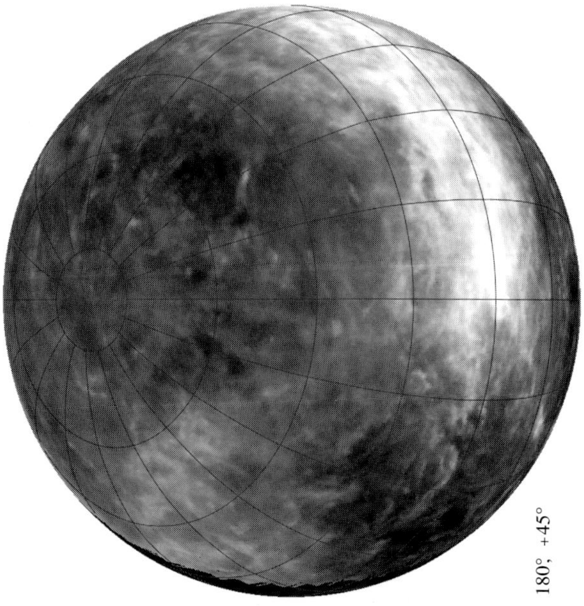

180°; +45°

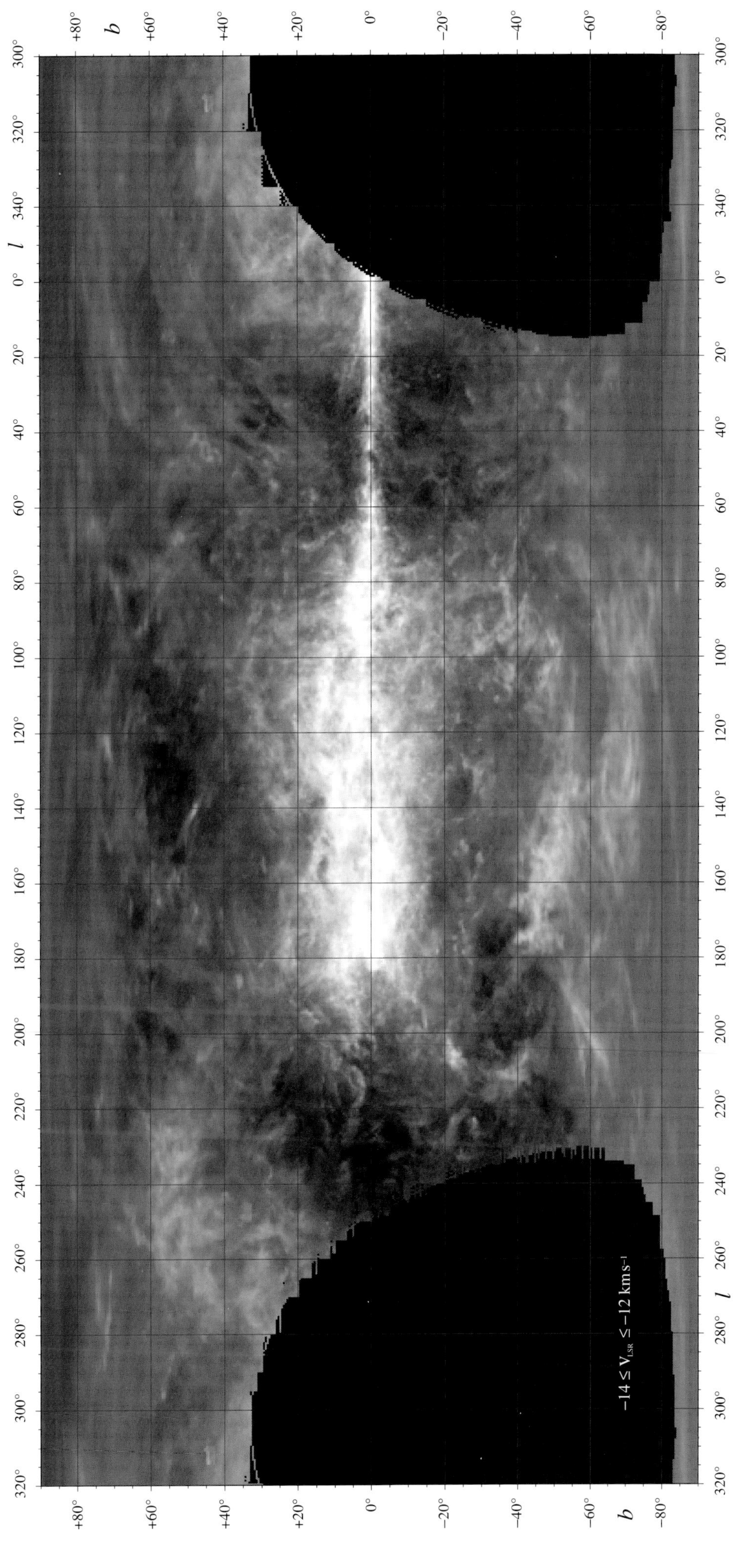
$-14 \leq V_{LSR} \leq -12$ km s^{-1}
l
b
+80°
+60°
+40°
+20°
0°
−20°
−40°
−60°
−80°
300°
320°
340°
0°
20°
40°
60°
80°
100°
120°
140°
160°
180°
200°
220°
240°
260°
280°
300°
320°

$-12 \le V_{LSR} \le -10$ km s^{-1}

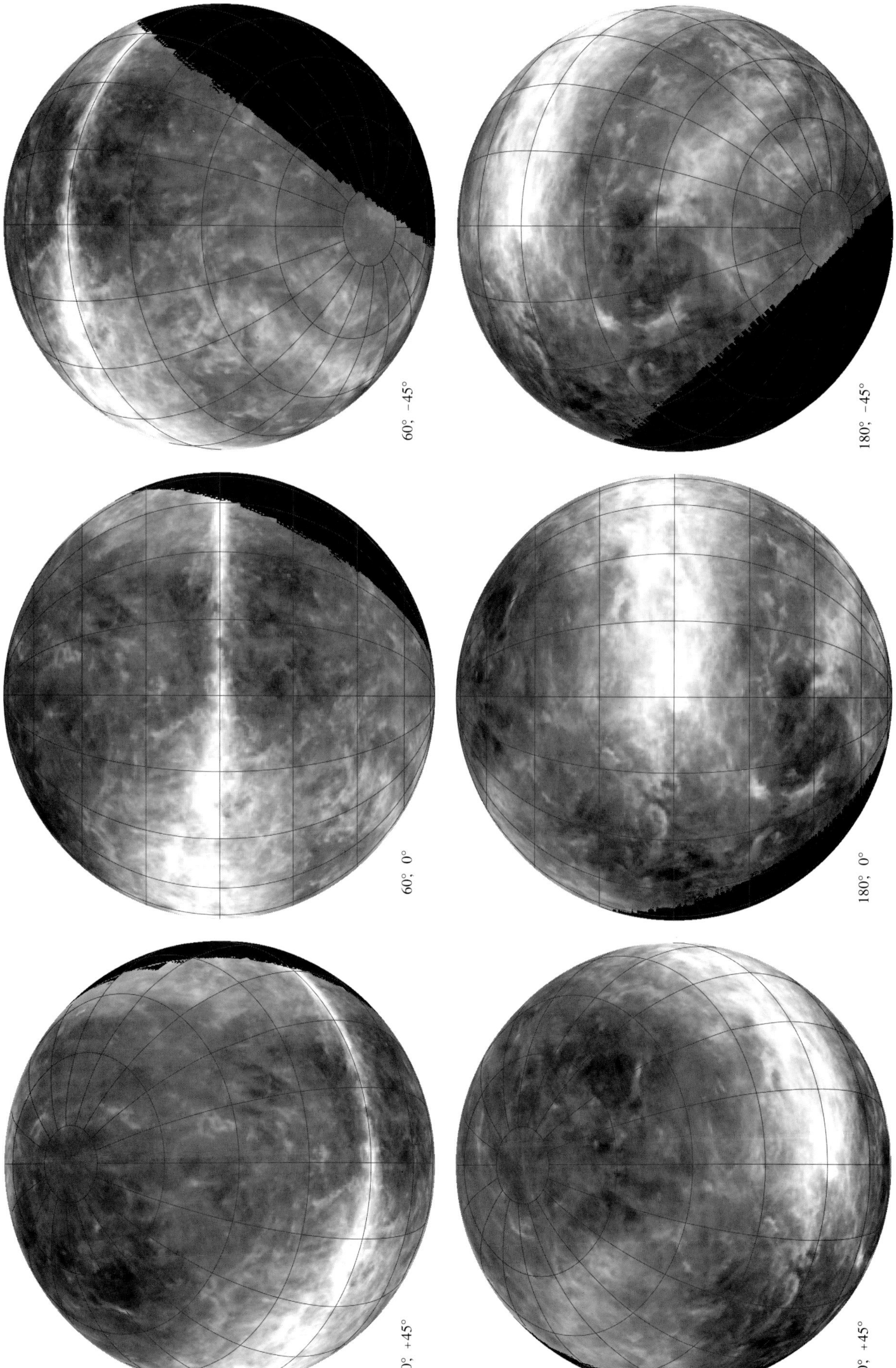

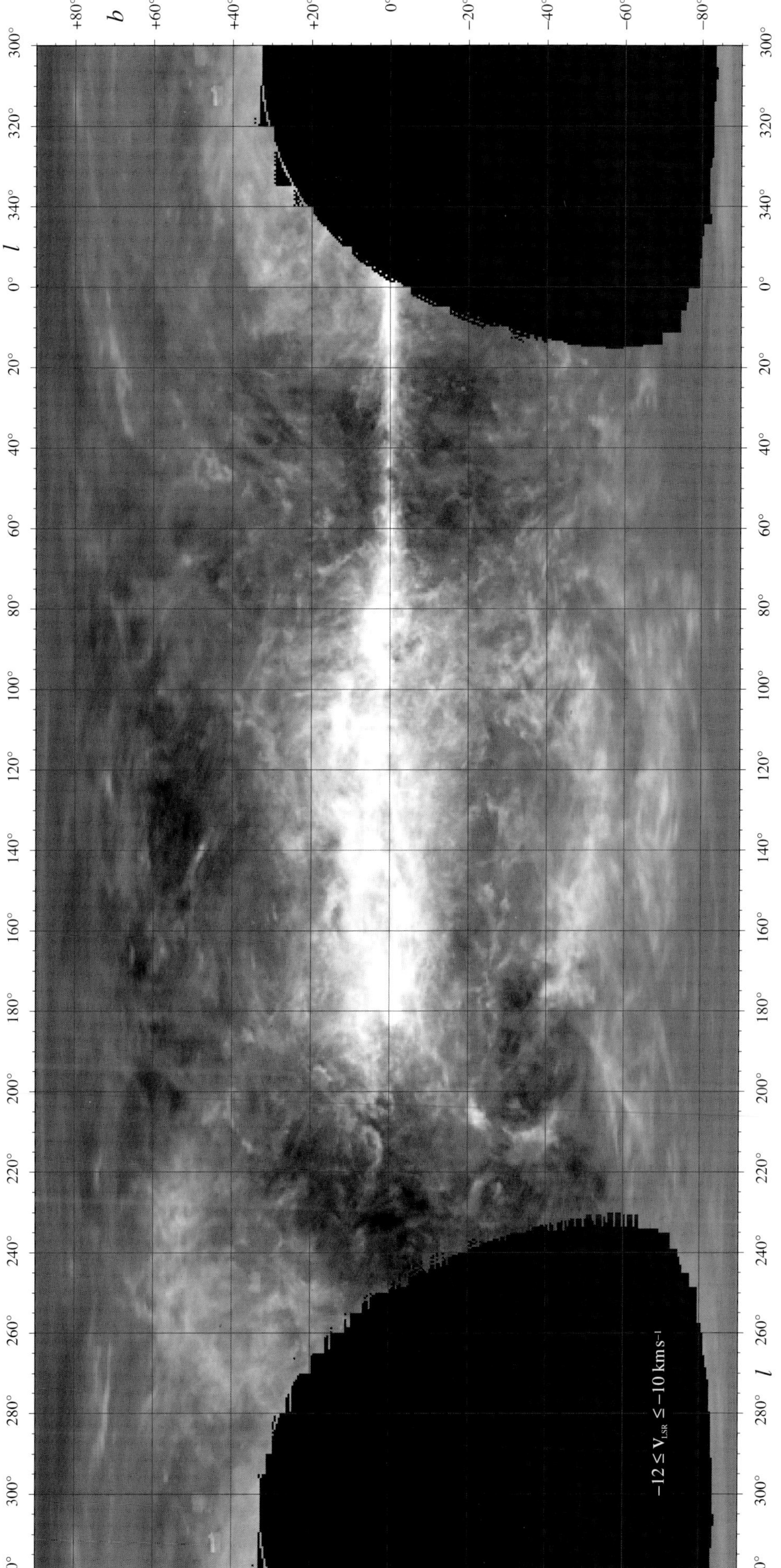
l
b
−12 ≤ V_LSR ≤ −10 km s^-1
300° 320° 340° 0° 20° 40° 60° 80° 100° 120° 140° 160° 180° 200° 220° 240° 260° 280° 300° 320°
+80° +60° +40° +20° 0° −20° −40° −60° −80°

$-10 \leq V_{LSR} \leq -8$ km s^{-1}

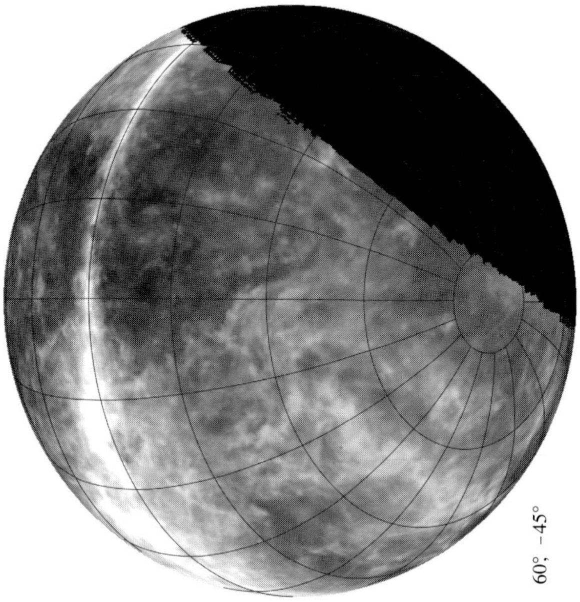

60°, −45°

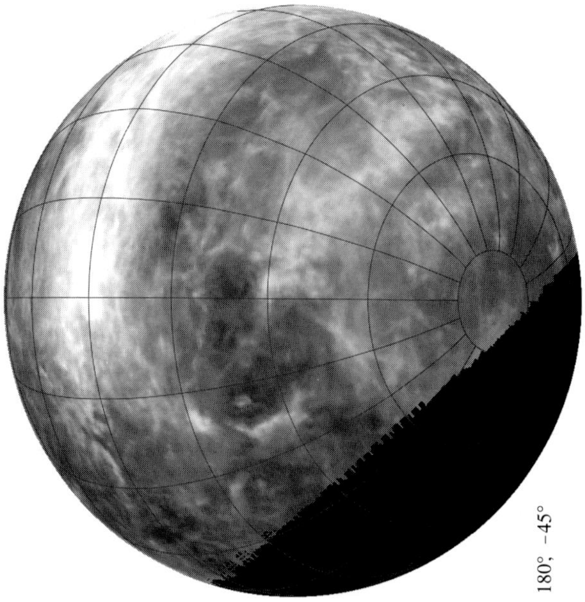

180°, −45°

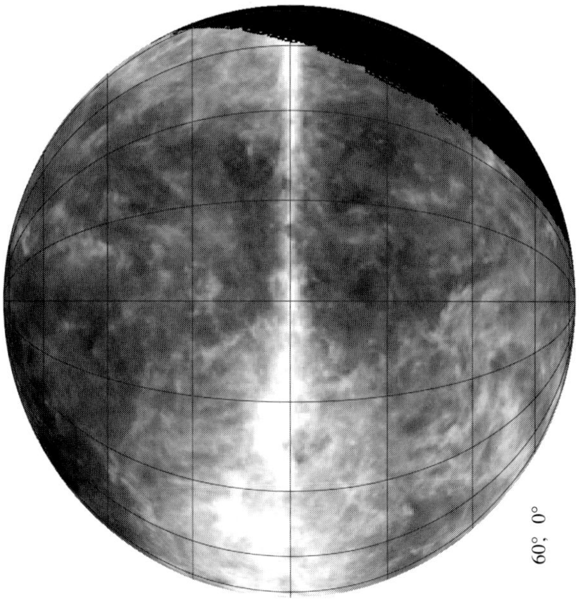

60°, 0°

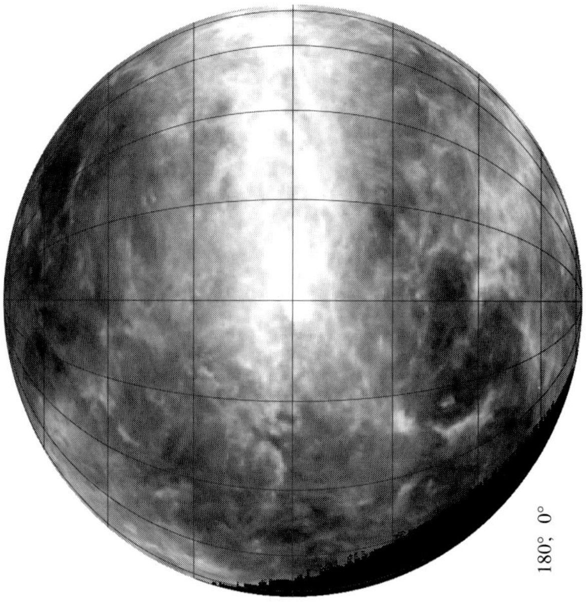

180°, 0°

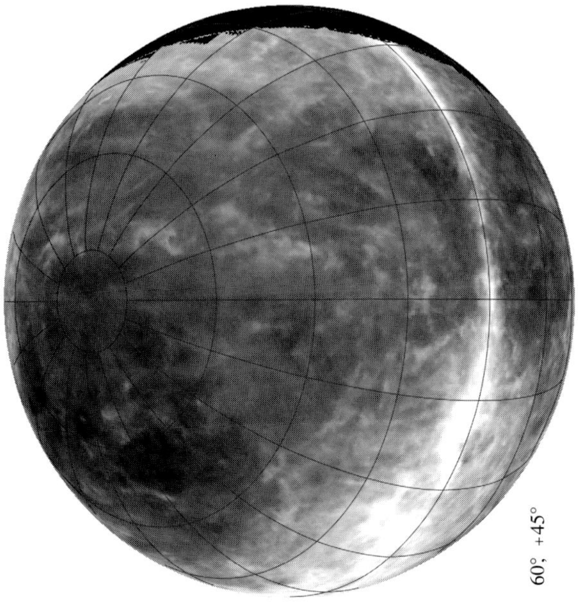

60°, +45°

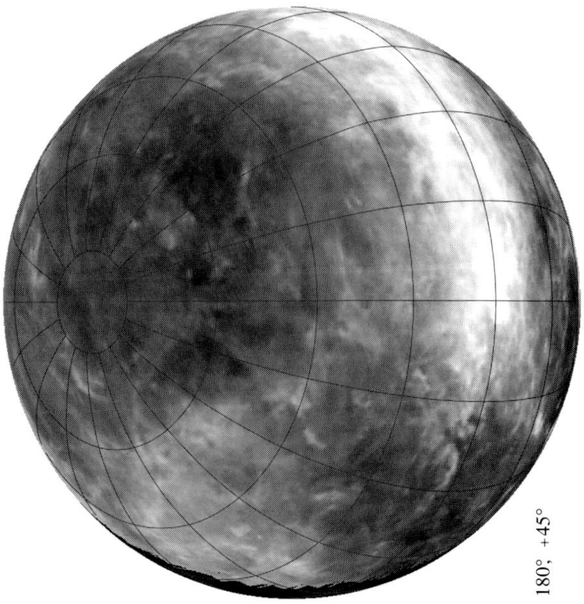

180°, +45°

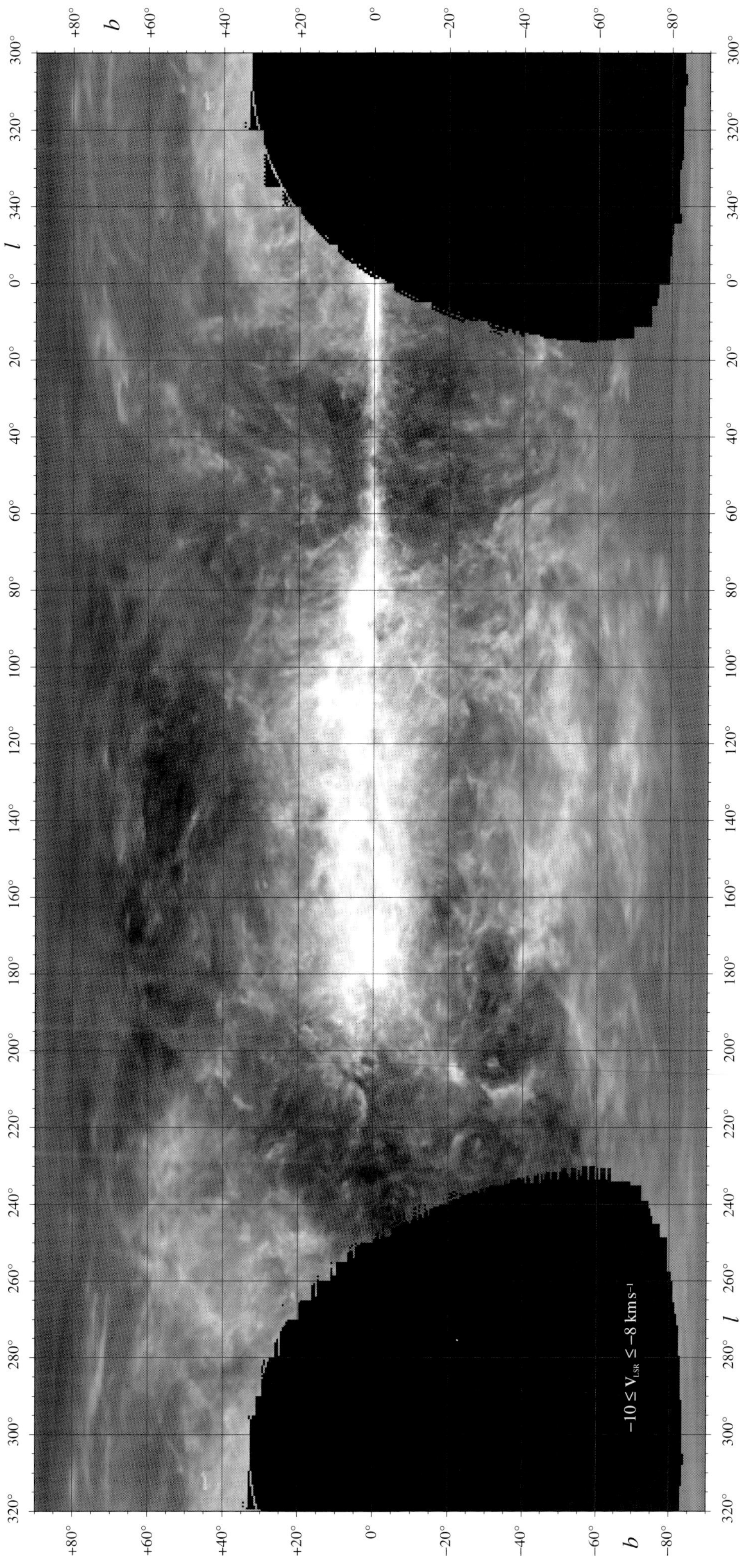
$-10 \leq V_{LSR} \leq -8$ km s^{-1}

$-8 \leq V_{LSR} \leq -6$ km s^{-1}

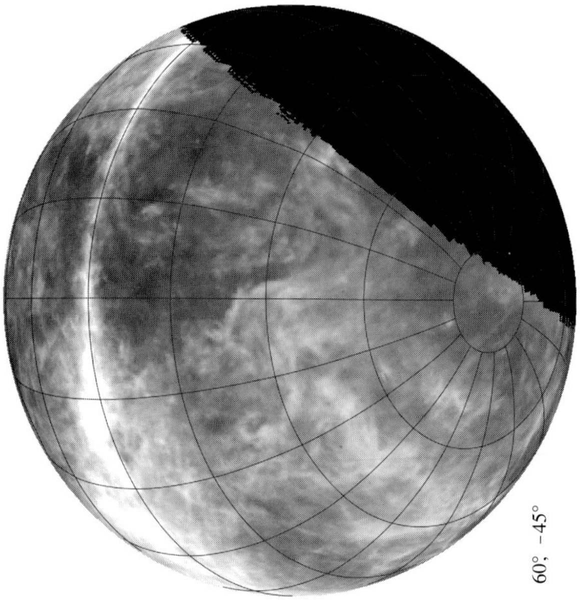

60°, −45°

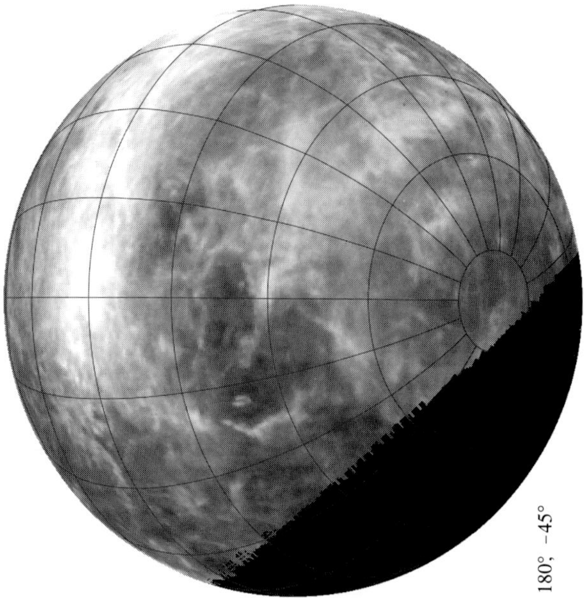

180°, −45°

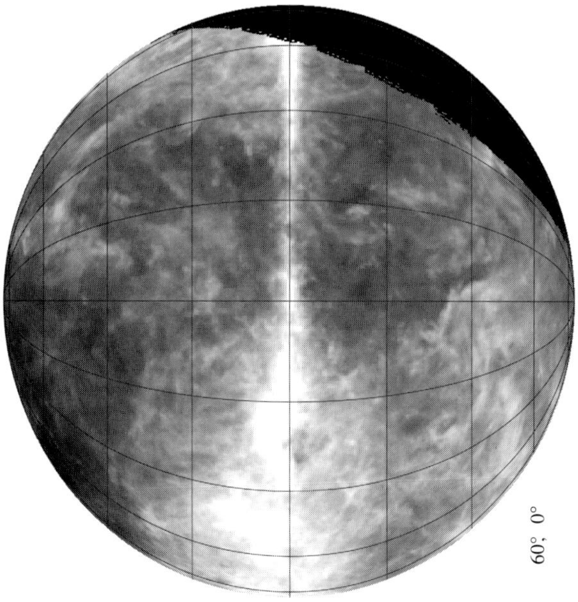

60°, 0°

180°, 0°

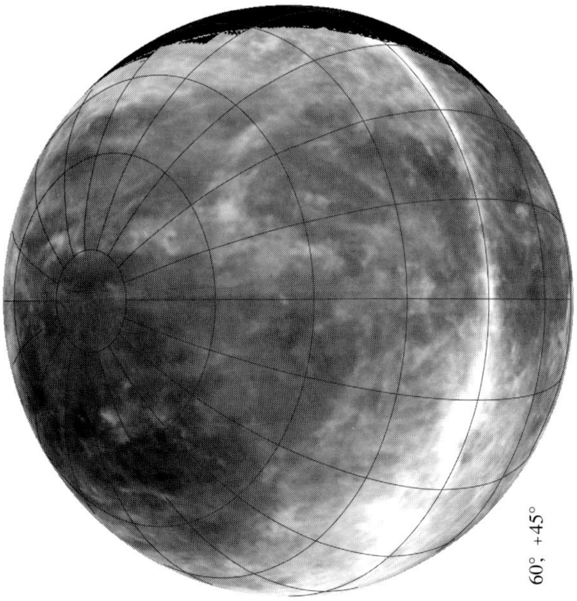

60°, +45°

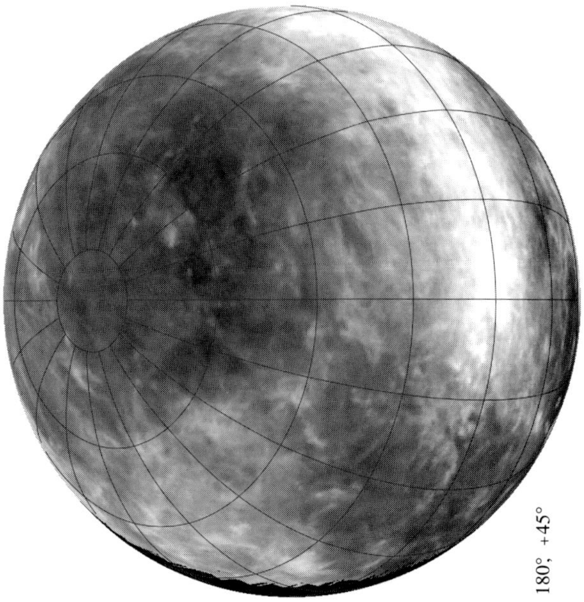

180°, +45°

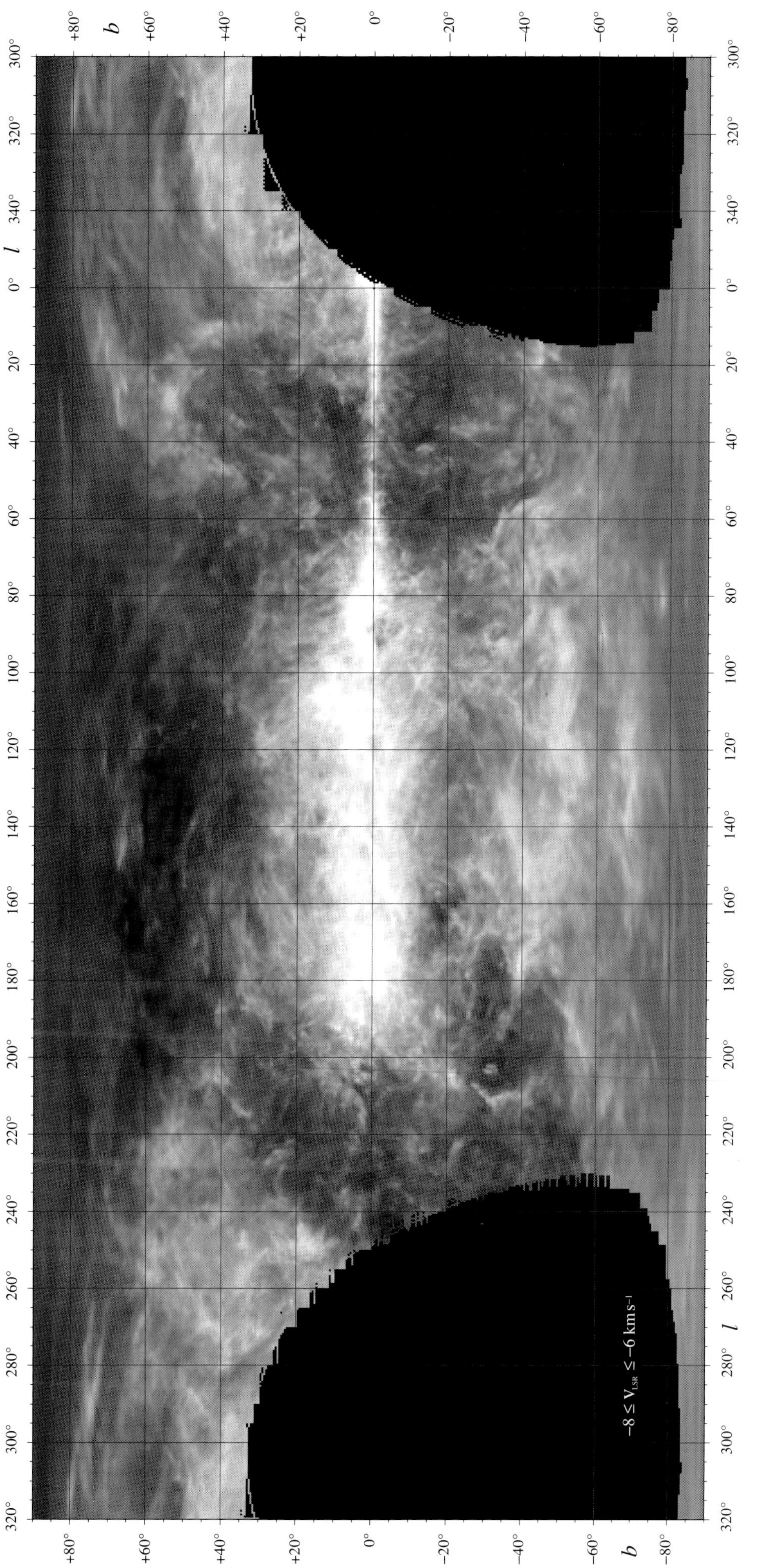

−8 ≤ V_LSR ≤ −6 km s⁻¹

$-6 \leq V_{LSR} \leq -4$ km s^{-1}

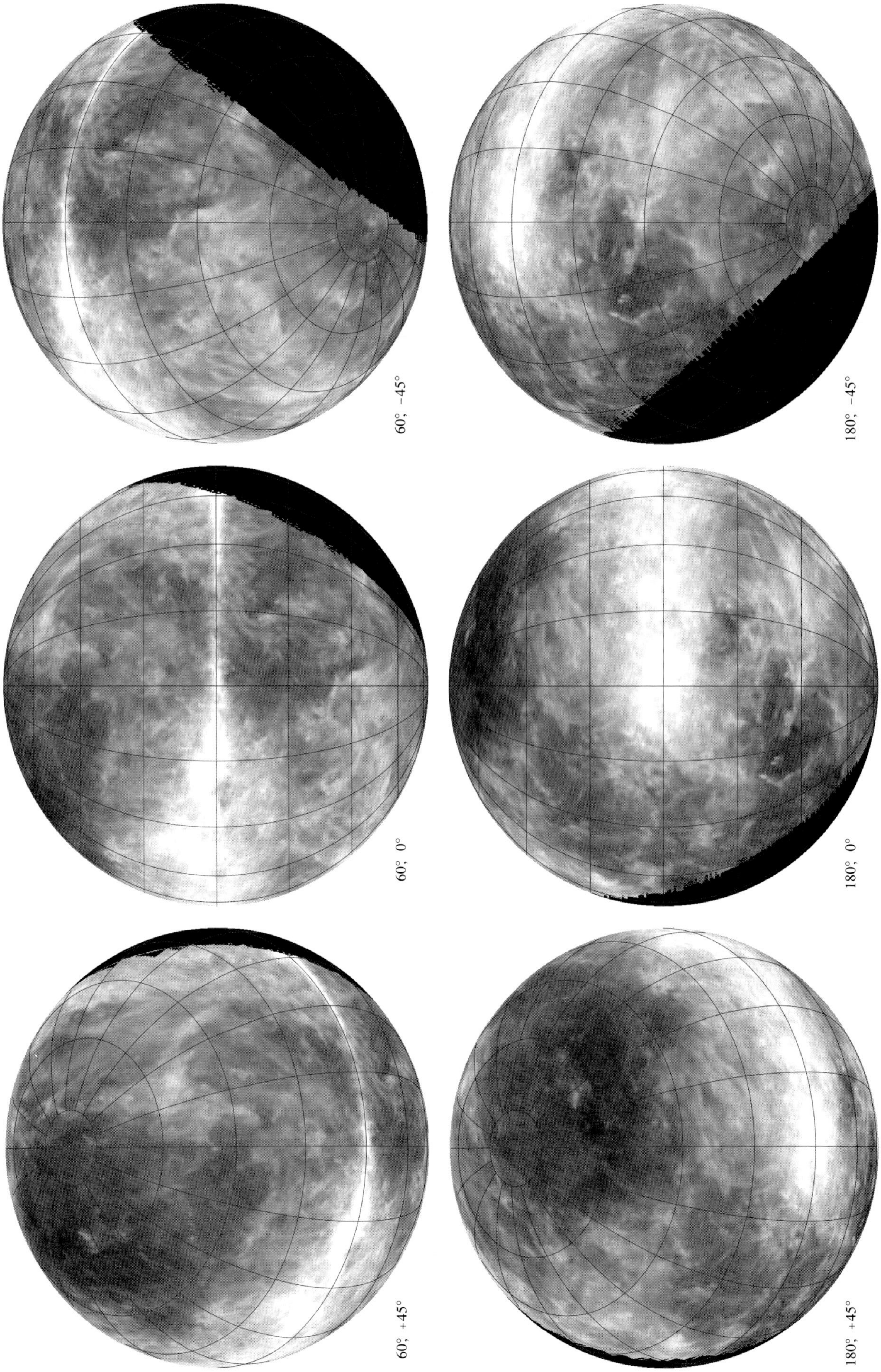

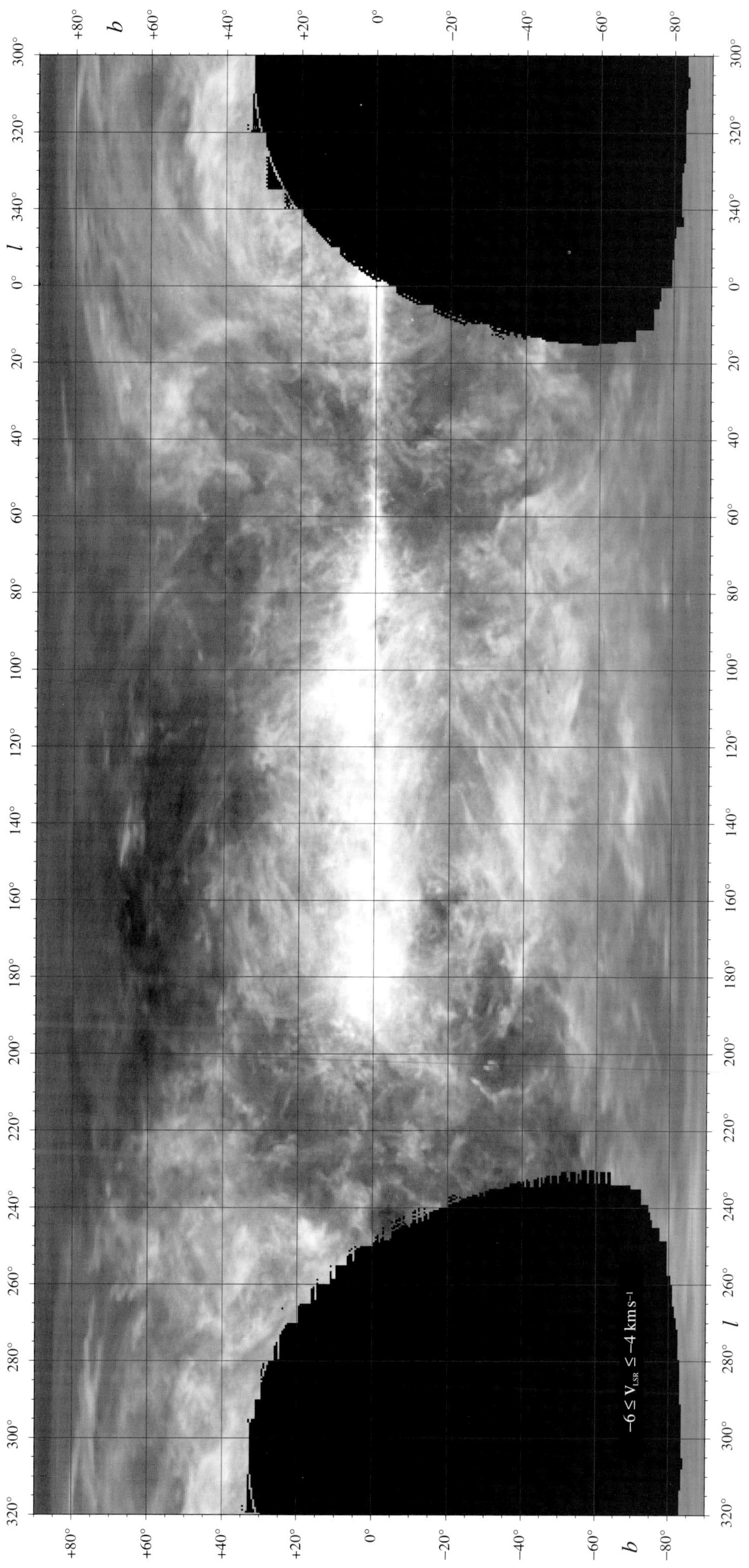

−6 ≤ V_LSR ≤ −4 km s−1
l
b

$-4 \leq V_{LSR} \leq -2 \text{ km s}^{-1}$

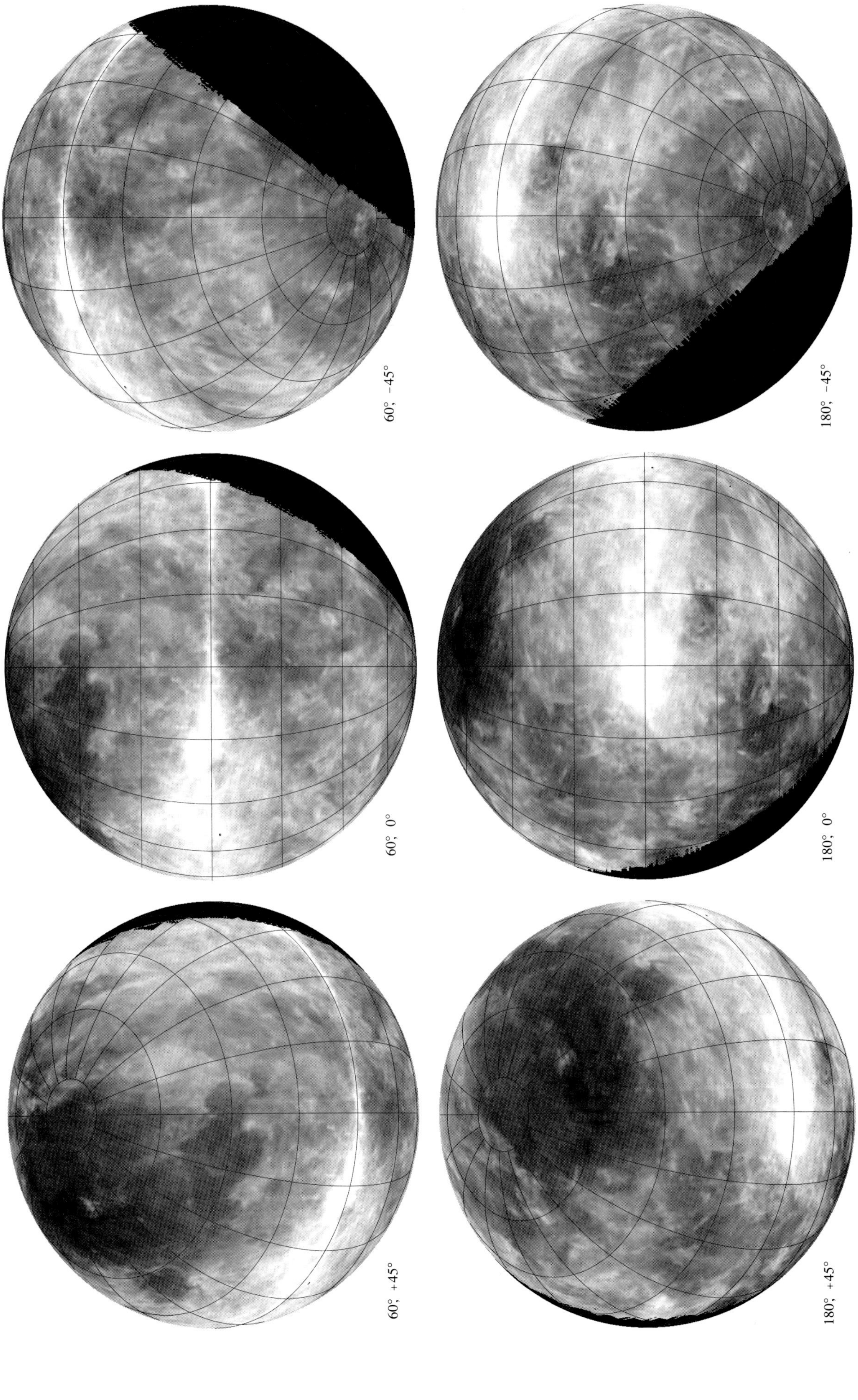

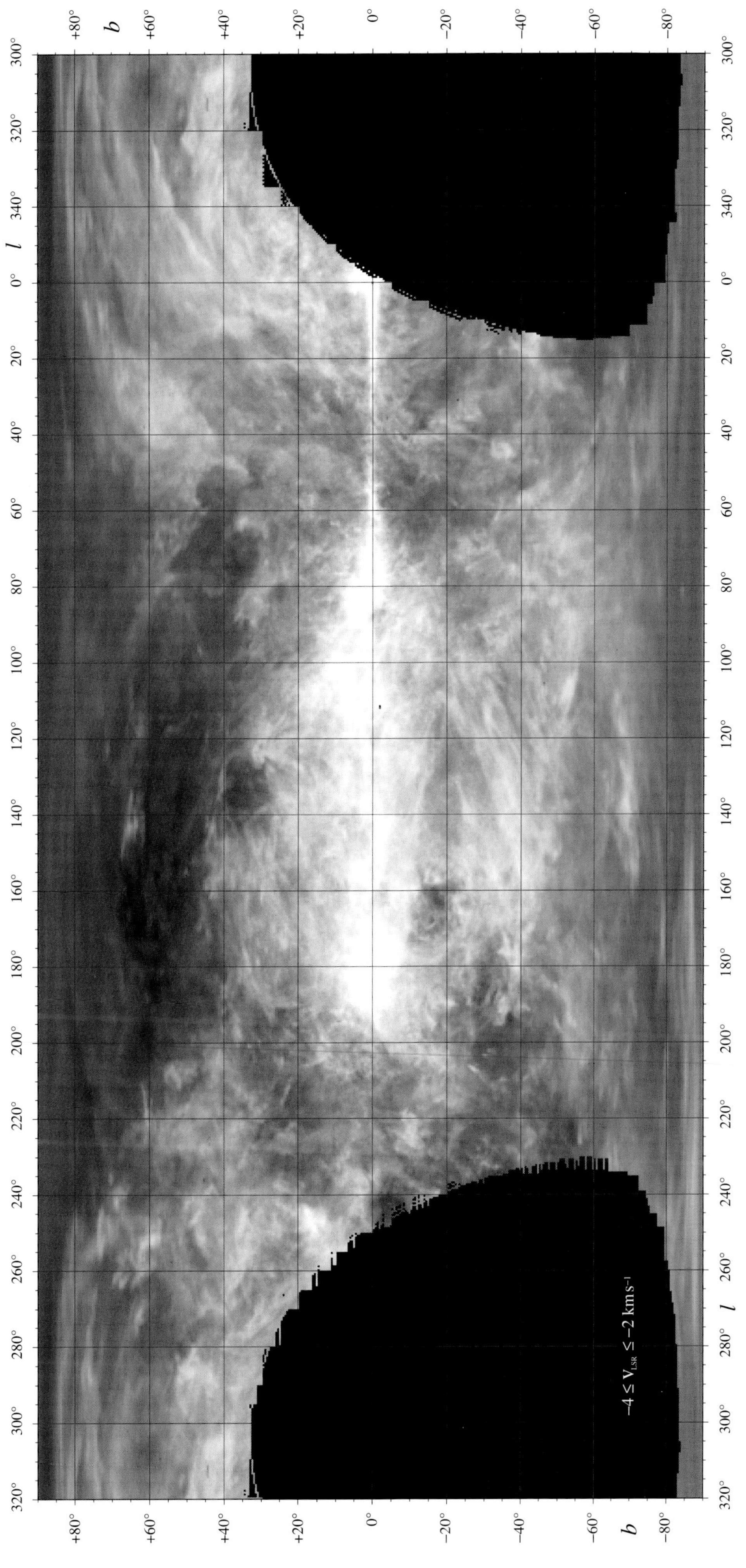
−4 ≤ V_LSR ≤ −2 km s⁻¹
l
b

$-2 \leq V_{LSR} \leq 0$ km s^{-1}

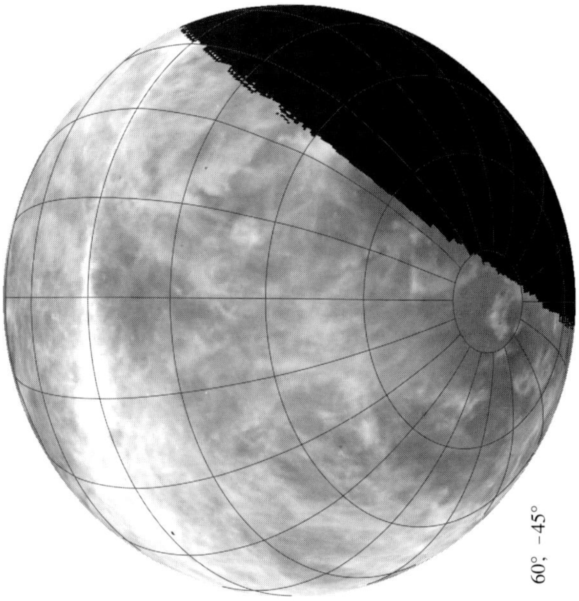

60°; −45°

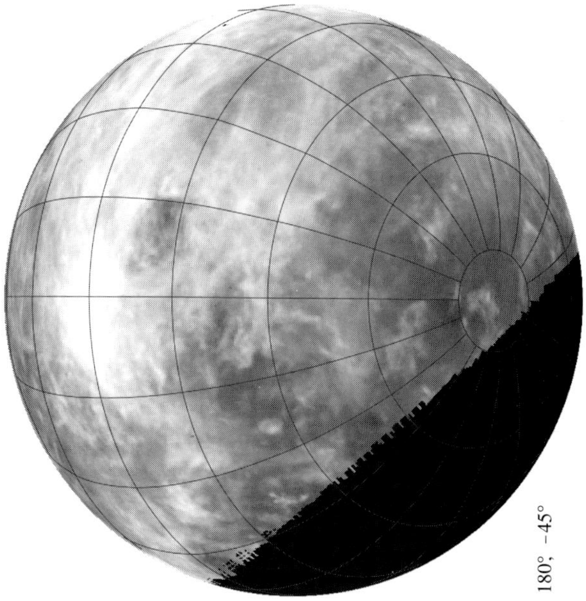

180°; −45°

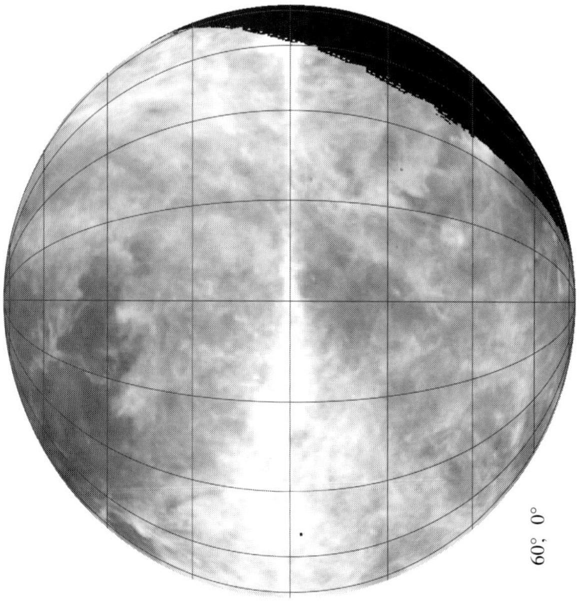

60°; 0°

180°; 0°

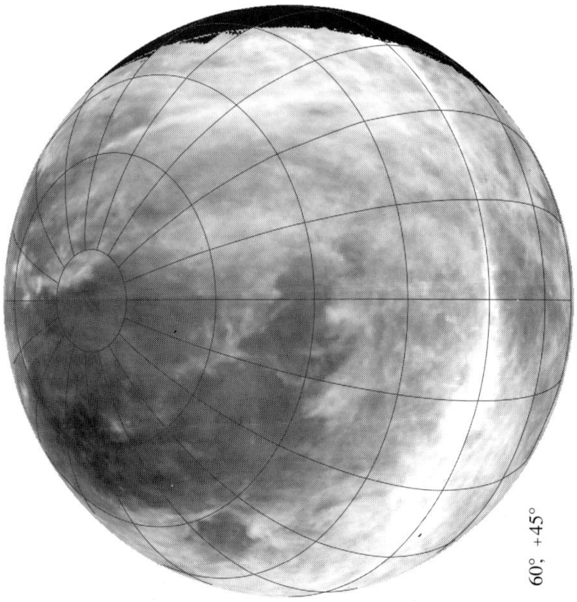

60°; +45°

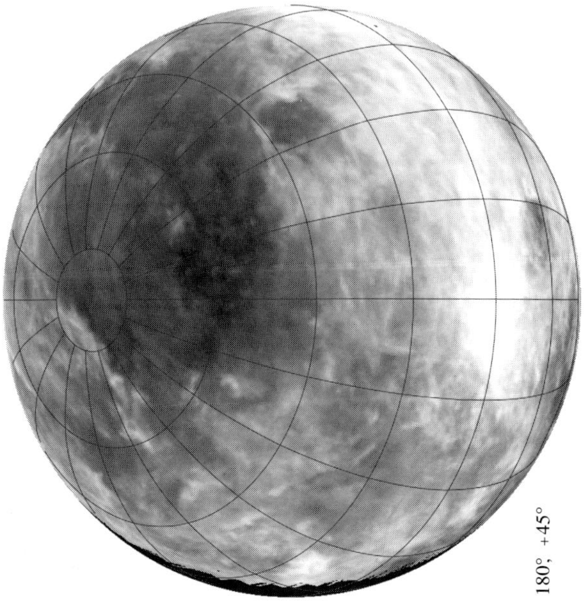

180°; +45°

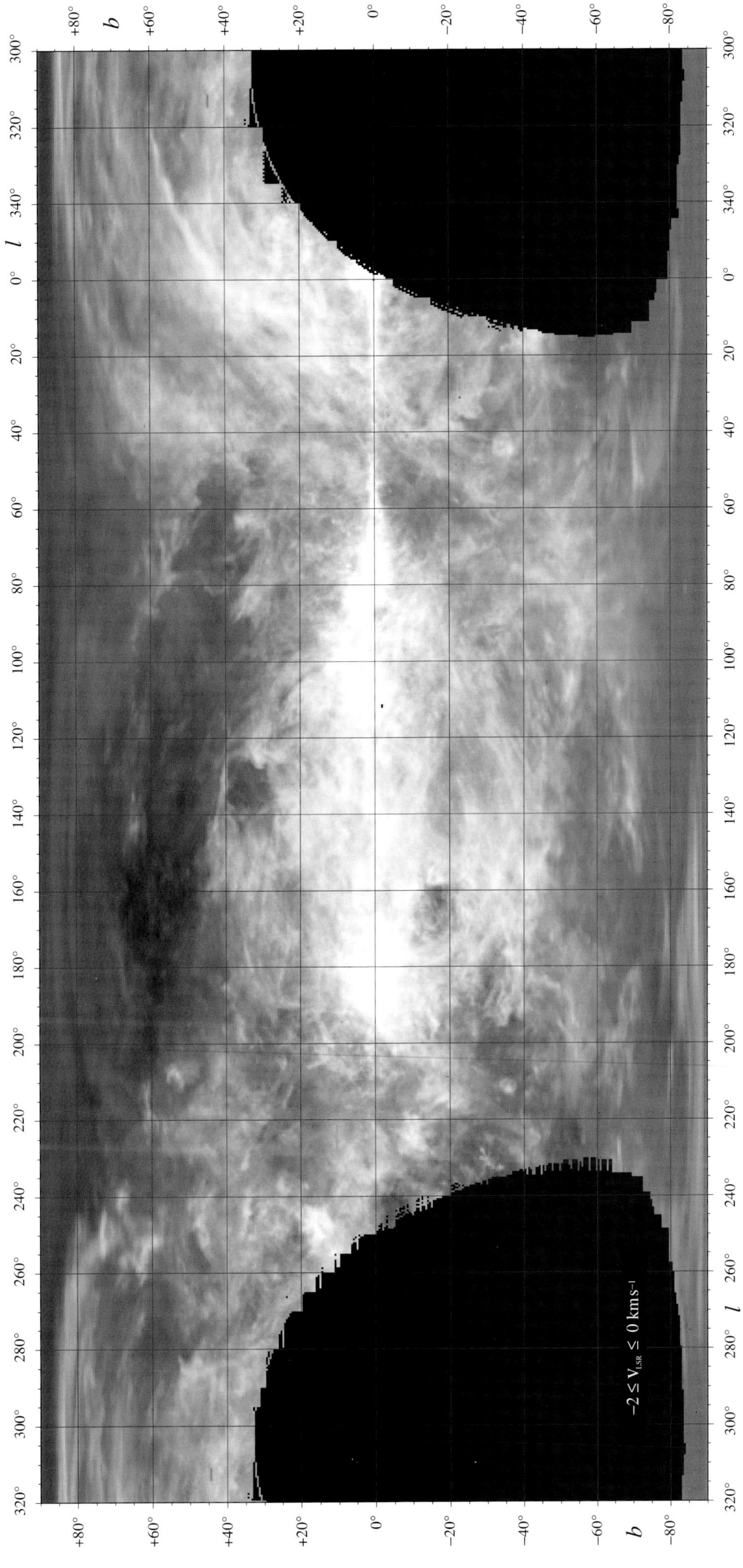
$-2 \leq V_{LSR} \leq 0$ kms^{-1}
l
b

$0 \leq V_{LSR} \leq +2$ km s^{-1}

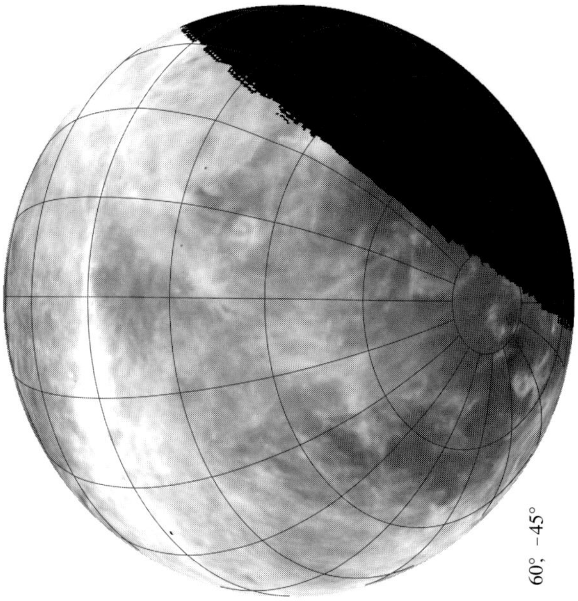

60°, −45°

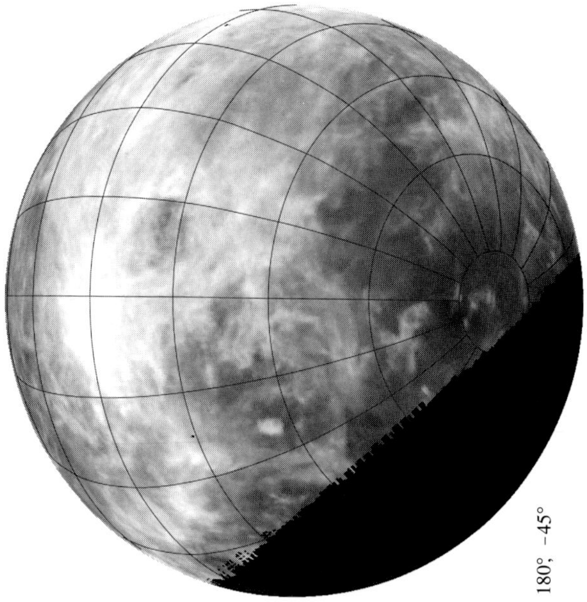

180°, −45°

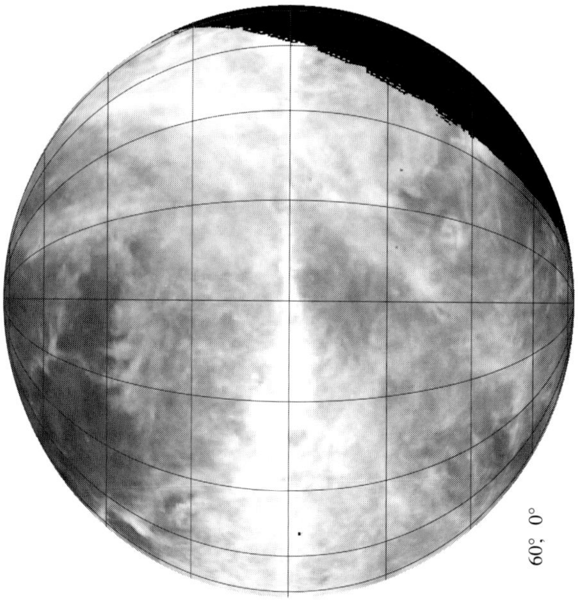

60°, 0°

180°, 0°

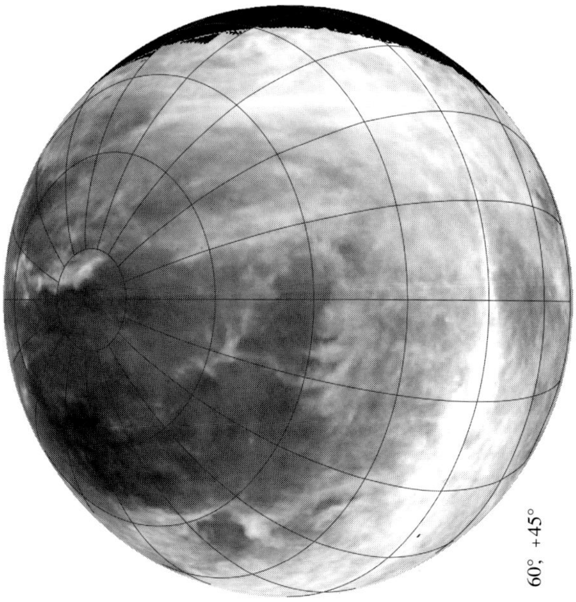

60°, +45°

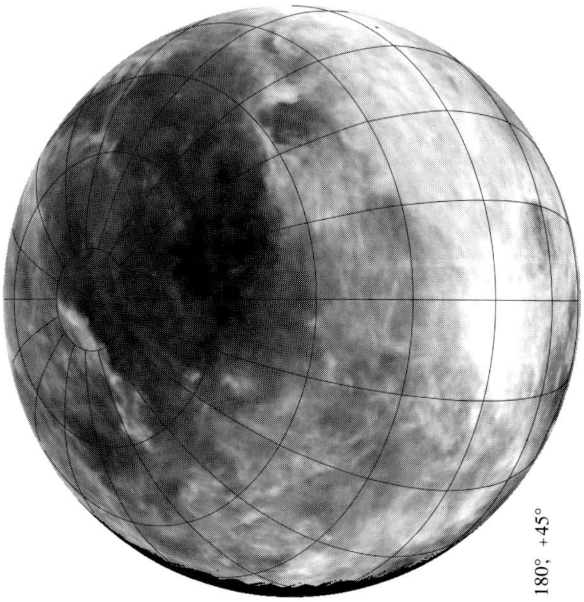

180°, +45°

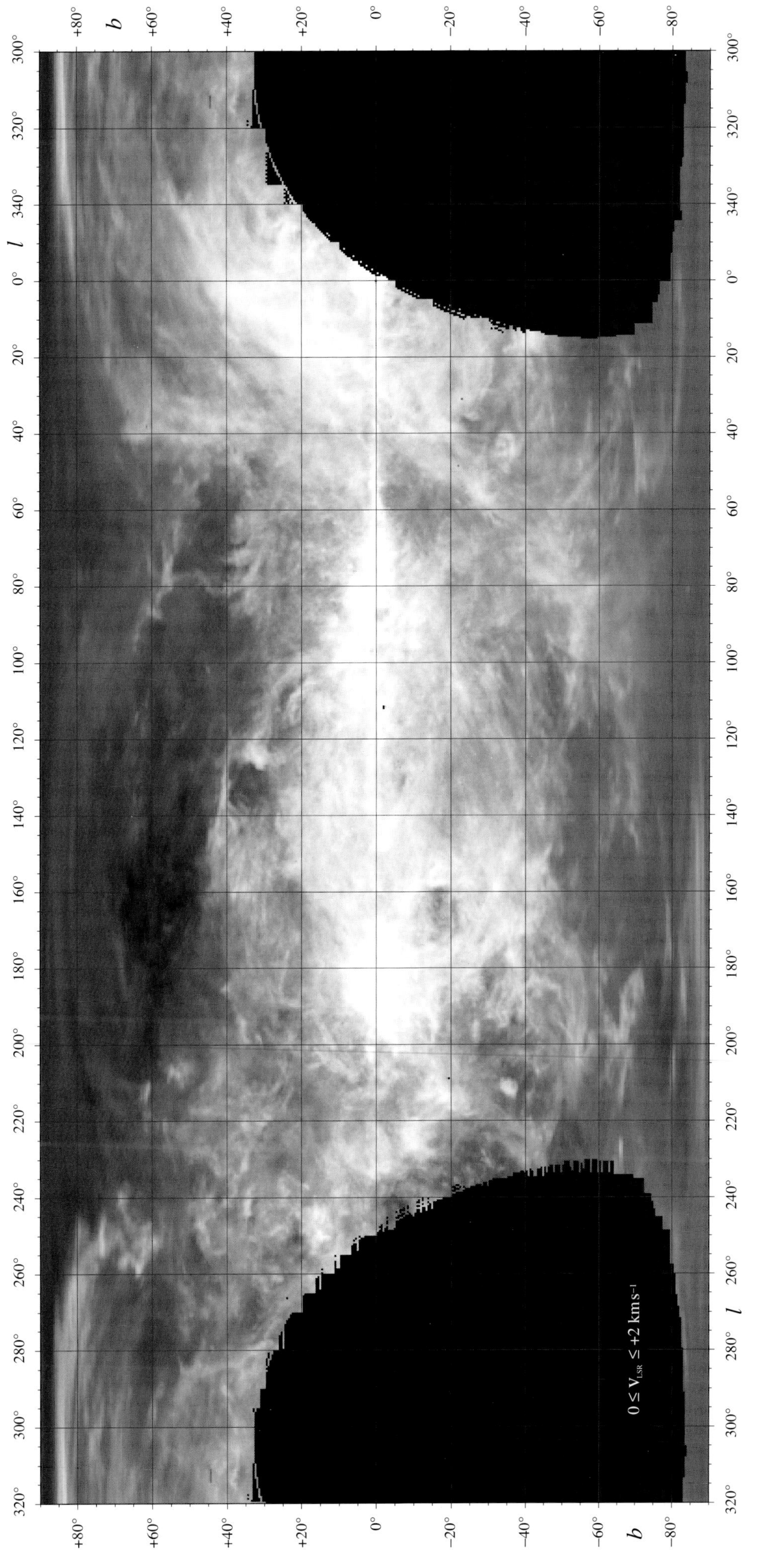
$0 \le V_{LSR} \le +2$ km s^{-1}
l
b

$+2 \leq V_{LSR} \leq +4$ km s^{-1}

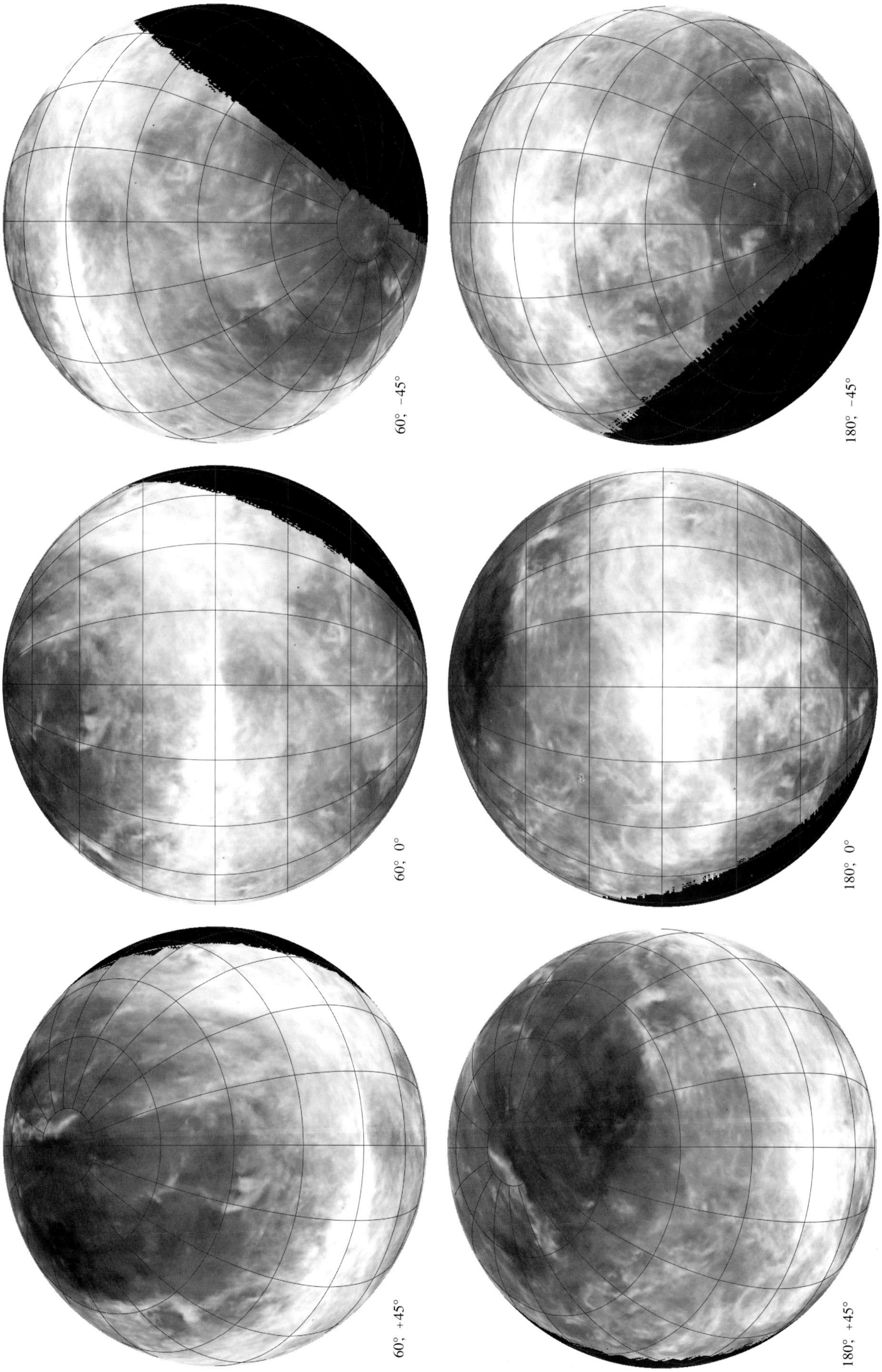

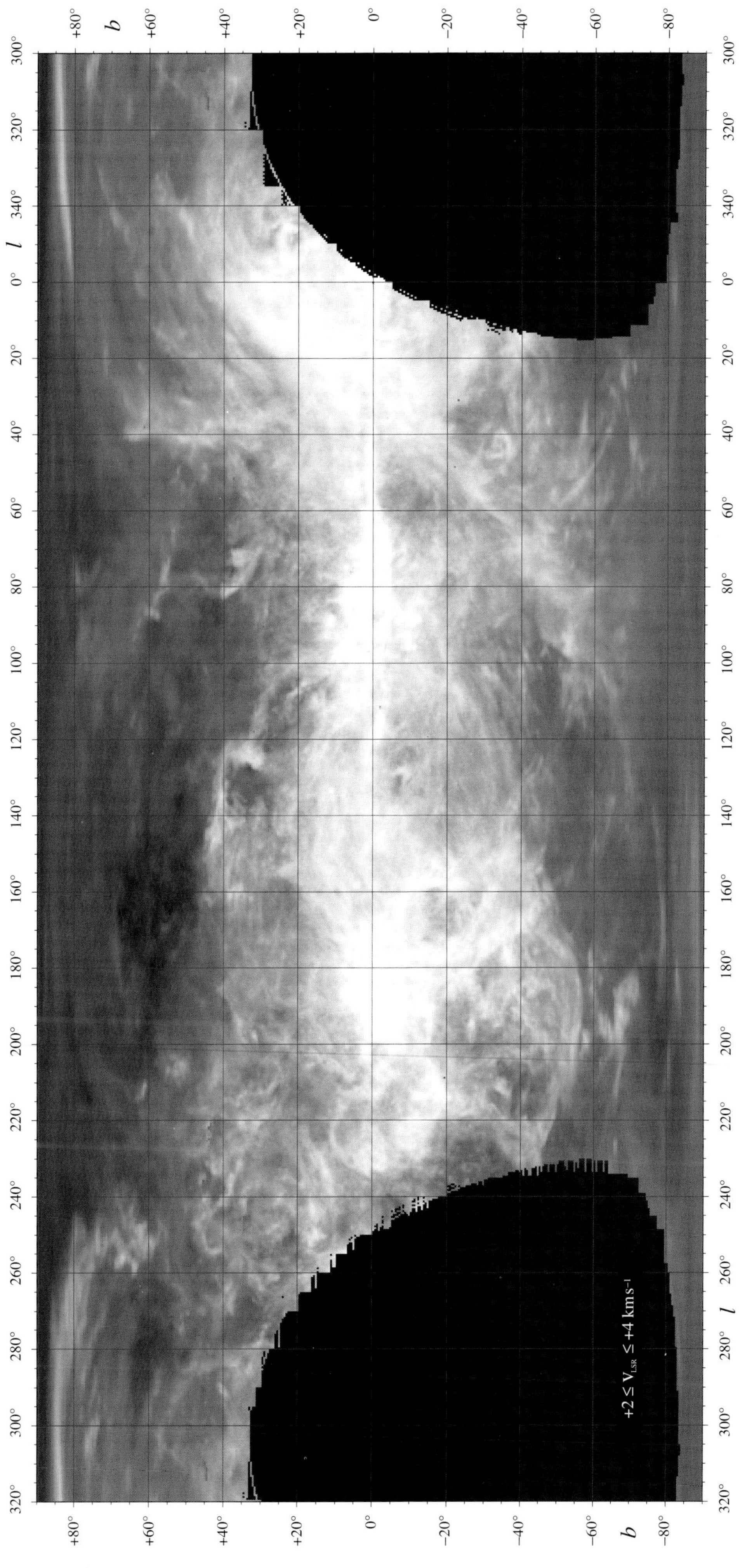
+2 ≤ V_LSR ≤ +4 kms⁻¹
l
b

$+4 \leq V_{LSR} \leq +6$ km s^{-1}

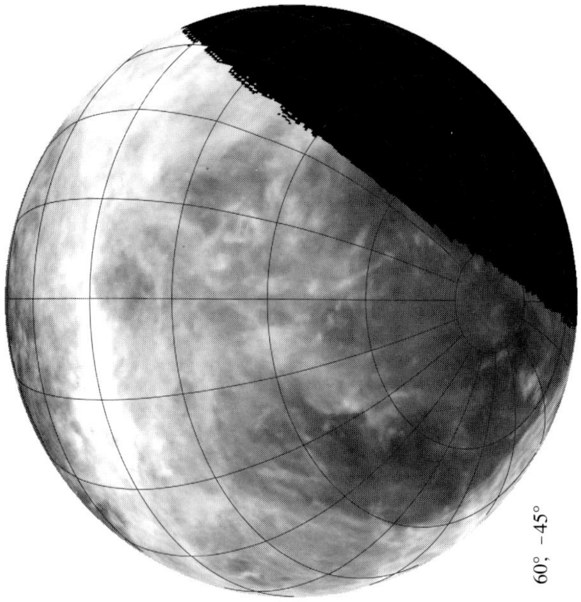

60°; −45°

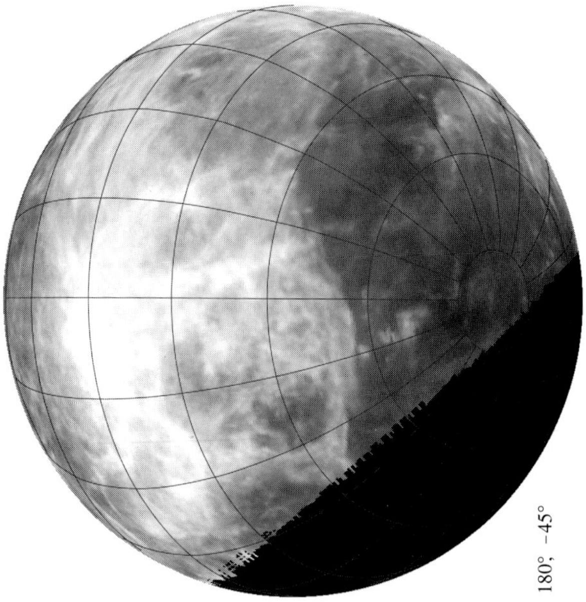

180°; −45°

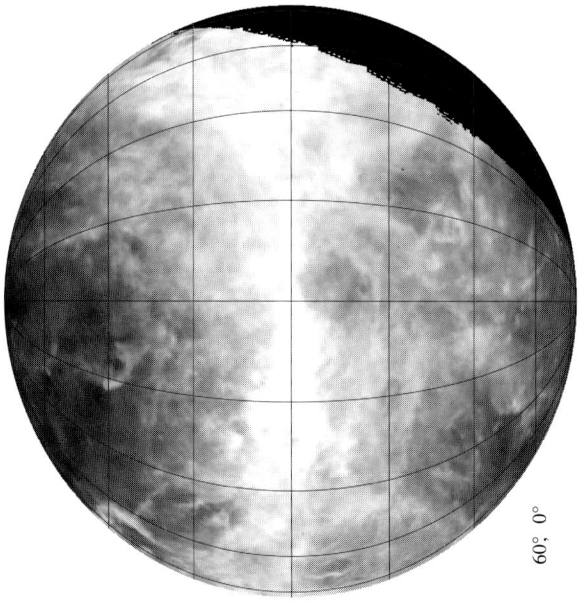

60°; 0°

180°; 0°

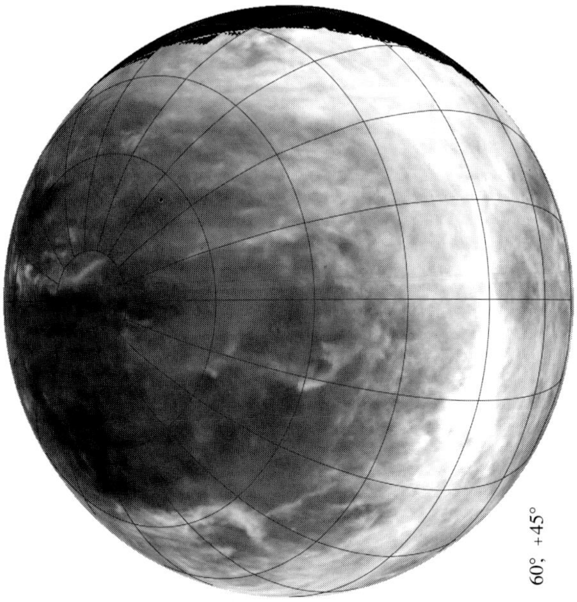

60°; +45°

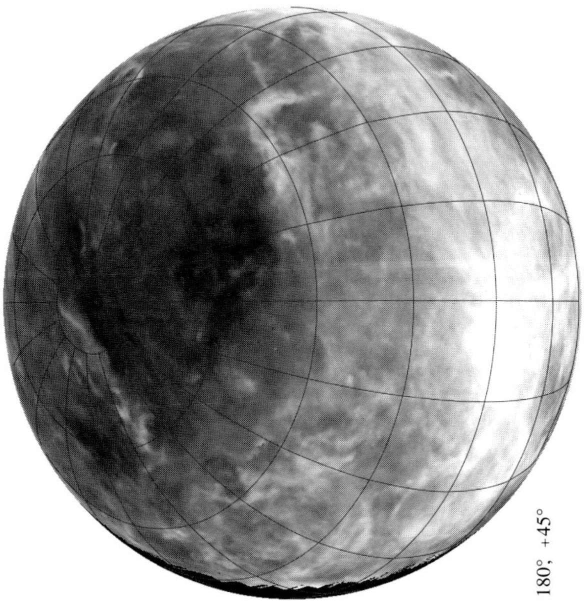

180°; +45°

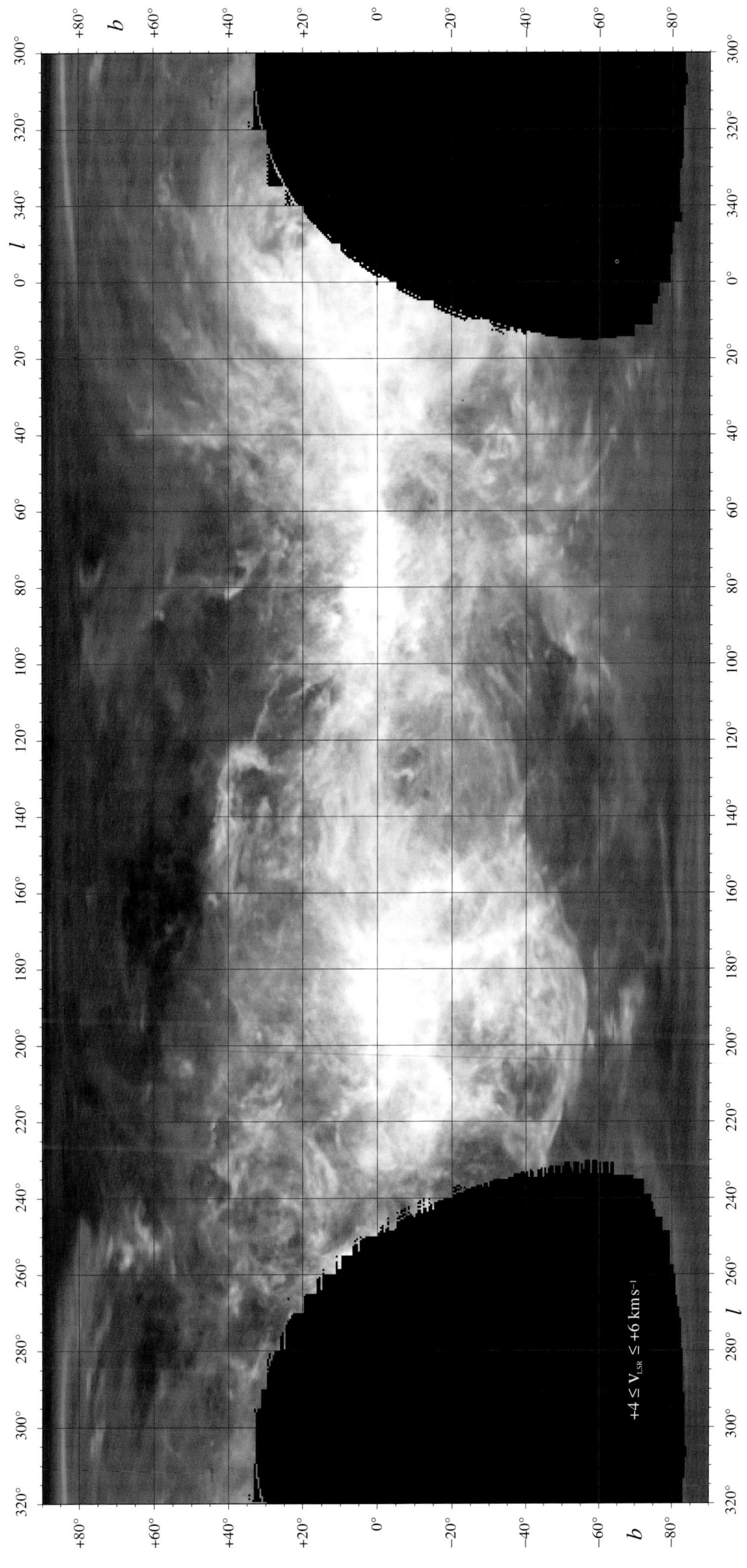
+4 ≤ V_LSR ≤ +6 km s⁻¹
l
b

$+6 \leq V_{LSR} \leq +8$ km s^{-1}

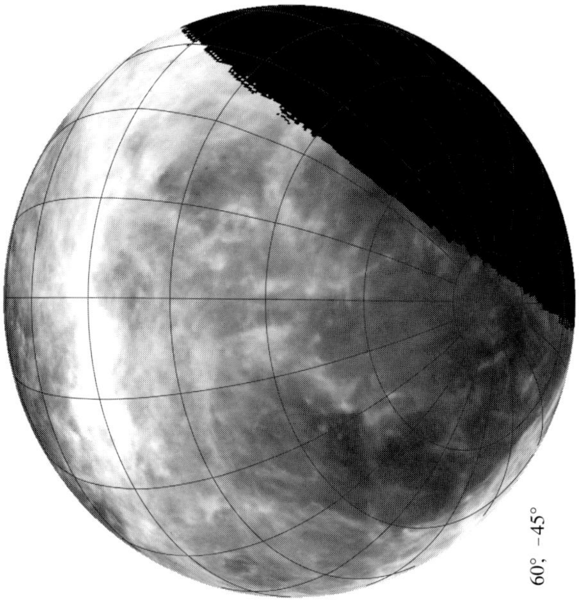

60°, −45°

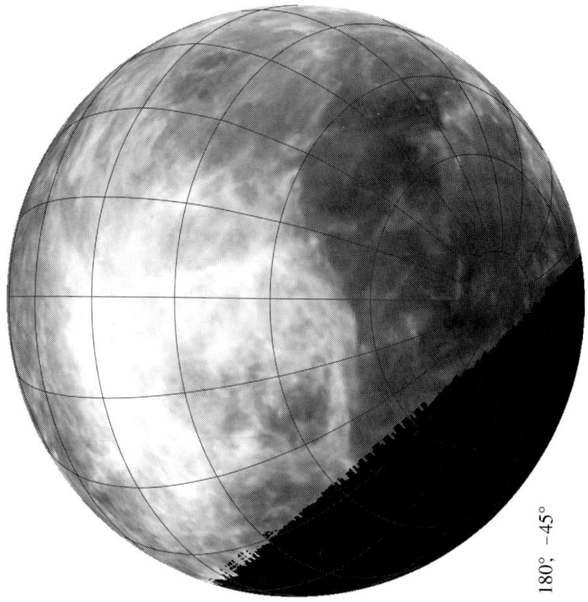

180°, −45°

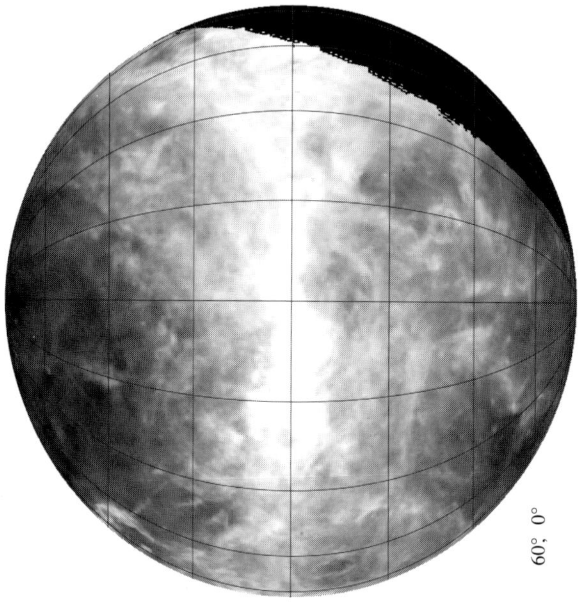

60°, 0°

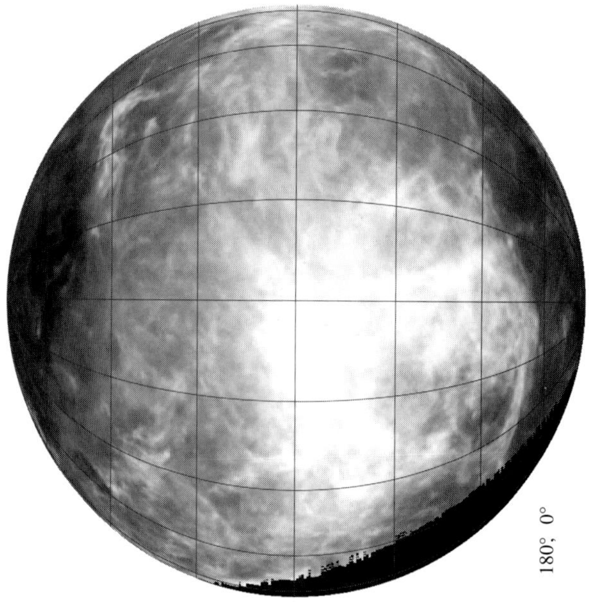

180°, 0°

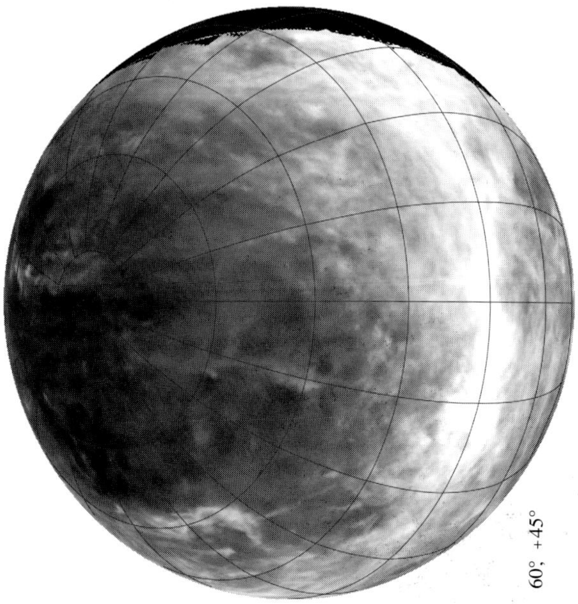

60°, +45°

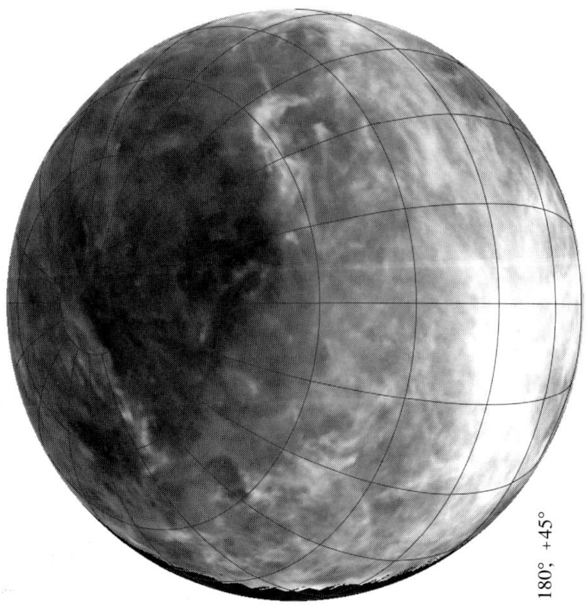

180°, +45°

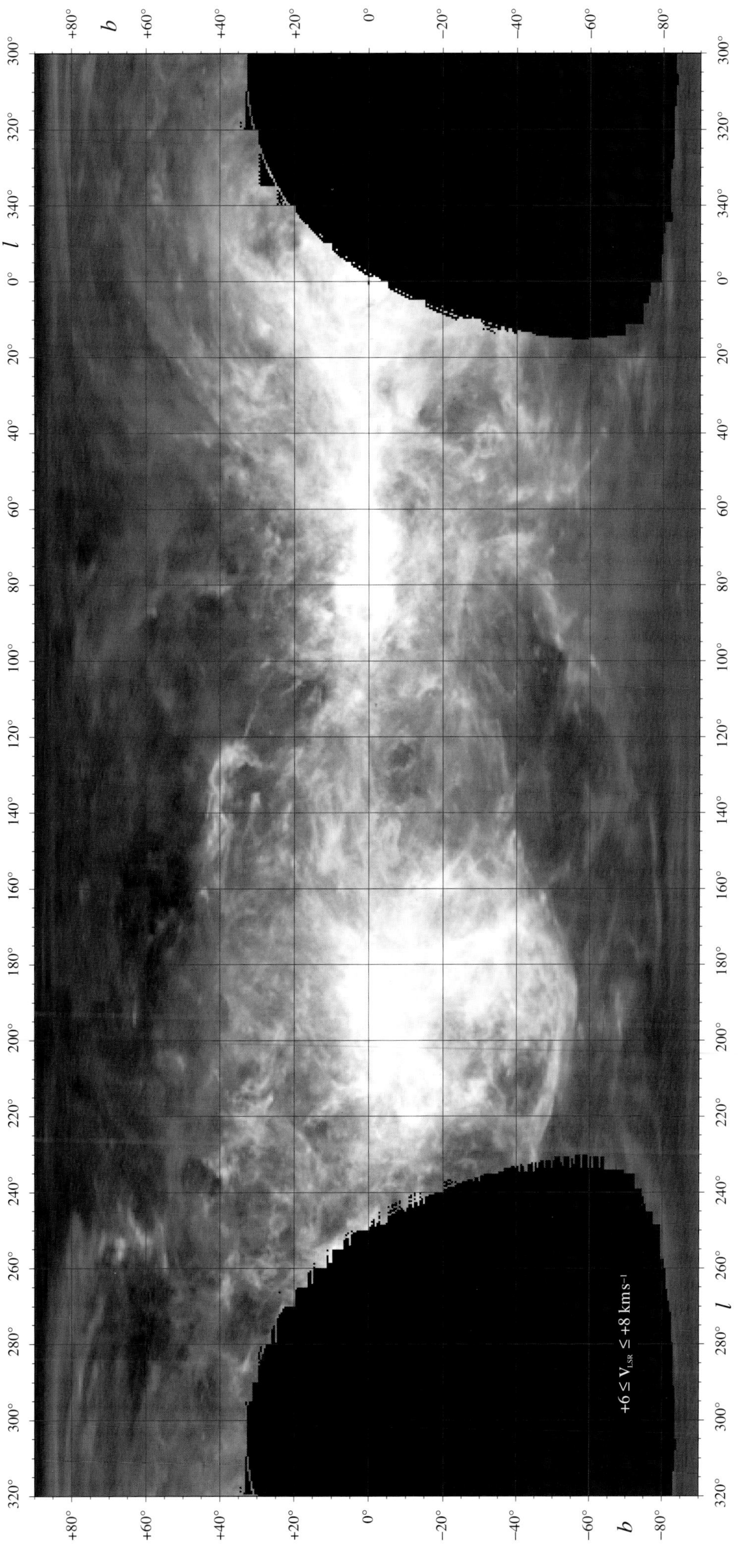
+6 ≤ V_LSR ≤ +8 km s⁻¹

$+8 \leq V_{LSR} \leq +10 \ \mathrm{km\,s^{-1}}$

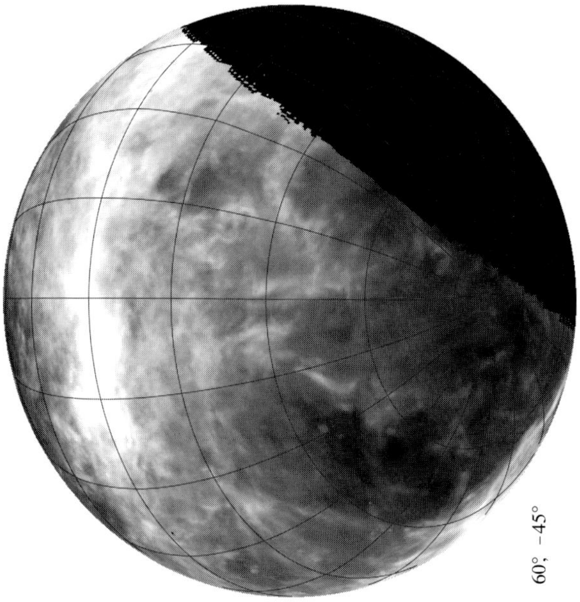

60°, −45°

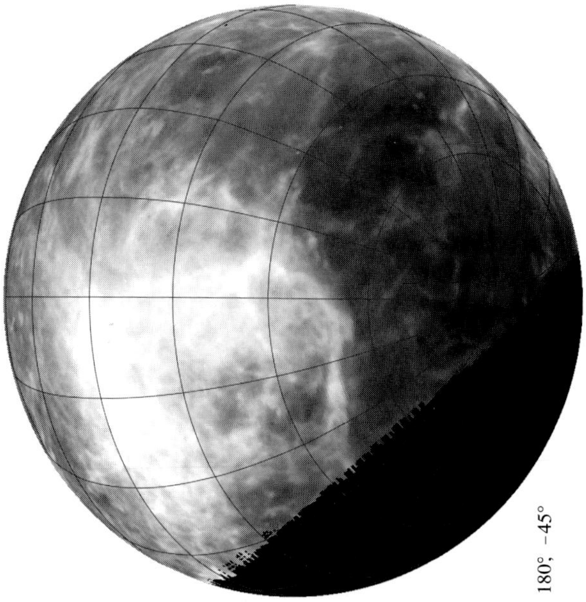

180°, −45°

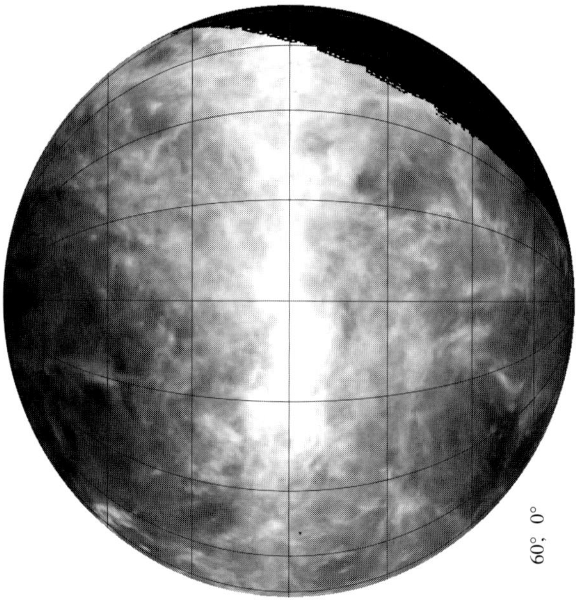

60°, 0°

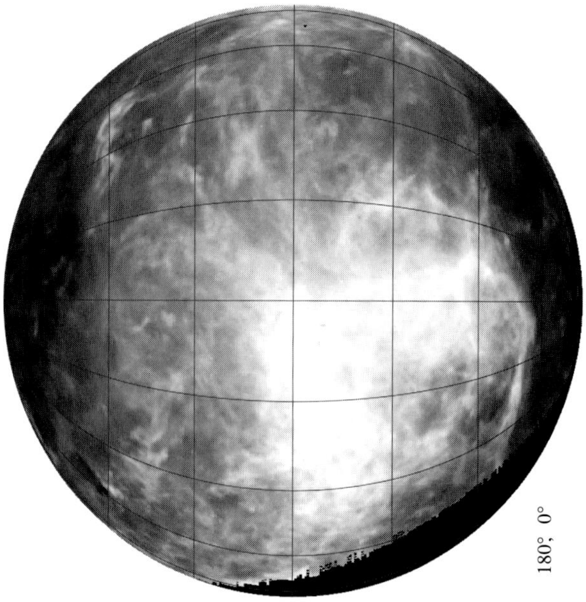

180°, 0°

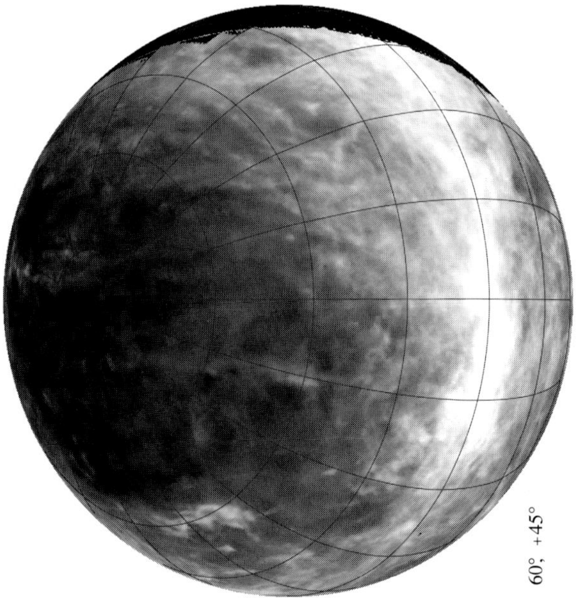

60°, +45°

180°, +45°

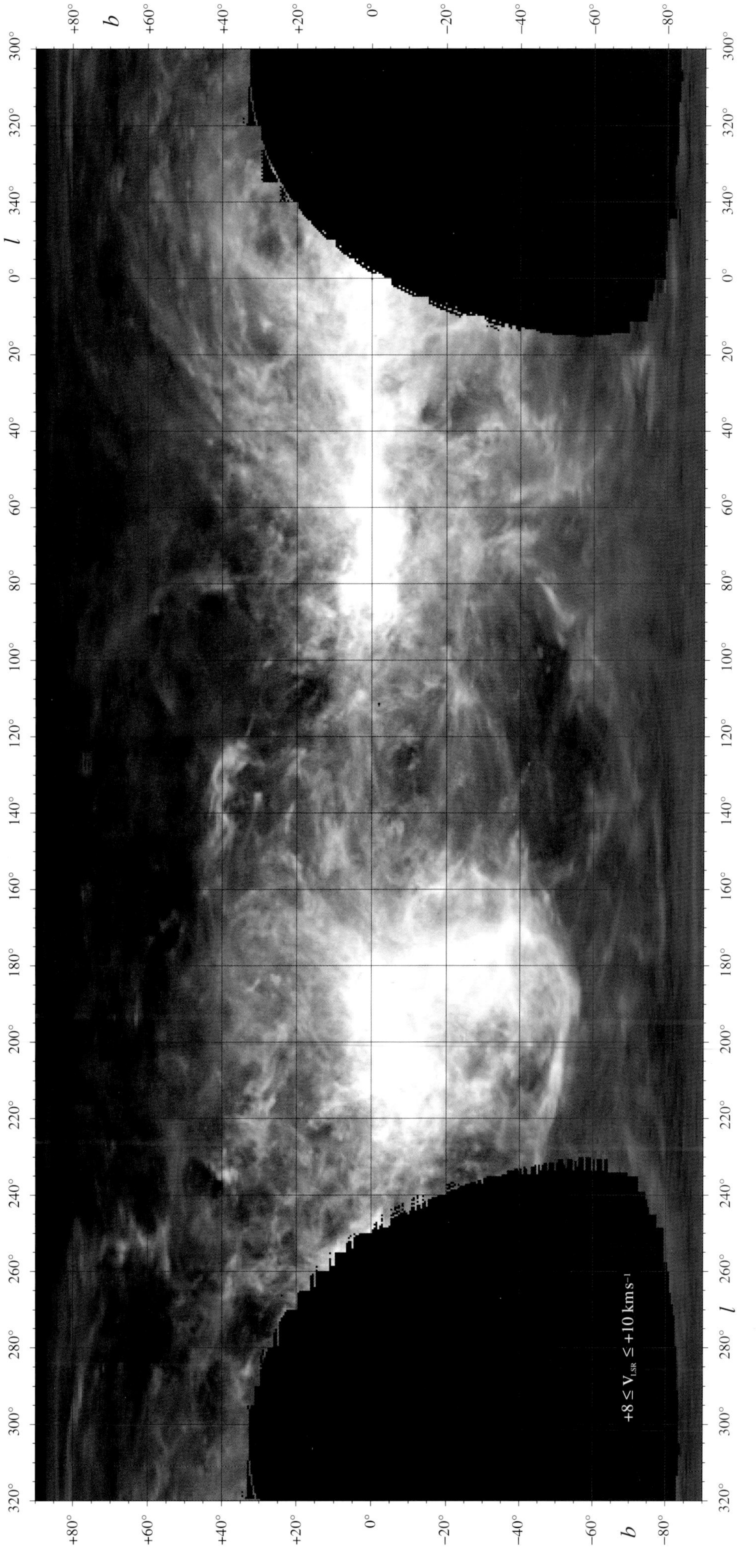
+8 ≤ V_LSR ≤ +10 kms⁻¹
l
b

$+10 \leq V_{LSR} \leq +12$ km s^{-1}

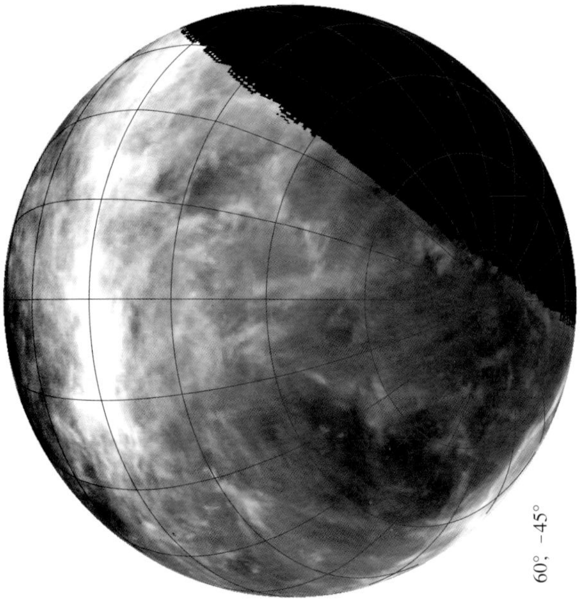

60°, −45°

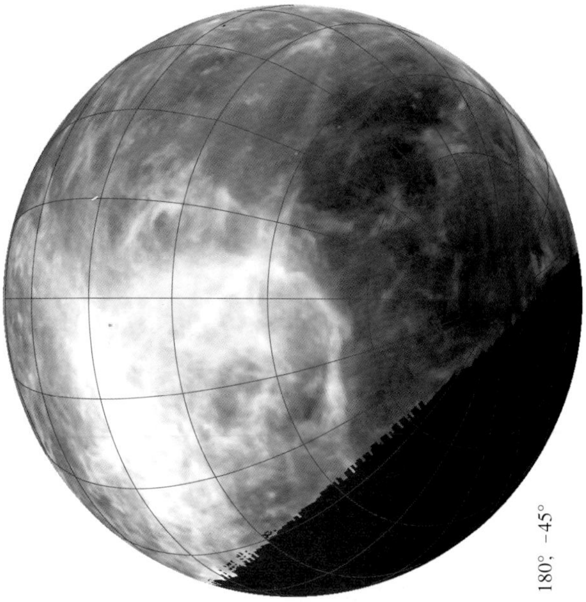

180°, −45°

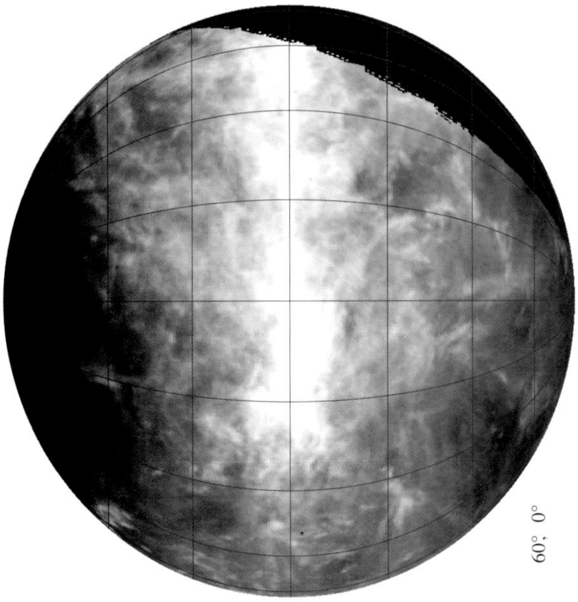

60°, 0°

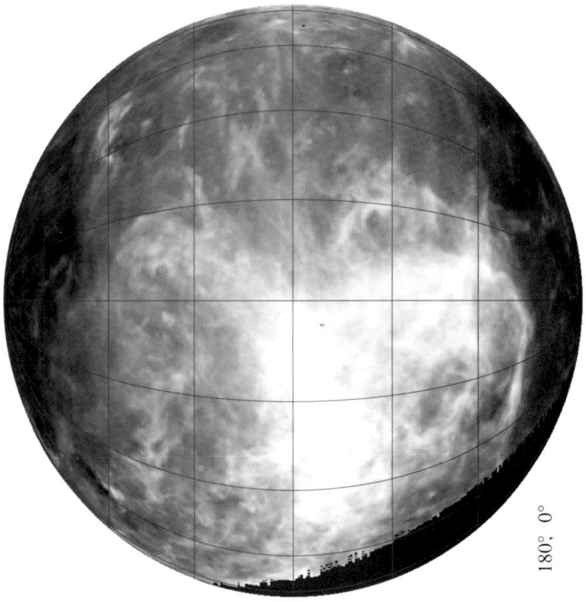

180°, 0°

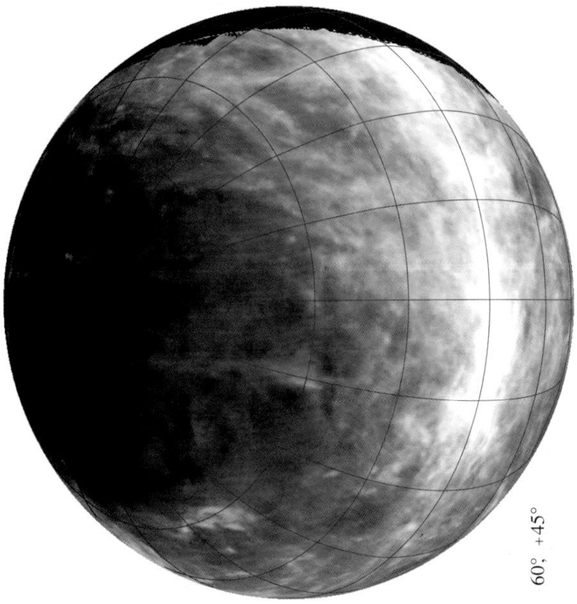

60°, +45°

180°, +45°

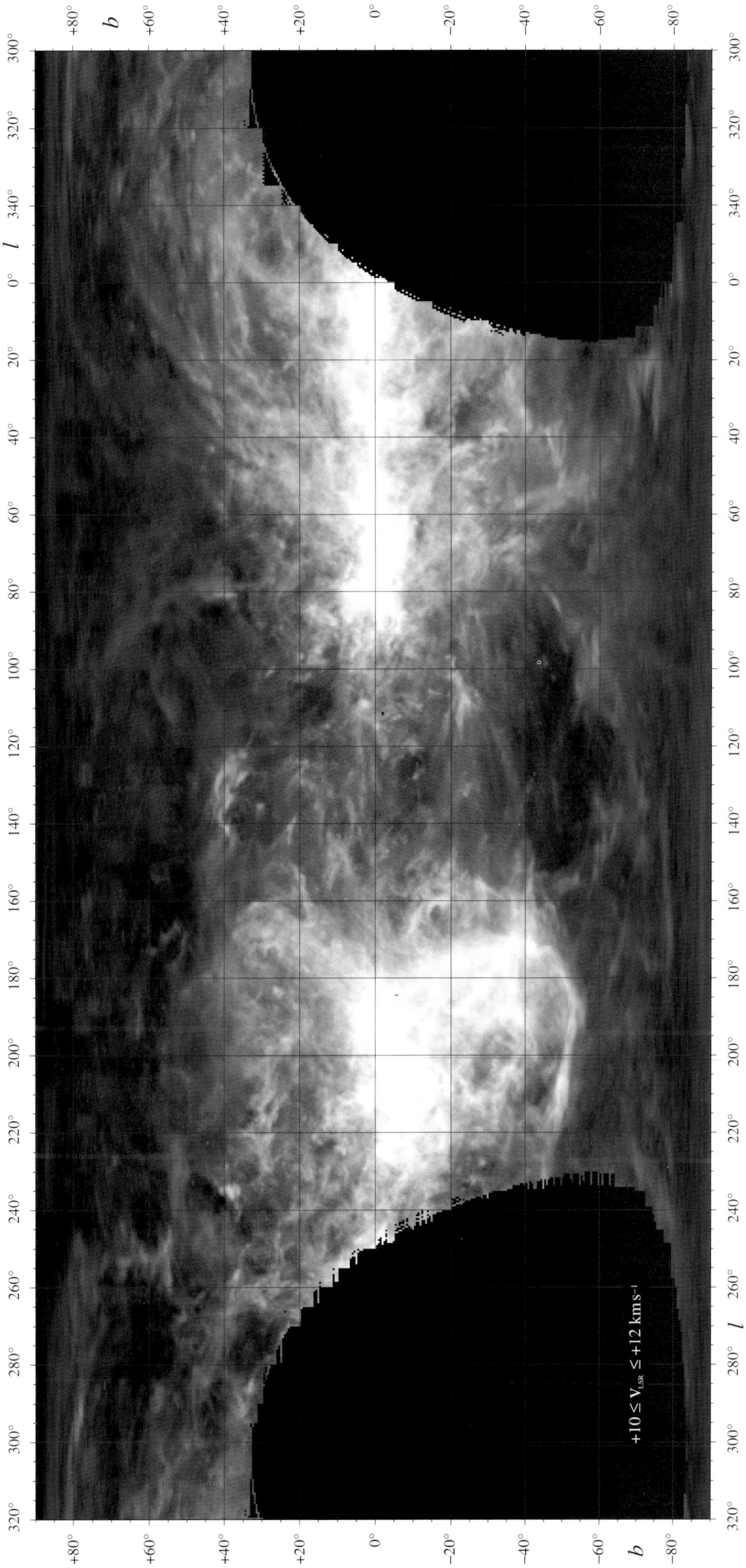
+10 ≤ V_LSR ≤ +12 kms⁻¹
l
b
300° 320° 340° 0° 20° 40° 60° 80° 100° 120° 140° 160° 180° 200° 220° 240° 260° 280° 300° 320°
+80° +60° +40° +20° 0° −20° −40° −60° −80°

$+12 \le V_{LSR} \le +14$ km s^{-1}

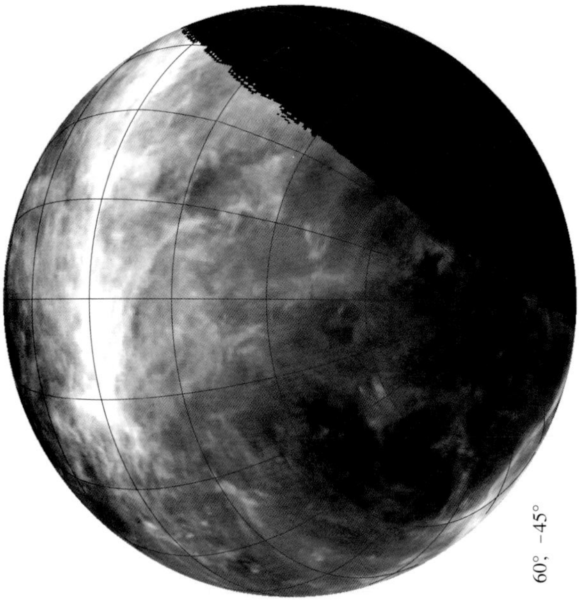

60°, −45°

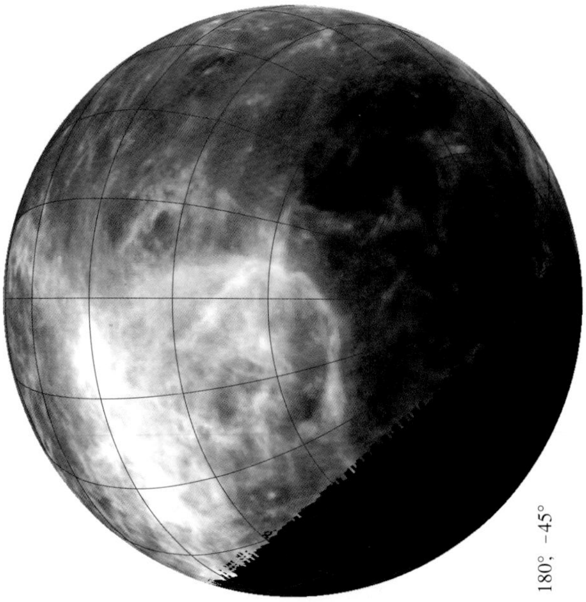

180°, −45°

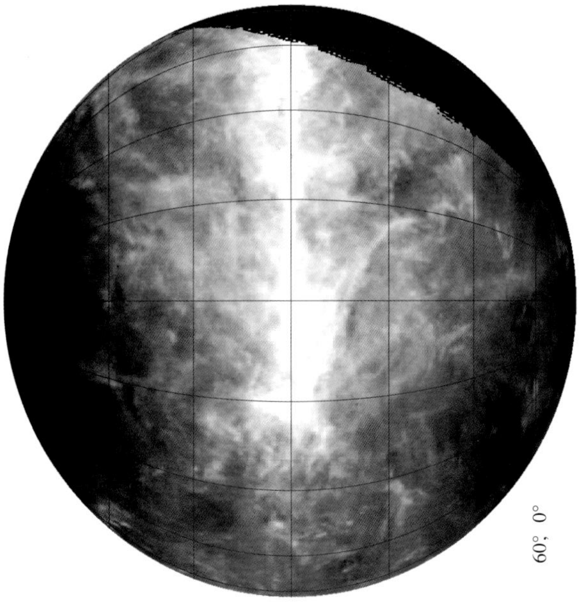

60°, 0°

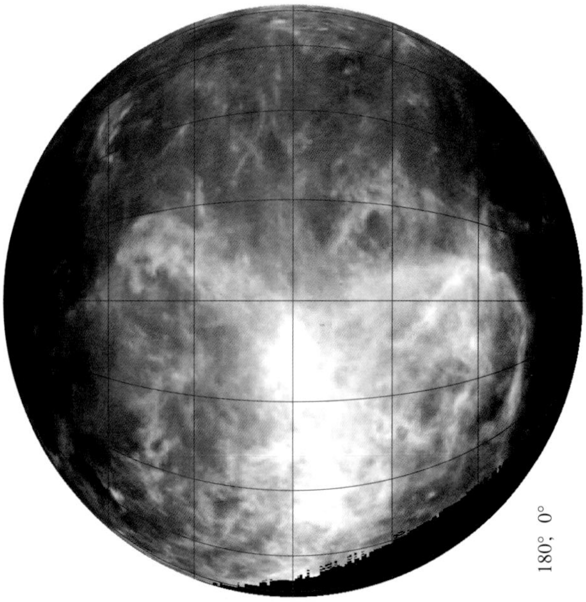

180°, 0°

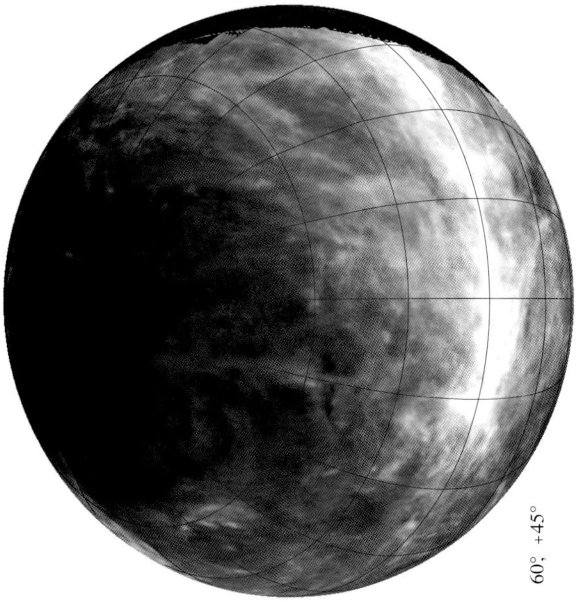

60°, +45°

180°, +45°

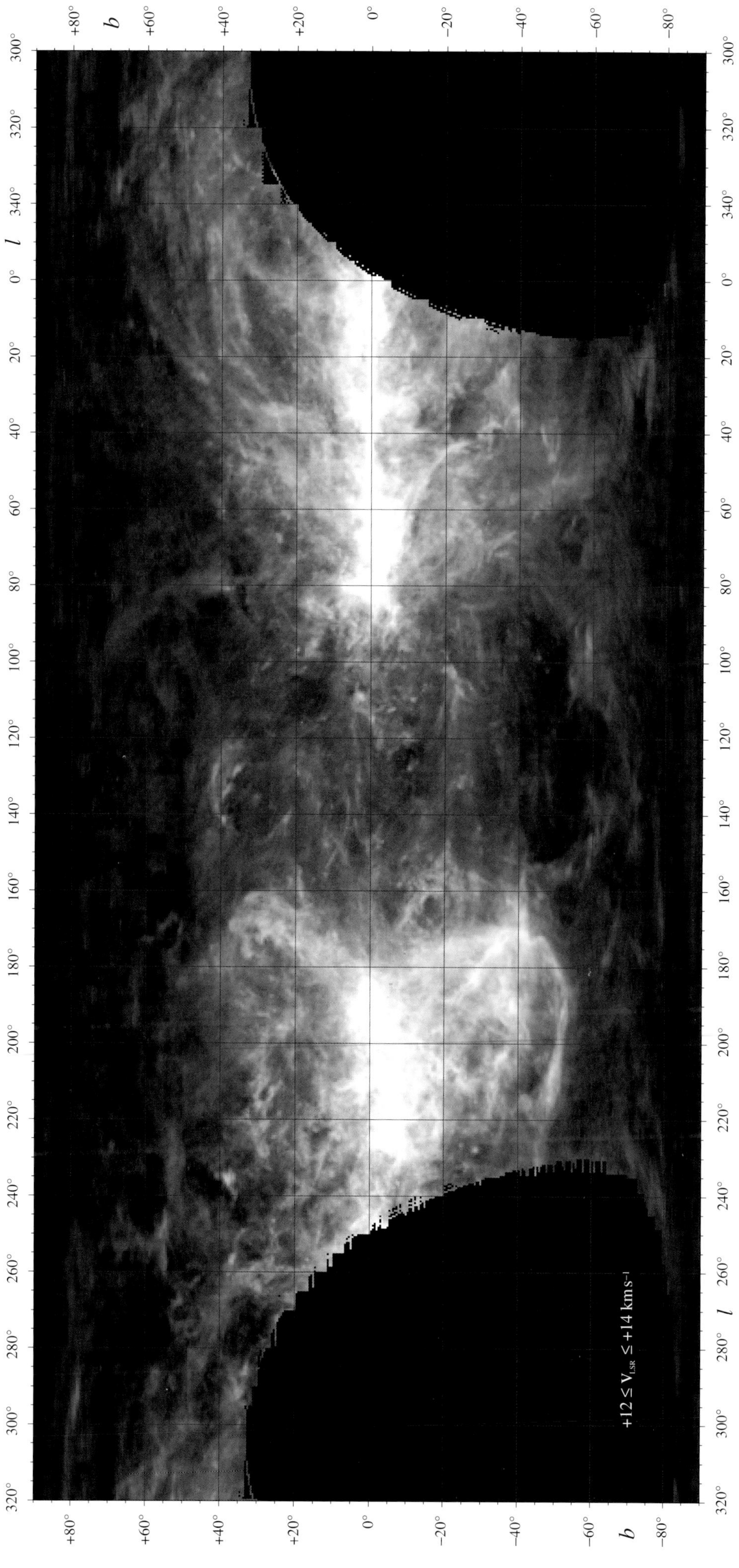

+12 ≤ V_LSR ≤ +14 km s⁻¹
l
b

$+14 \leq V_{LSR} \leq +16$ km s^{-1}

60°, −45°

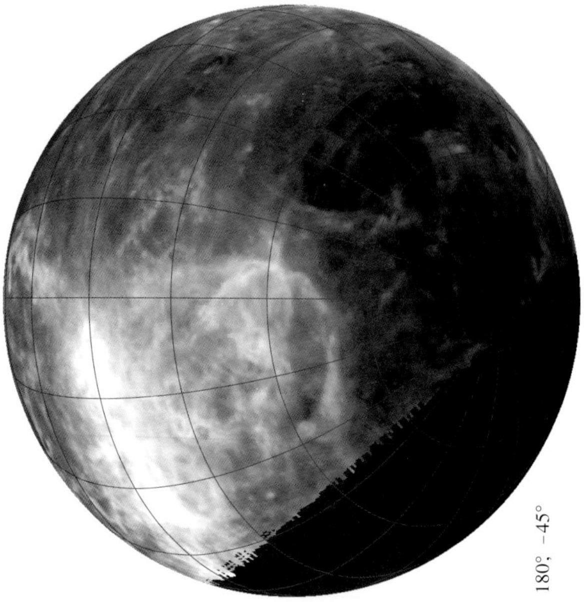

180°, −45°

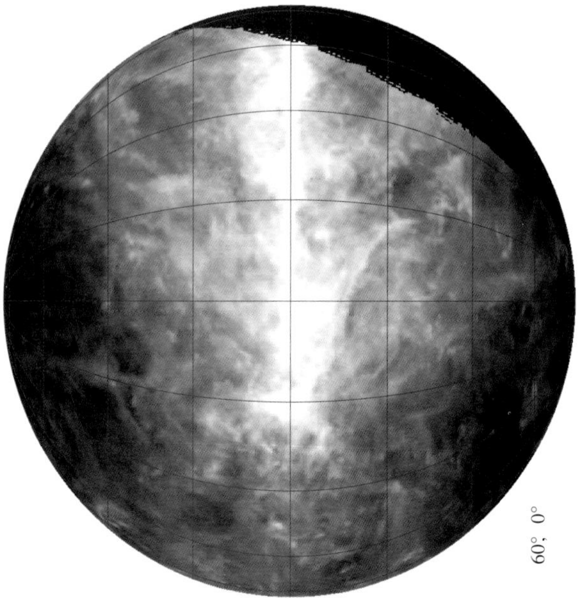

60°, 0°

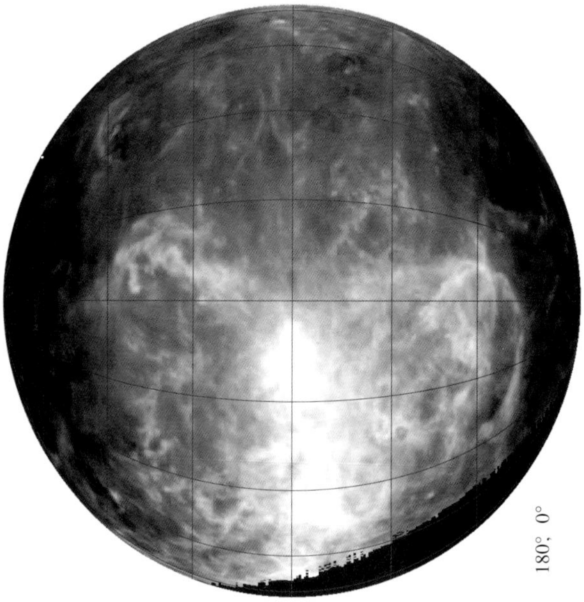

180°, 0°

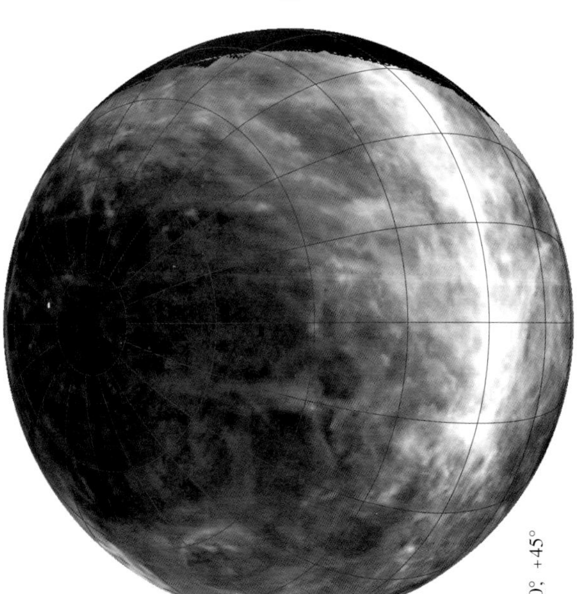

60°, +45°

180°, +45°

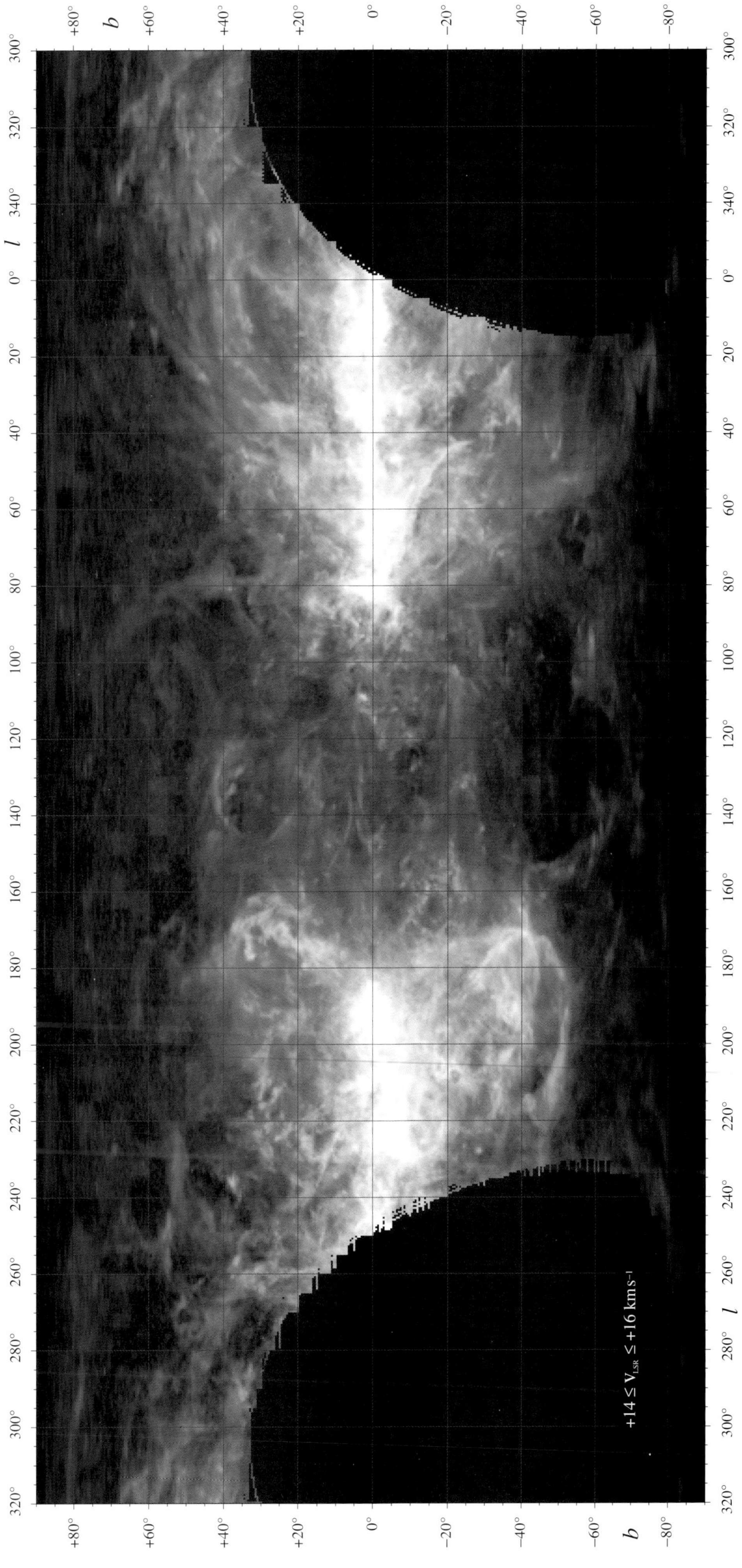
$+14 \le V_{LSR} \le +16$ km s^{-1}
l
b
300° 320° 340° 0° 20° 40° 60° 80° 100° 120° 140° 160° 180° 200° 220° 240° 260° 280° 300° 320°
+80° +60° +40° +20° 0° −20° −40° −60° −80°

$+16 \leq V_{LSR} \leq +18$ km s^{-1}

60°, −45°

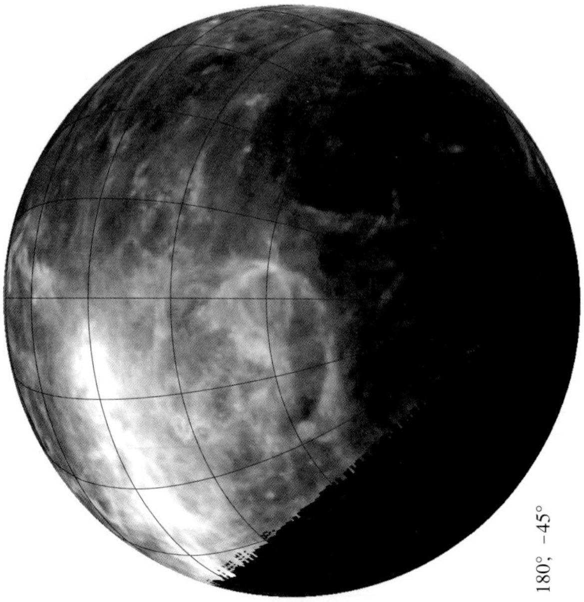

180°, −45°

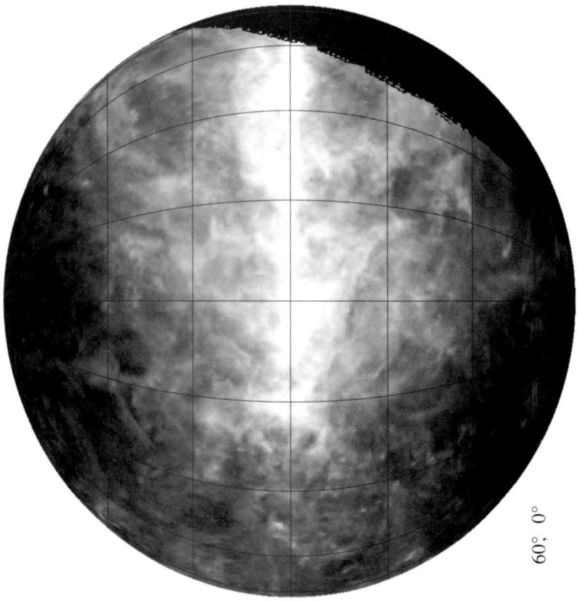

60°, 0°

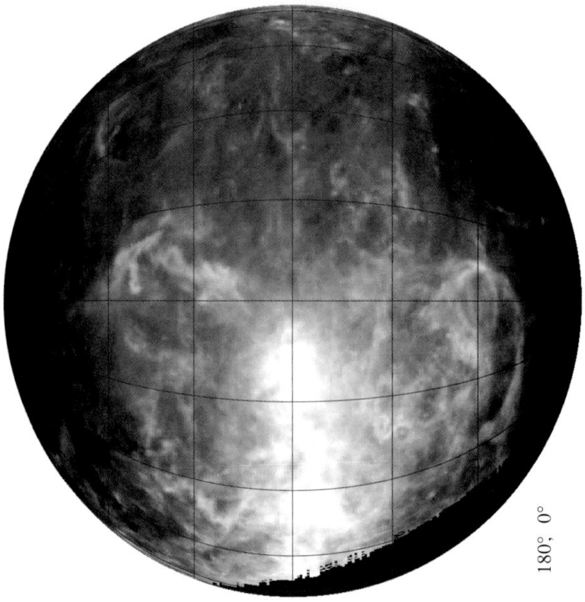

180°, 0°

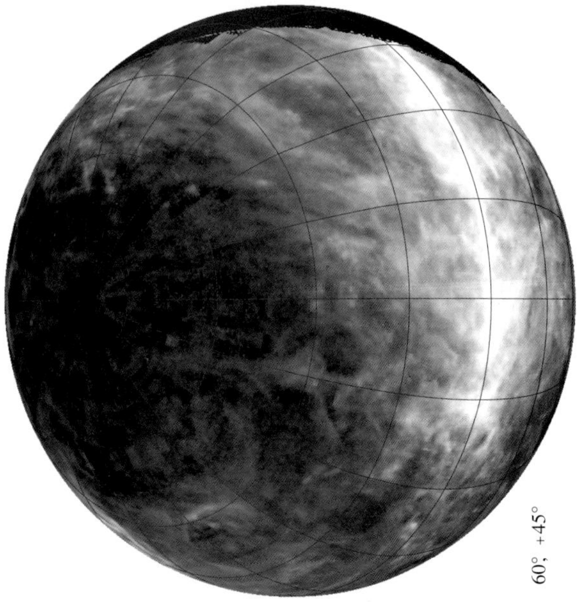

60°, +45°

180°, +45°

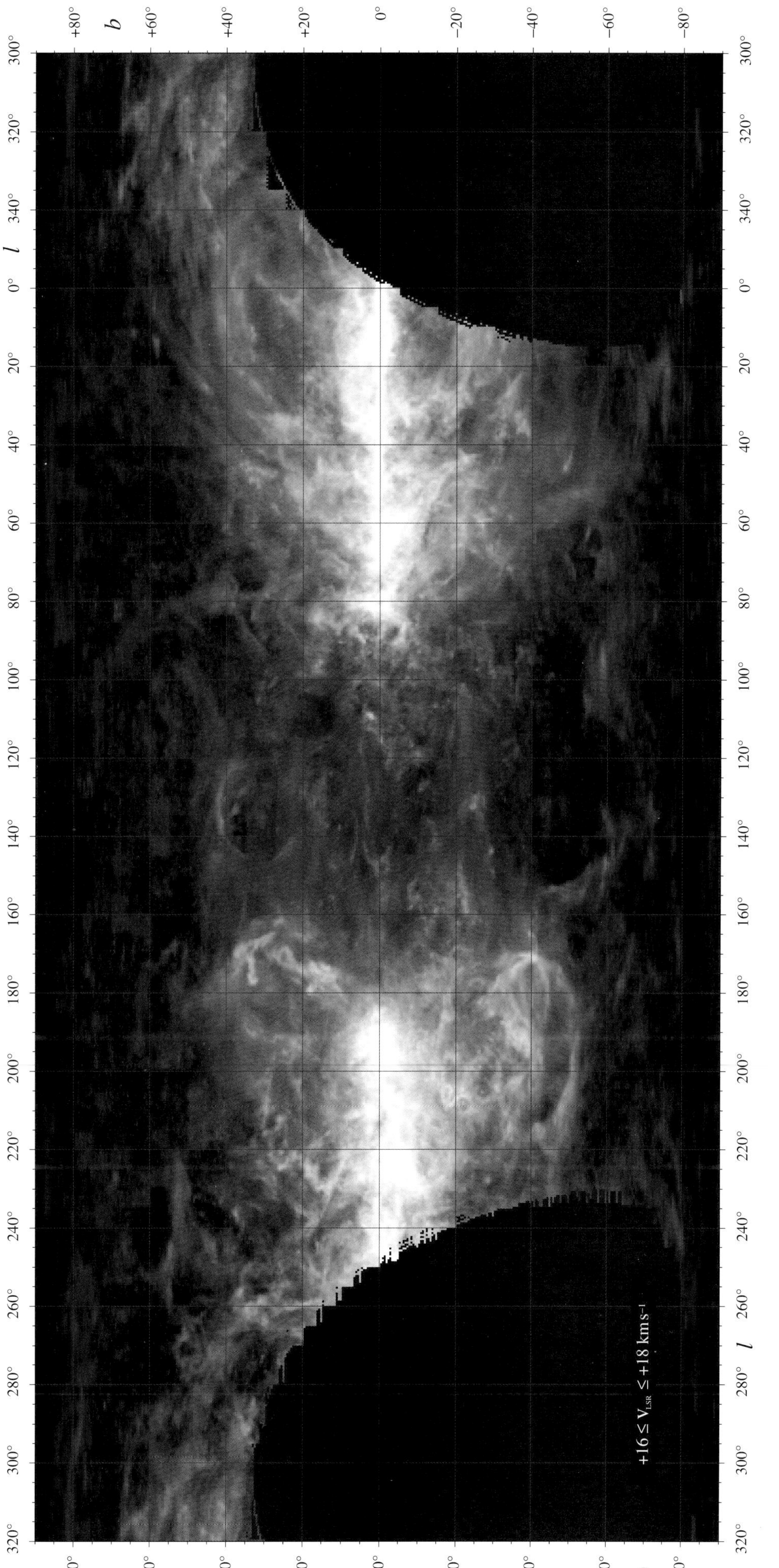
$+16 \leq V_{LSR} \leq +18$ kms^{-1}
l
b

$+18 \leq V_{LSR} \leq +20$ km s^{-1}

60°, −45°

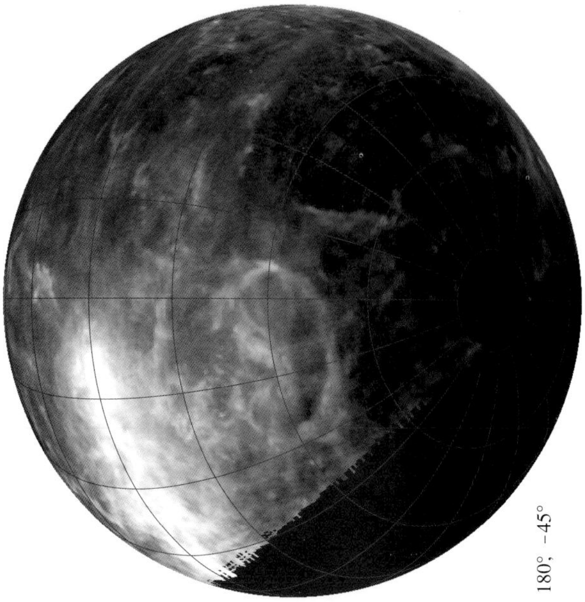

180°, −45°

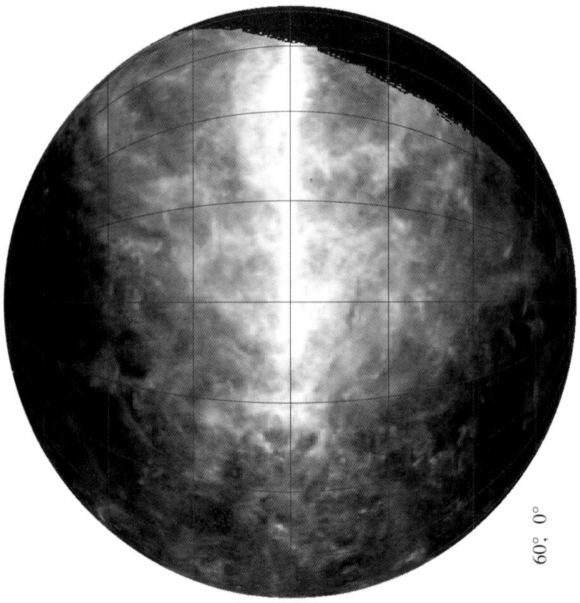

60°, 0°

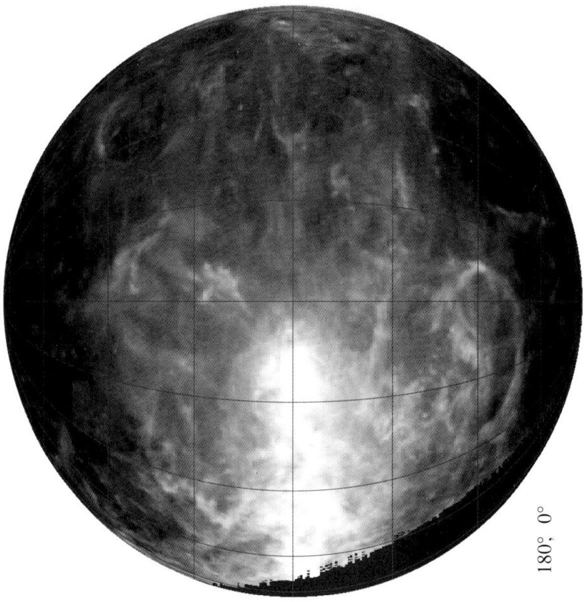

180°, 0°

60°, +45°

180°, +45°

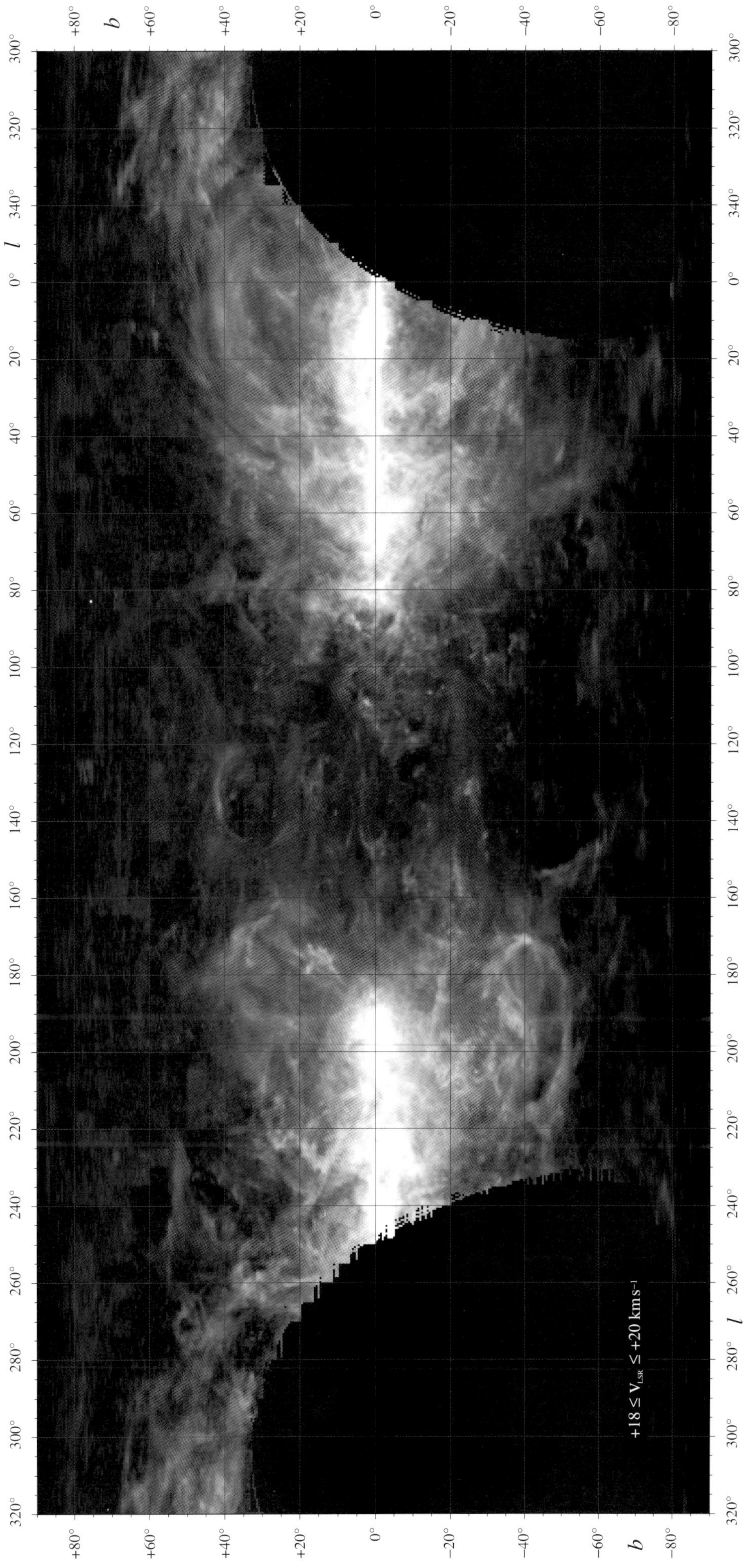
$+18 \le V_{LSR} \le +20$ km s^{-1}

$+20 \leq V_{LSR} \leq +22\ km\,s^{-1}$

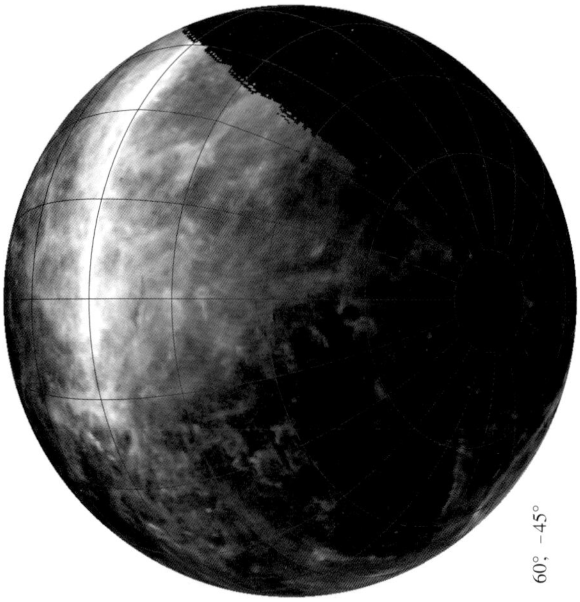

60°, −45°

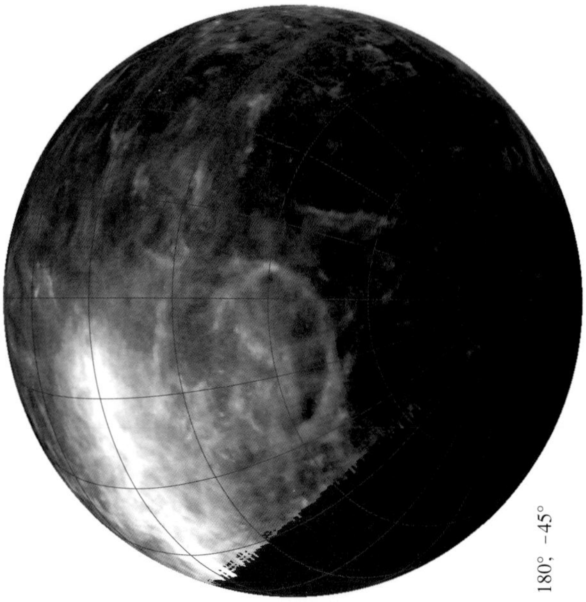

180°, −45°

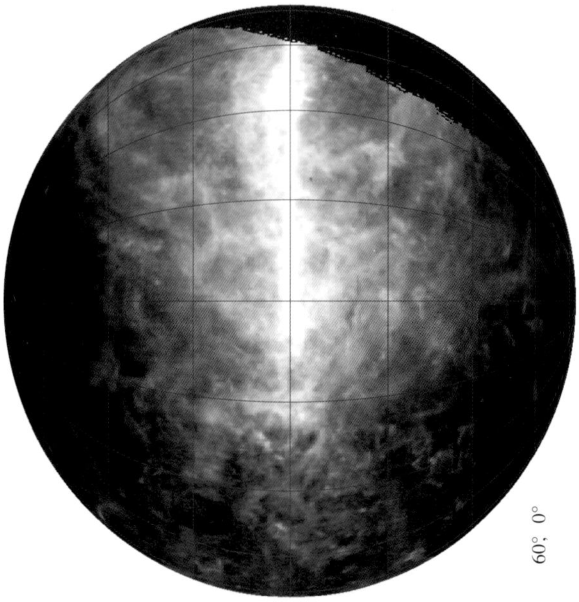

60°, 0°

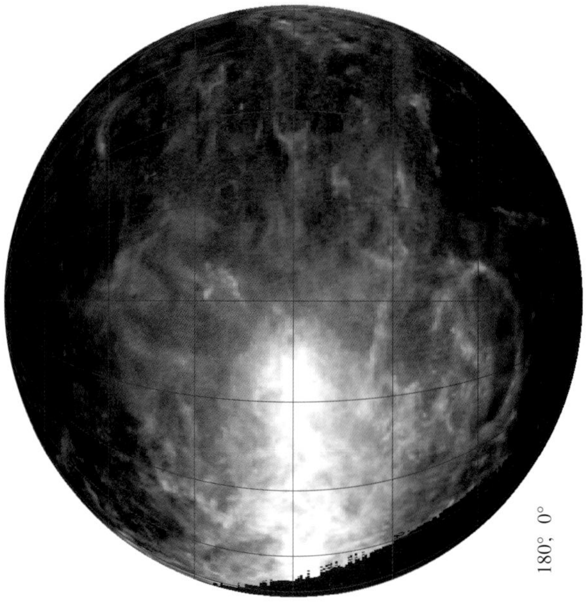

180°, 0°

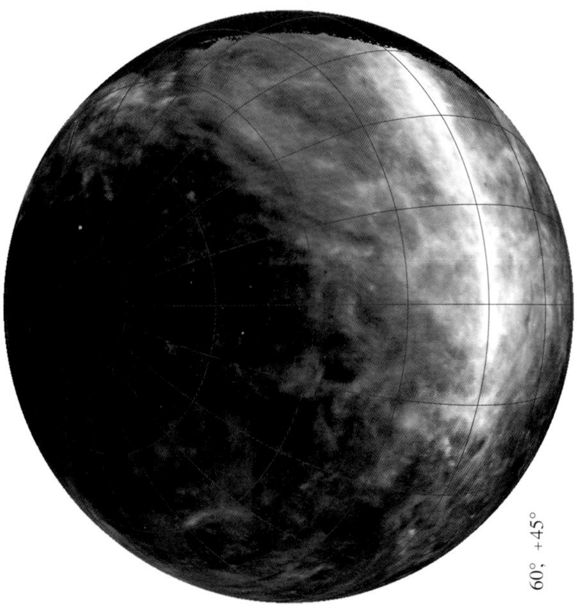

60°, +45°

180°, +45°

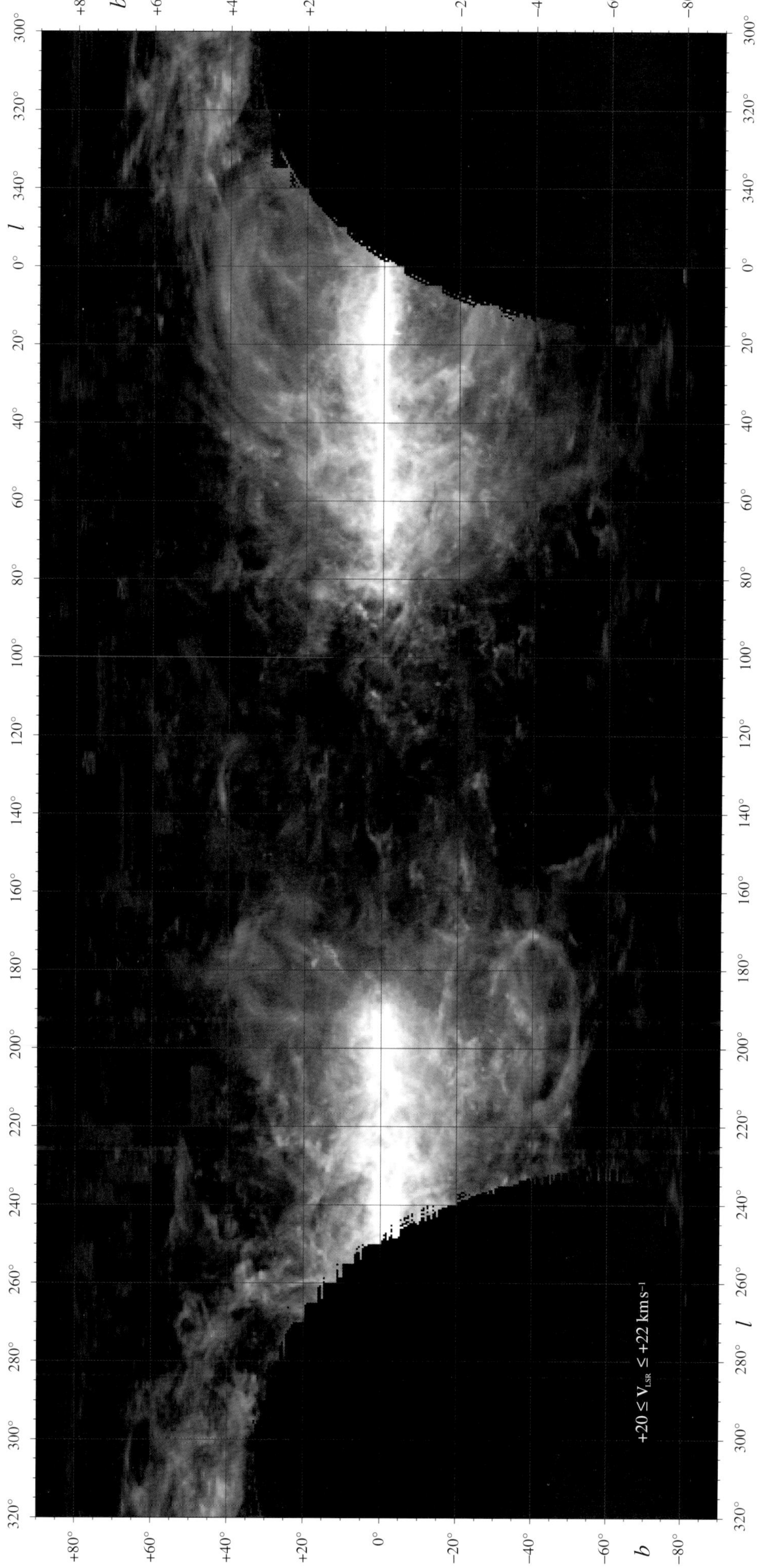
+20 ≤ V_LSR ≤ +22 km s^-1
l
b

$+22 \leq V_{LSR} \leq +24\ \mathrm{km\,s^{-1}}$

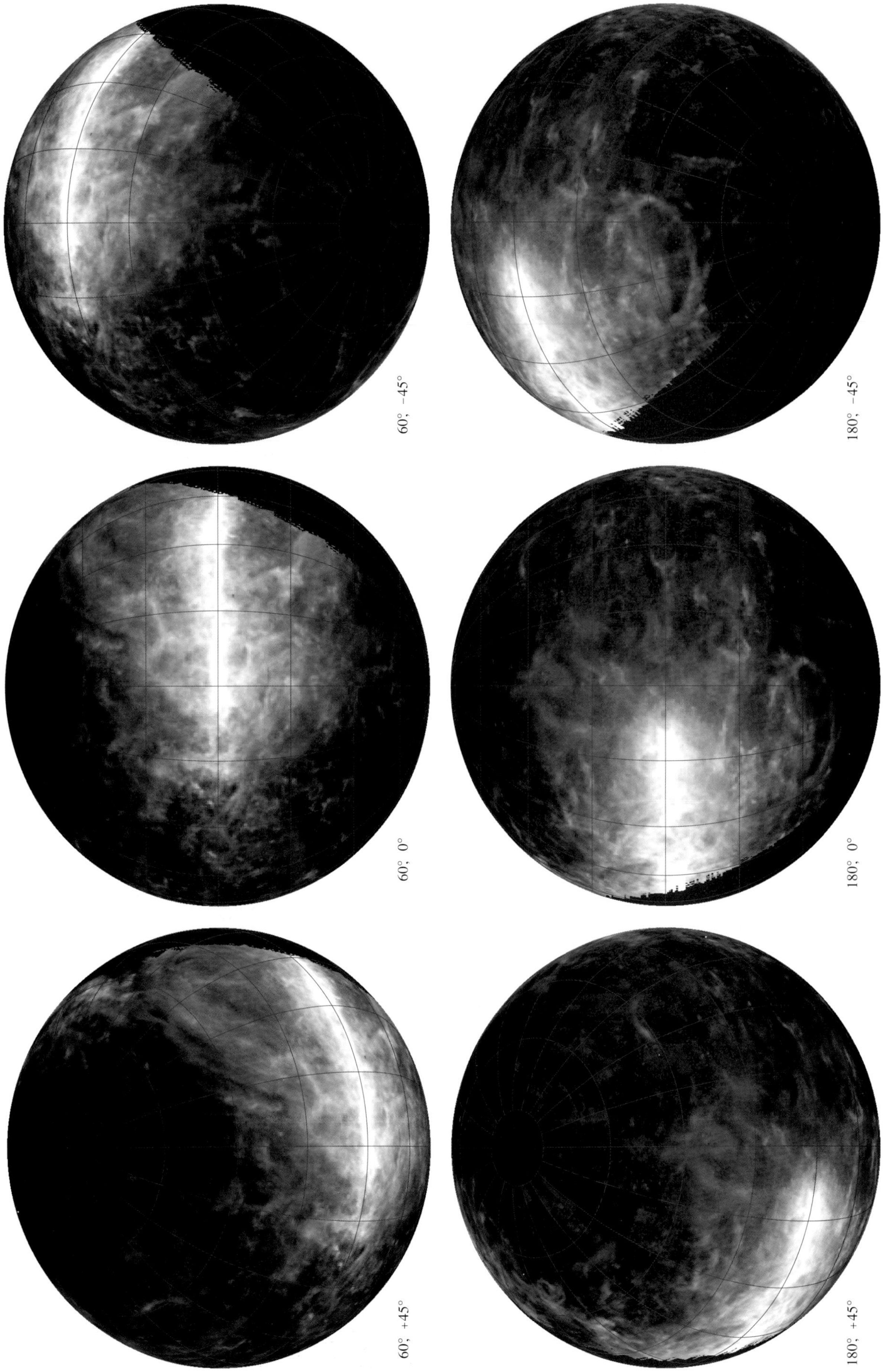

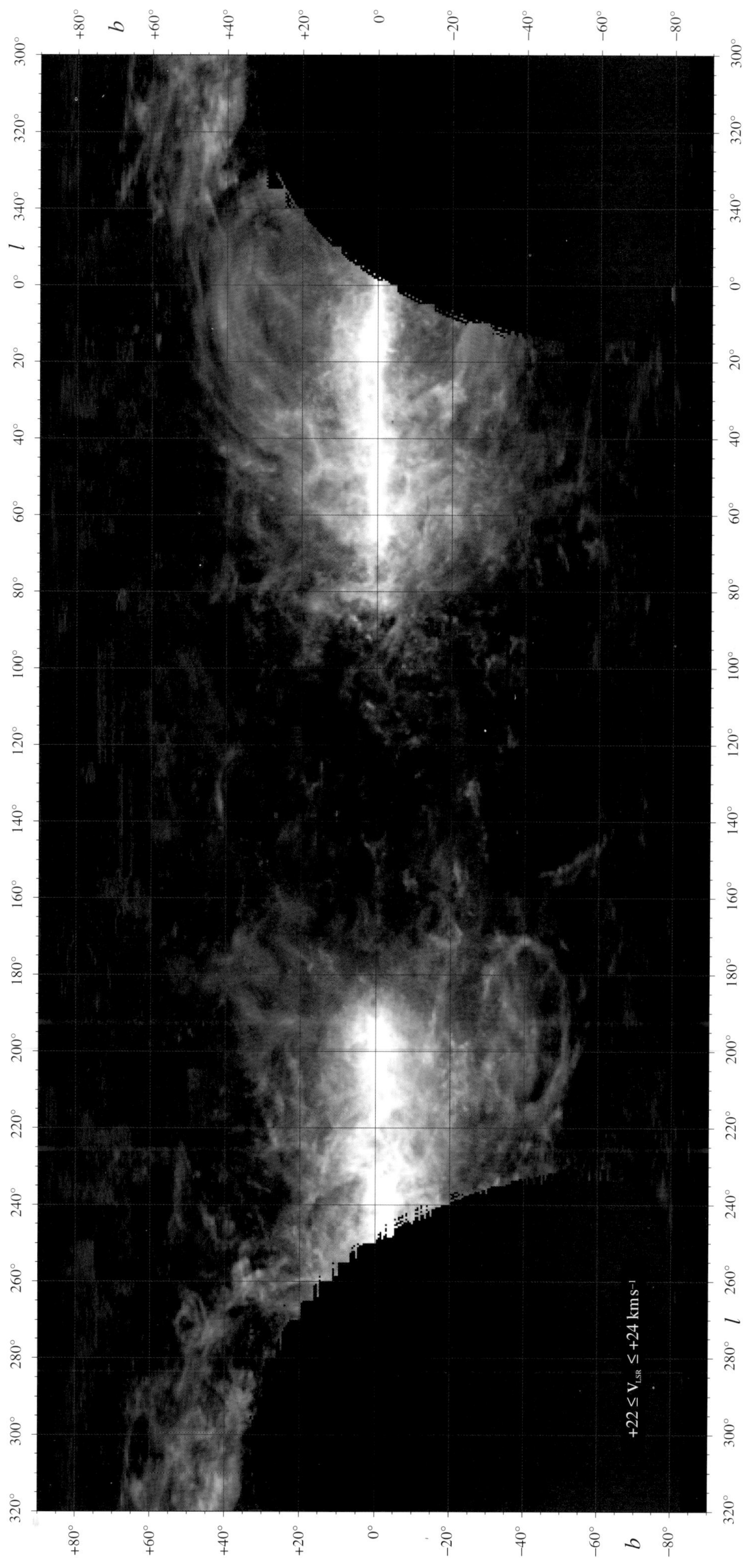
$+22 \leq V_{LSR} \leq +24$ km s^{-1}
l
b

$+24 \leq V_{LSR} \leq +26$ km s^{-1}

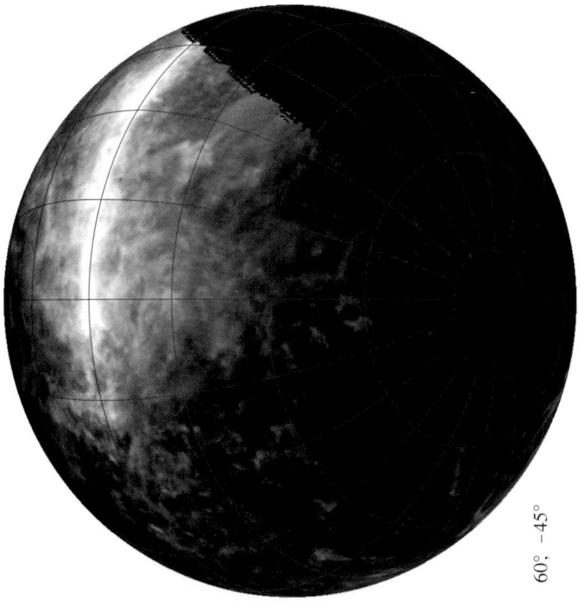

60°; −45°

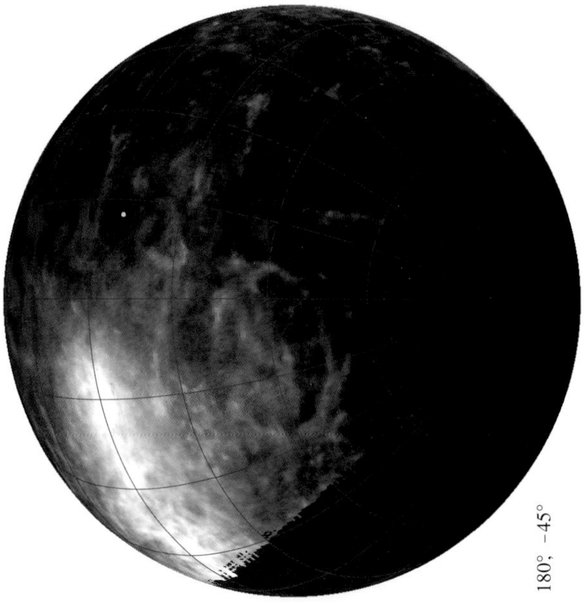

180°; −45°

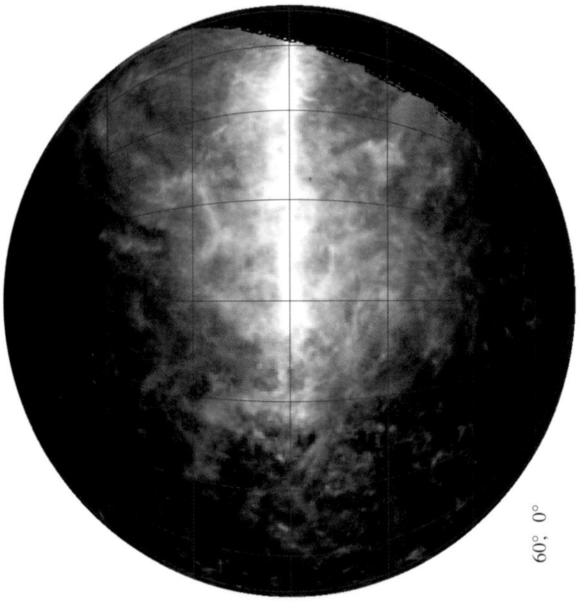

60°; 0°

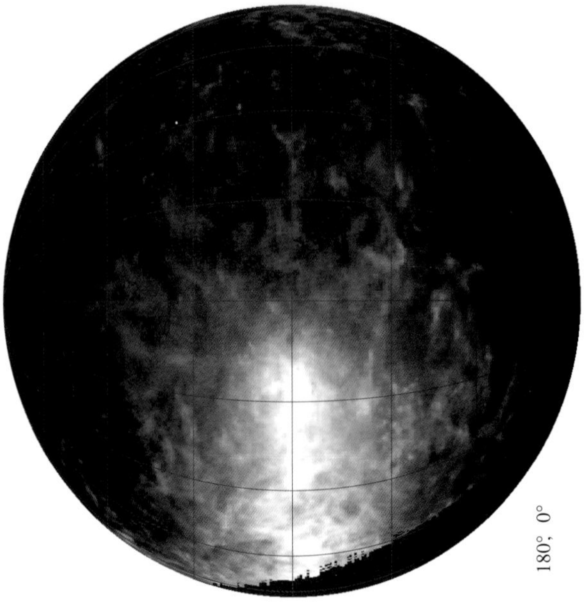

180°; 0°

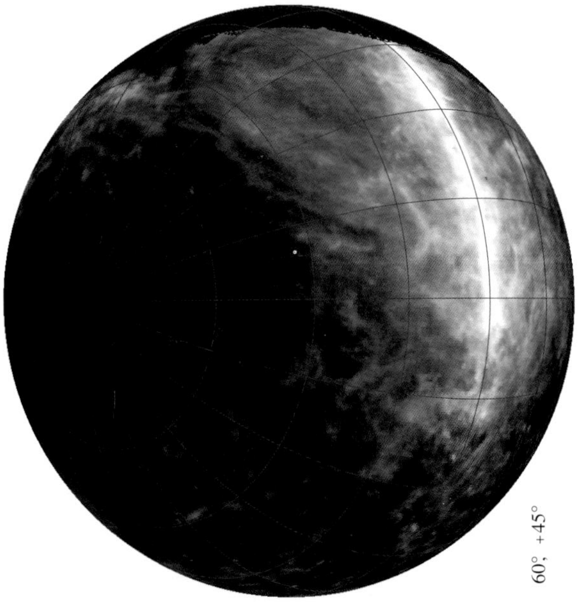

60°; +45°

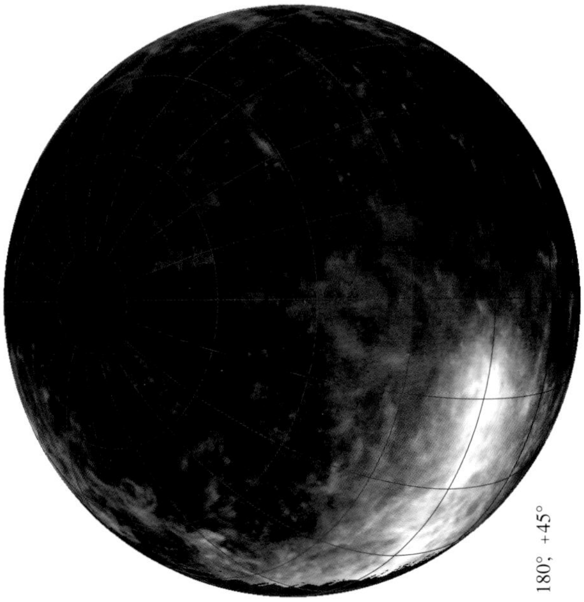

180°; +45°

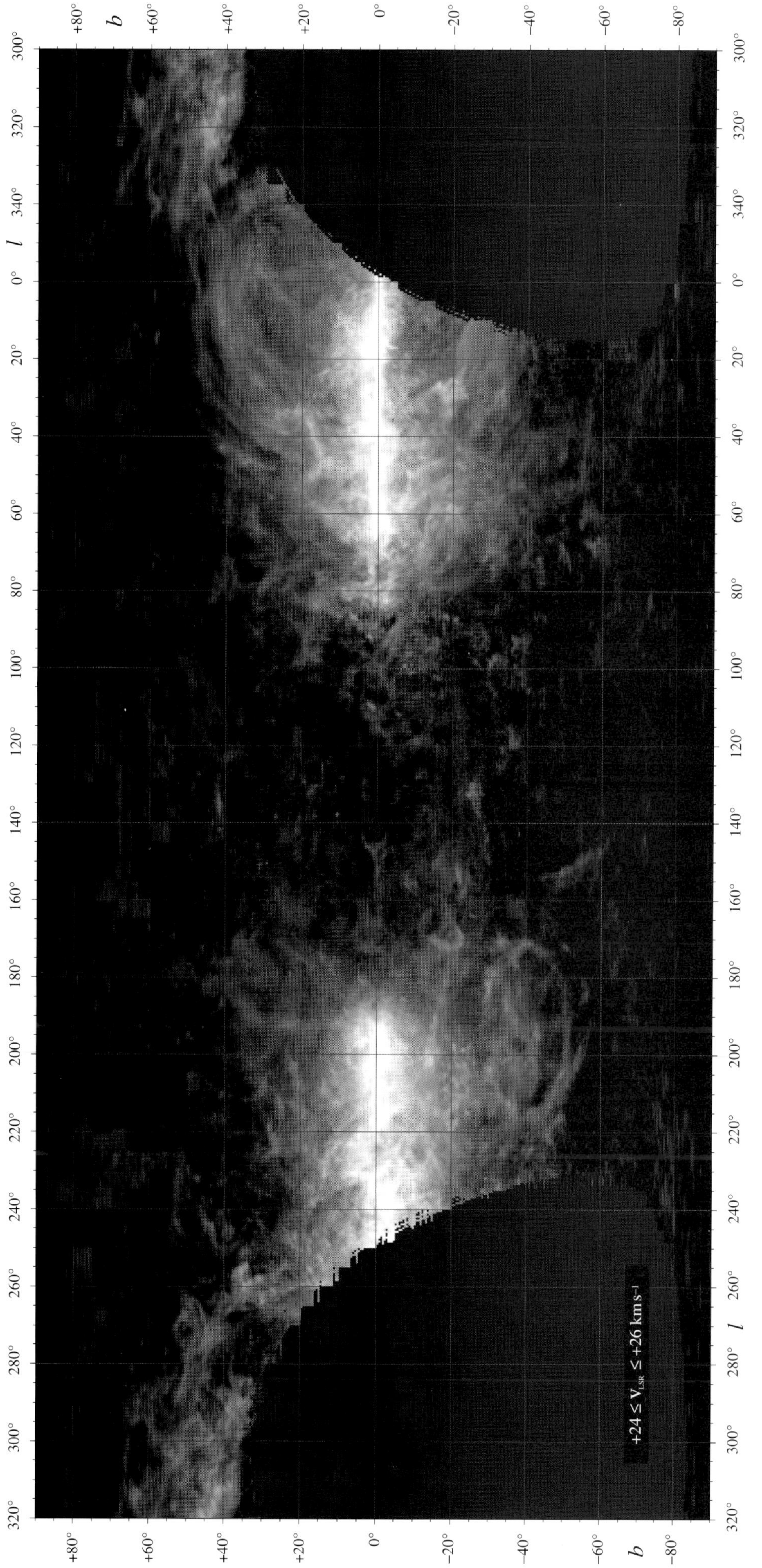
+24 ≤ V_LSR ≤ +26 kms-1

$+26 \leq V_{LSR} \leq +28$ km s^{-1}

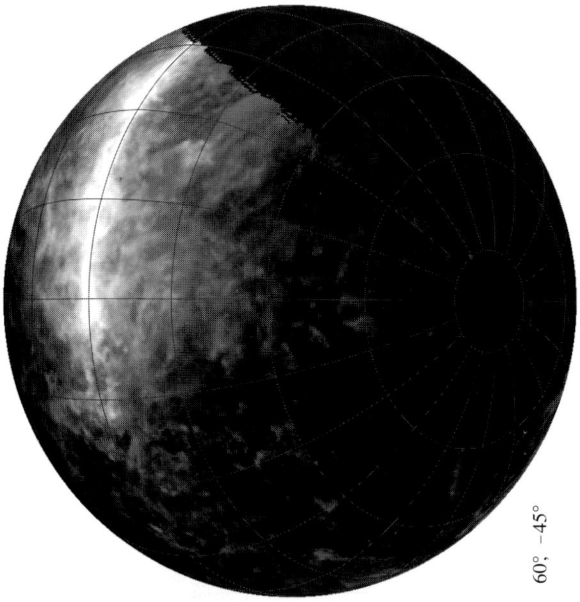

60°, −45°

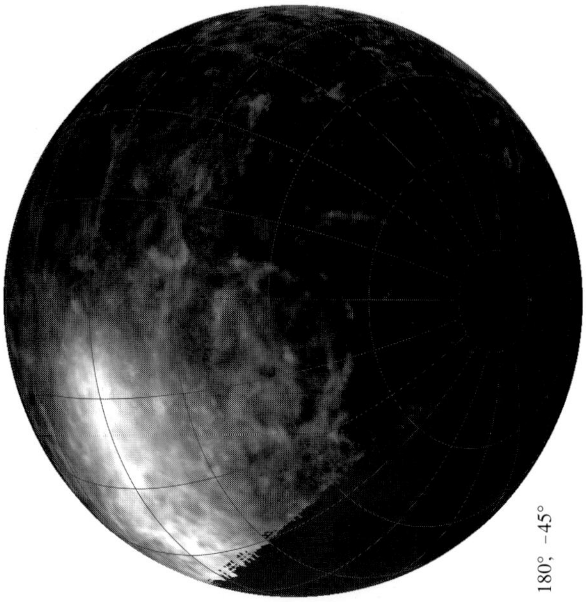

180°, −45°

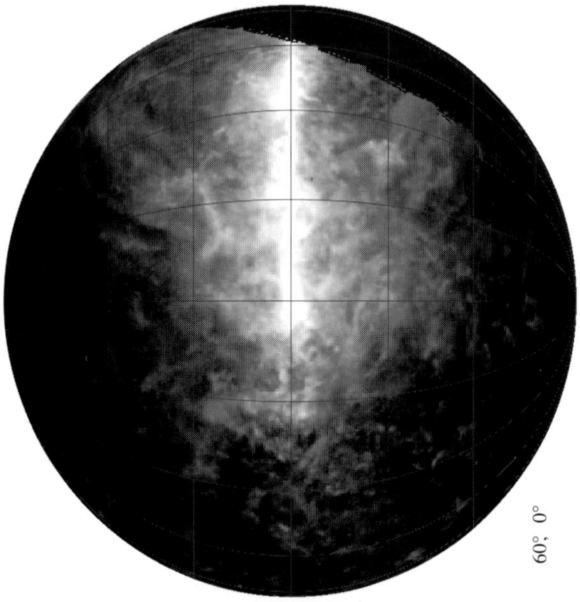

60°, 0°

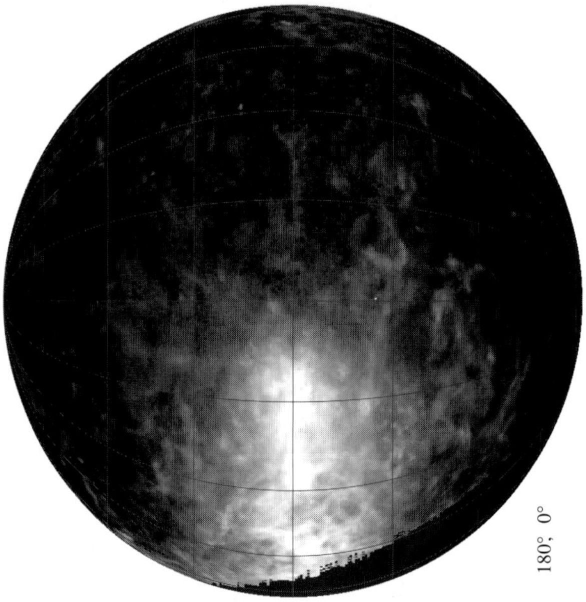

180°, 0°

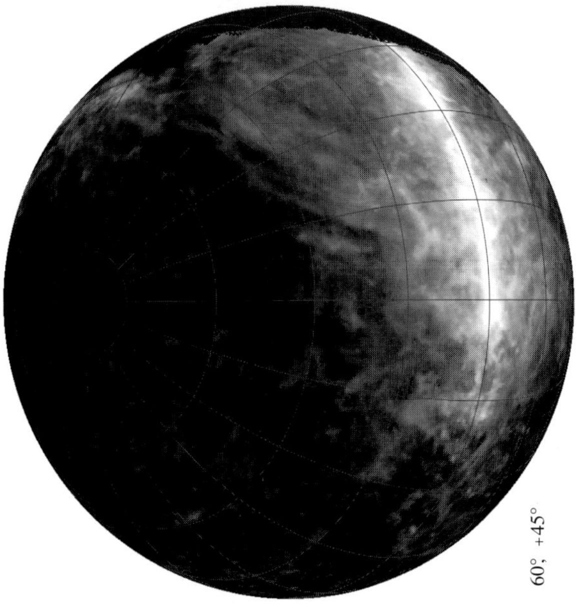

60°, +45°

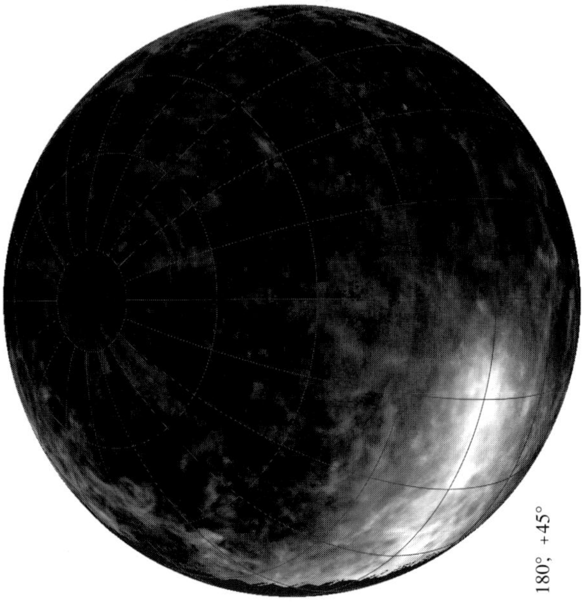

180°, +45°

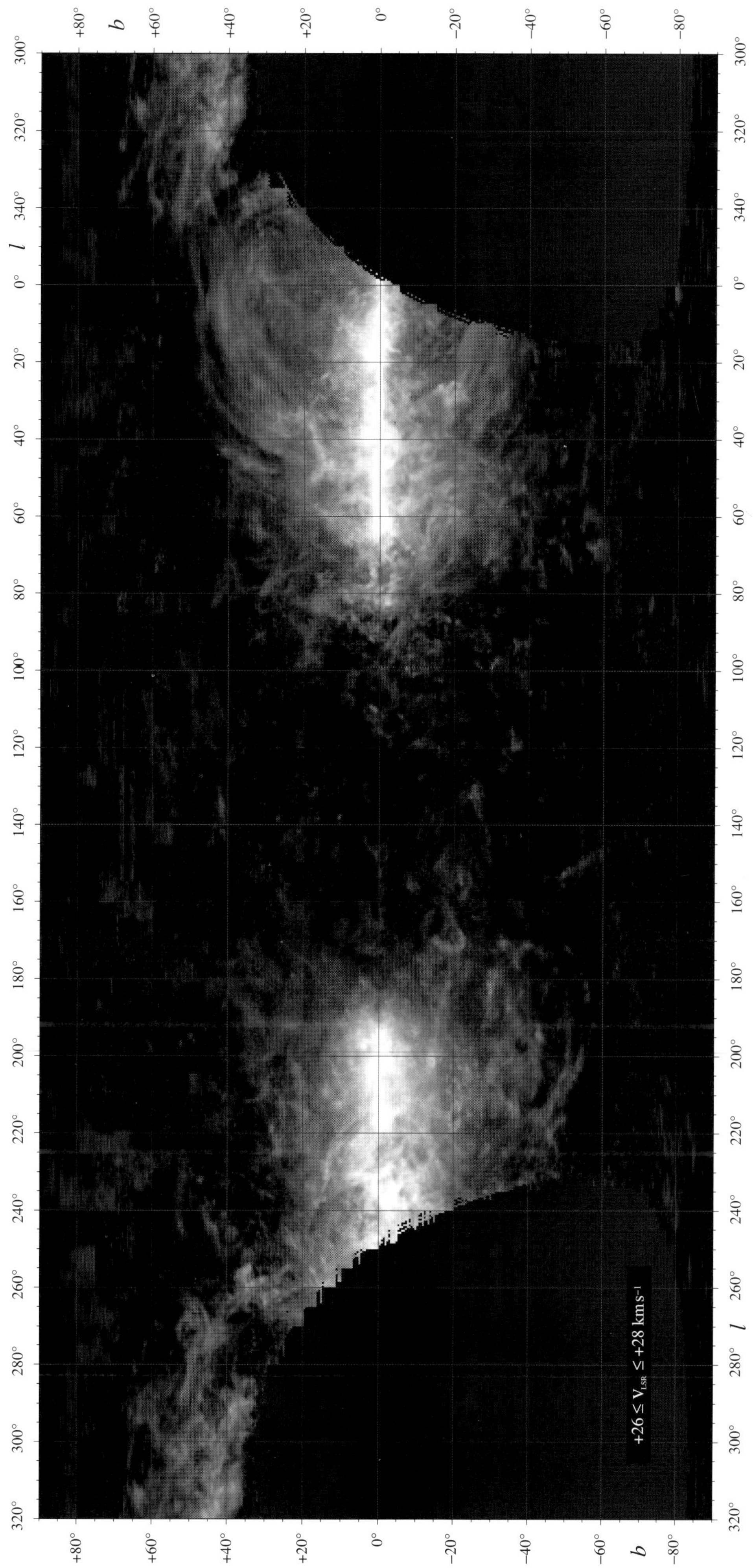
+26 ≤ V_LSR ≤ +28 km s⁻¹

$+28 \leq V_{LSR} \leq +30$ km s^{-1}

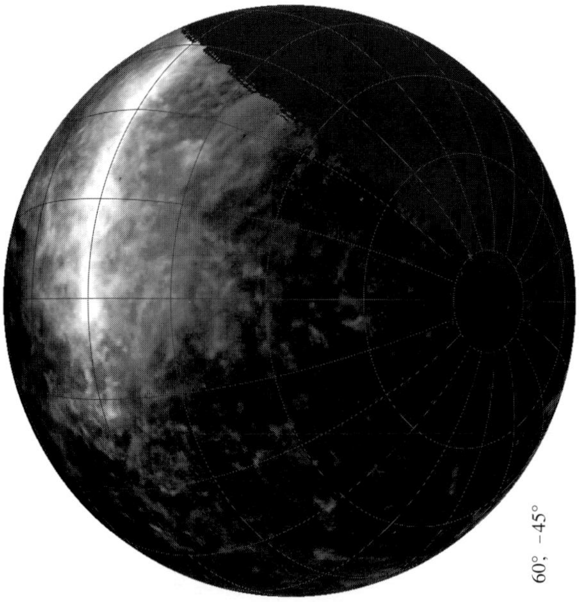

60°, −45°

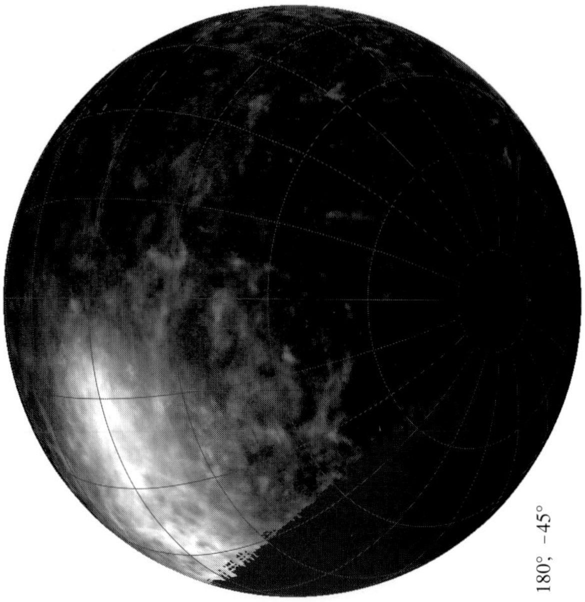

180°, −45°

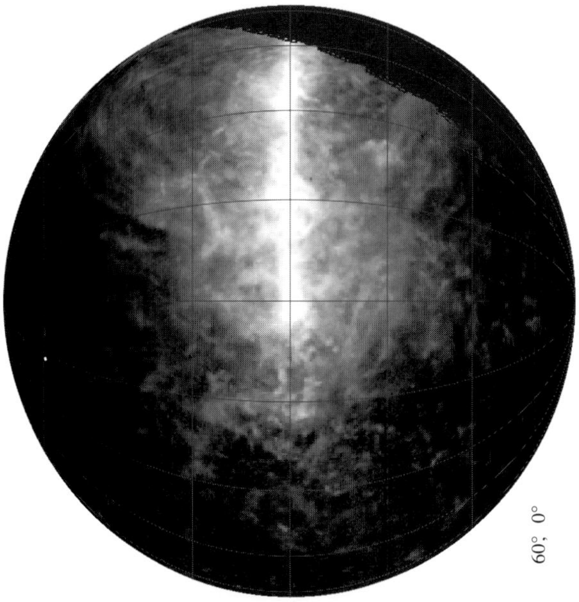

60°, 0°

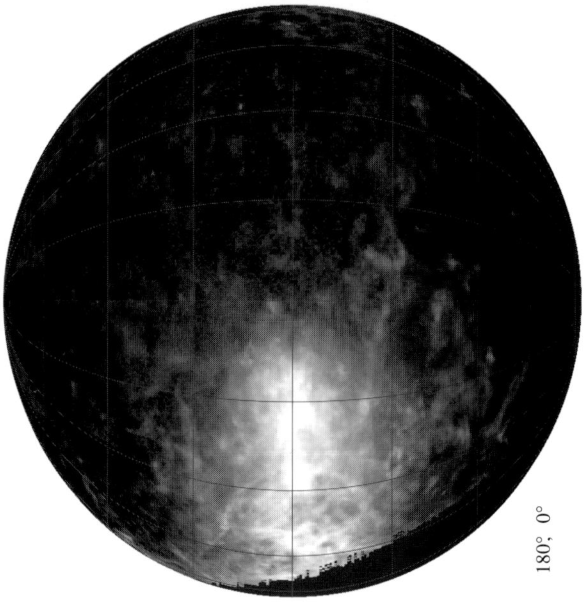

180°, 0°

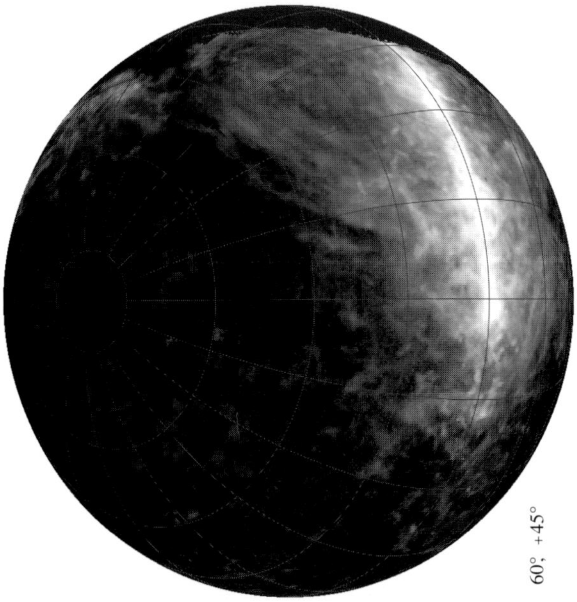

60°, +45°

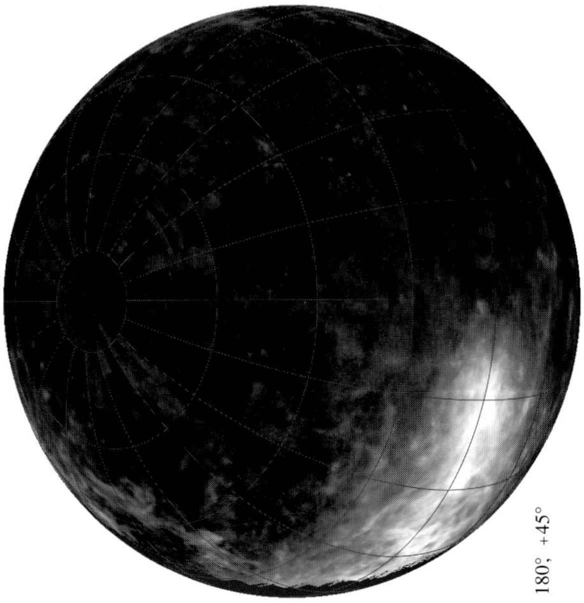

180°, +45°

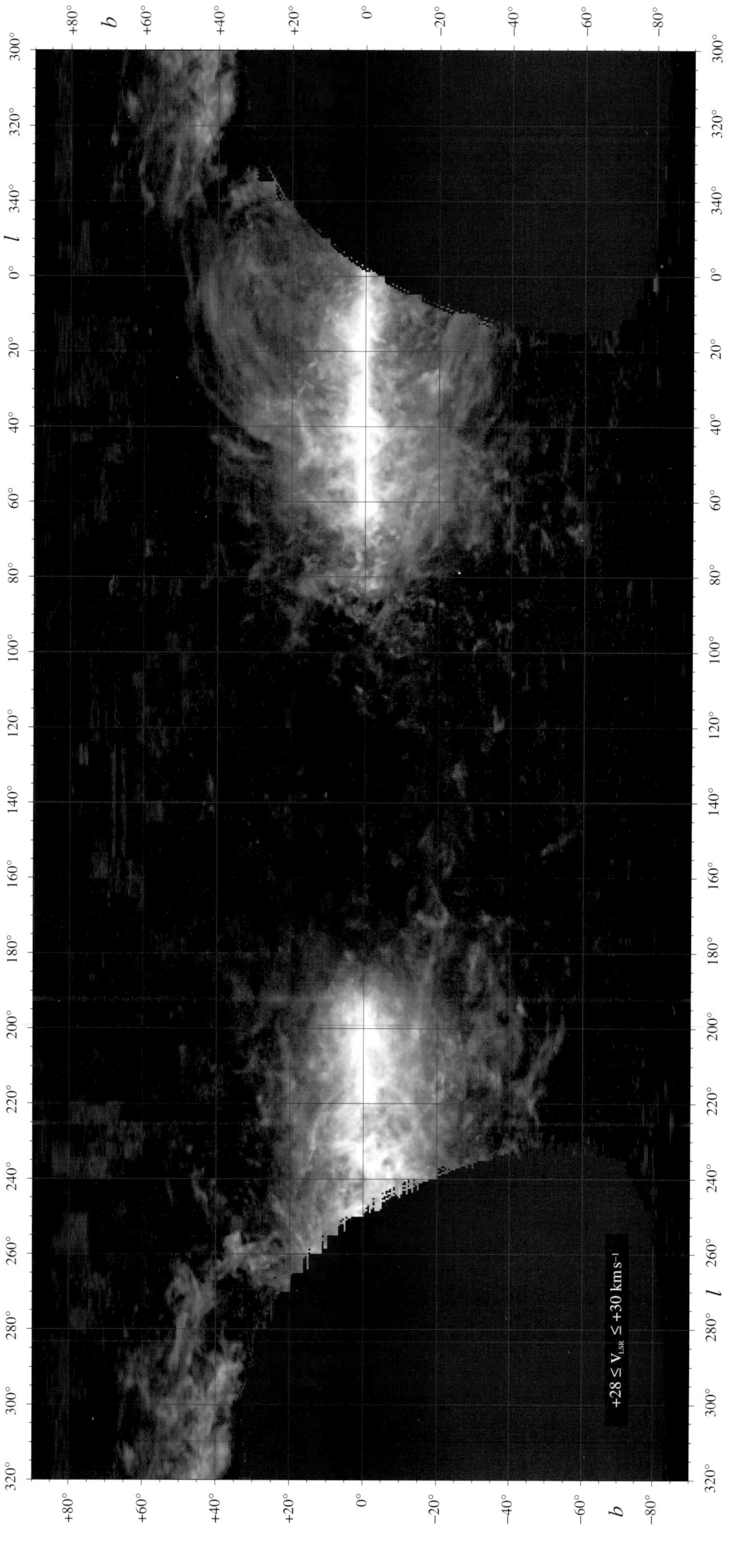
+28 ≤ V_LSR ≤ +30 kms-1
l
b

$+30 \leq V_{LSR} \leq +32$ km s^{-1}

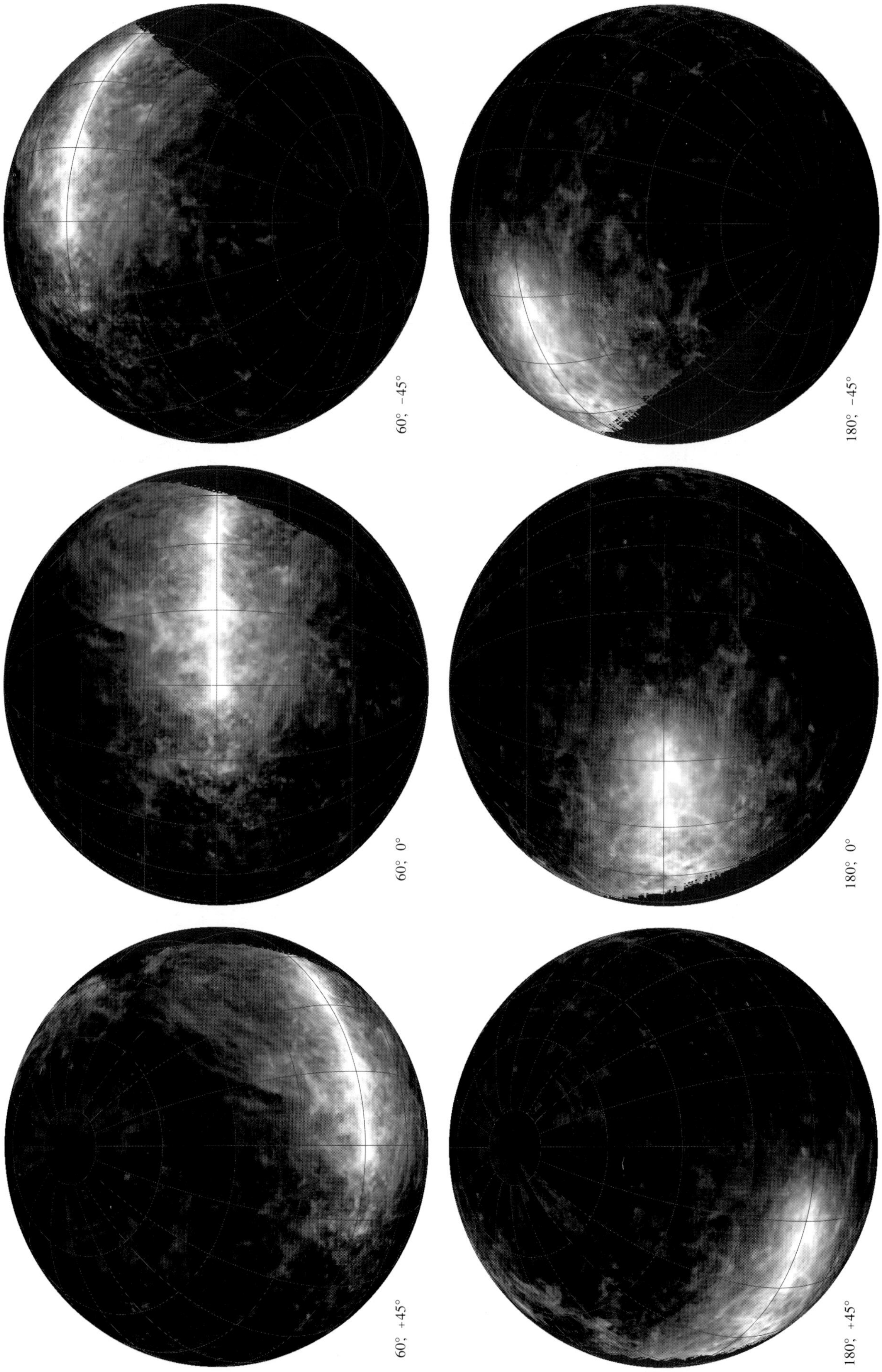

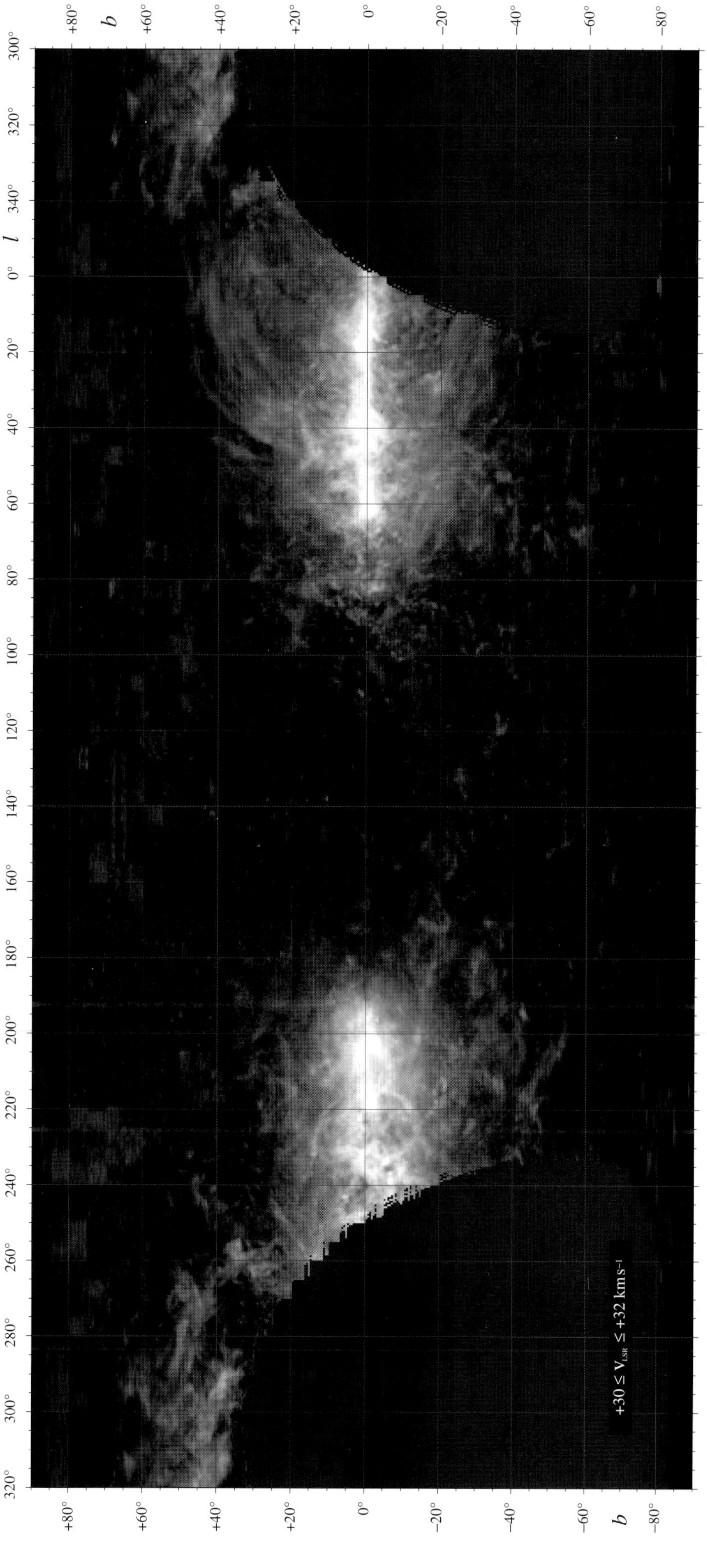
+30 ≤ V_LSR ≤ +32 km s⁻¹
l
b
300° 320° 340° 0° 20° 40° 60° 80° 100° 120° 140° 160° 180° 200° 220° 240° 260° 280° 300° 320°
+80° +60° +40° +20° 0° −20° −40° −60° −80°

$+32 \leq V_{LSR} \leq +34$ km s^{-1}

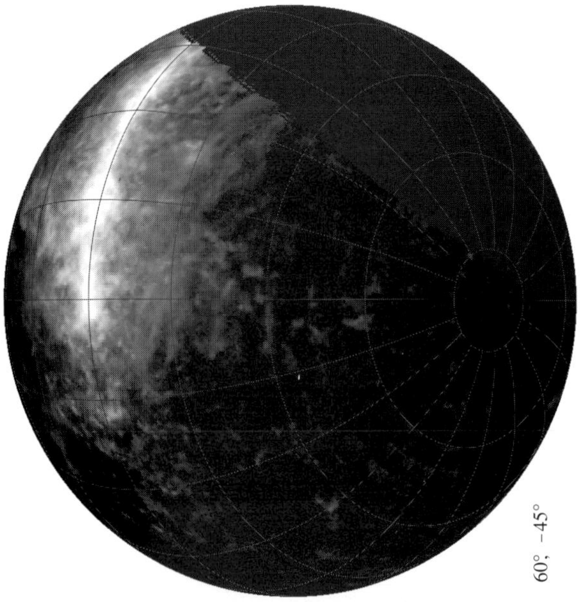

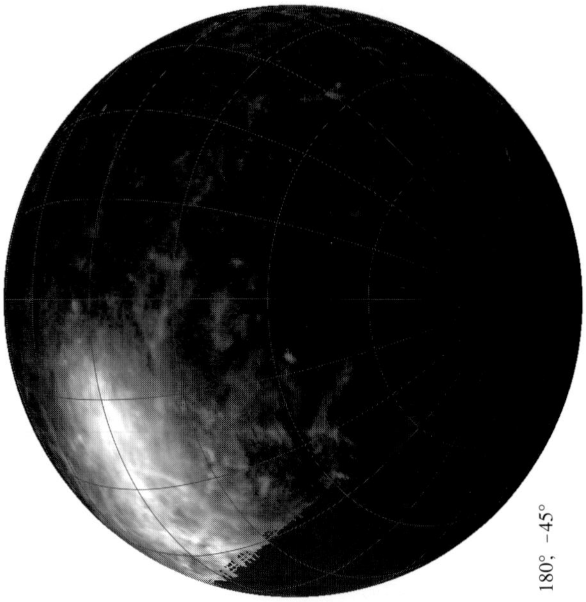

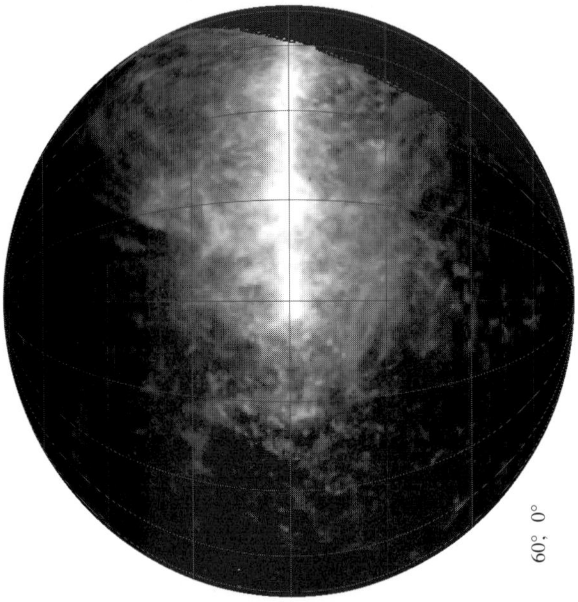

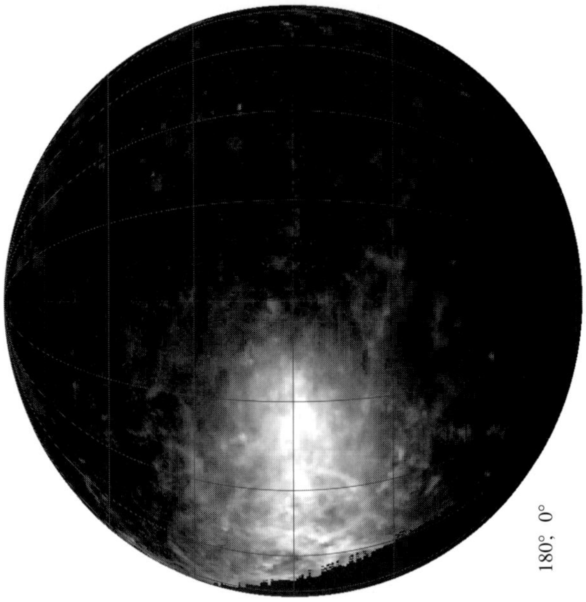

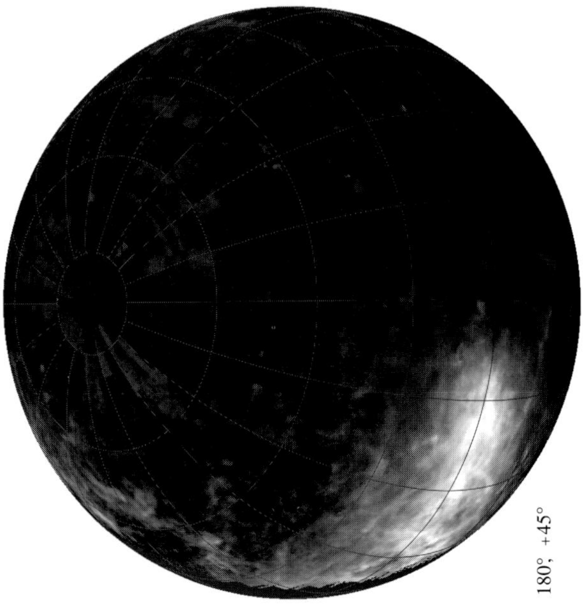

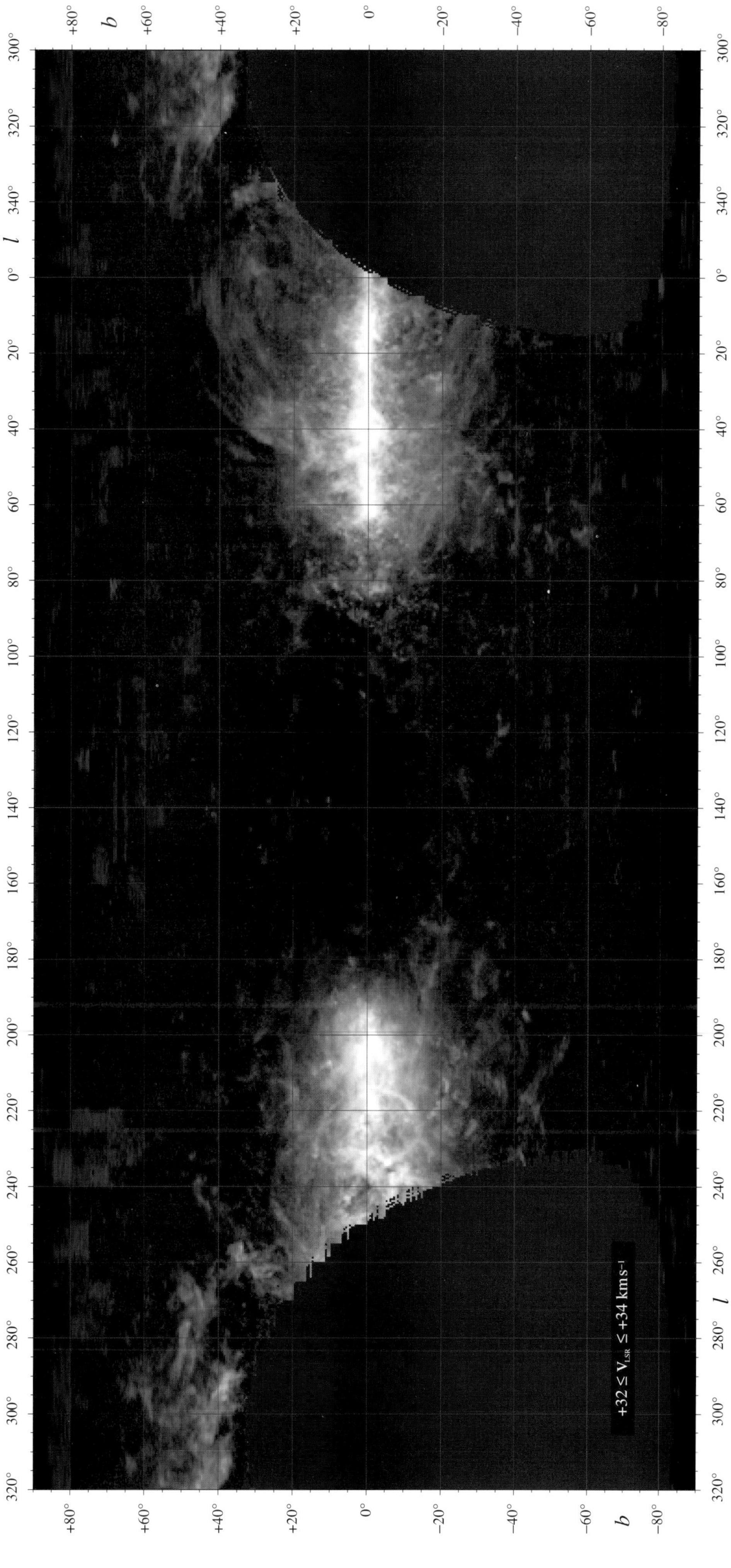
+32 ≤ V_LSR ≤ +34 kms⁻¹
l
b
300° 320° 340° 0° 20° 40° 60° 80° 100° 120° 140° 160° 180° 200° 220° 240° 260° 280° 300° 320°
+80° +60° +40° +20° 0° −20° −40° −60° −80°

$+34 \leq V_{LSR} \leq +36$ km s^{-1}

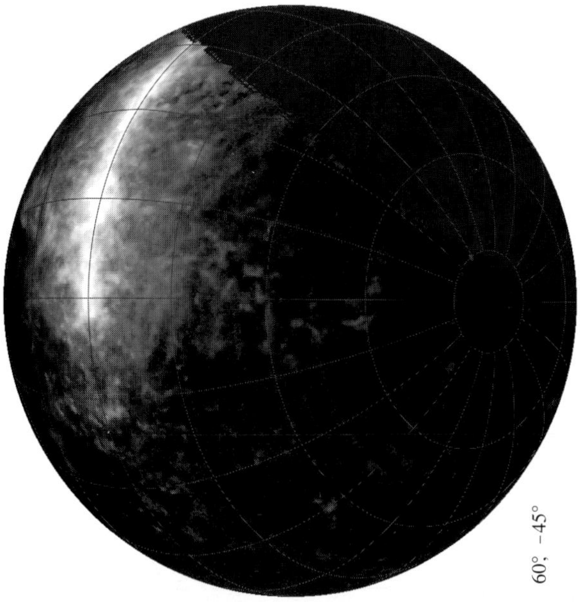

60°, −45°

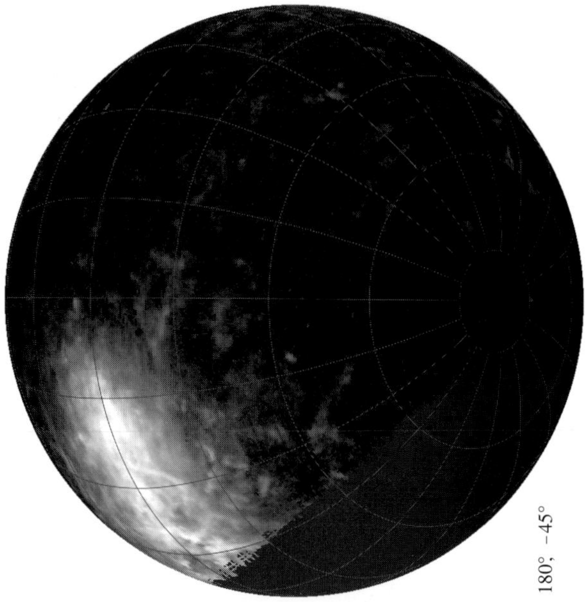

180°, −45°

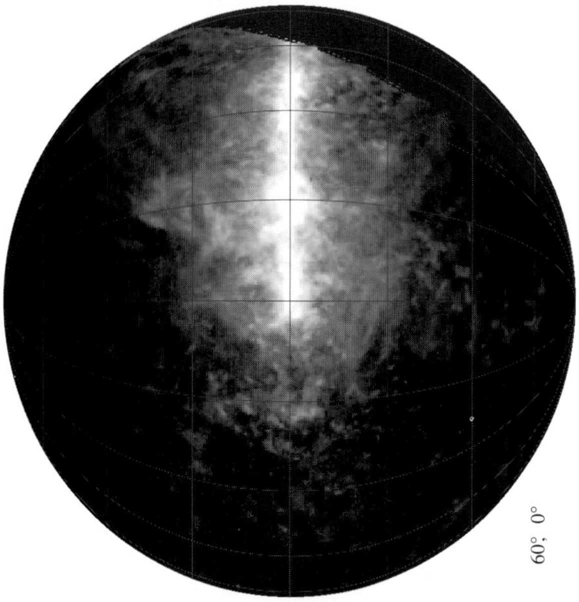

60°, 0°

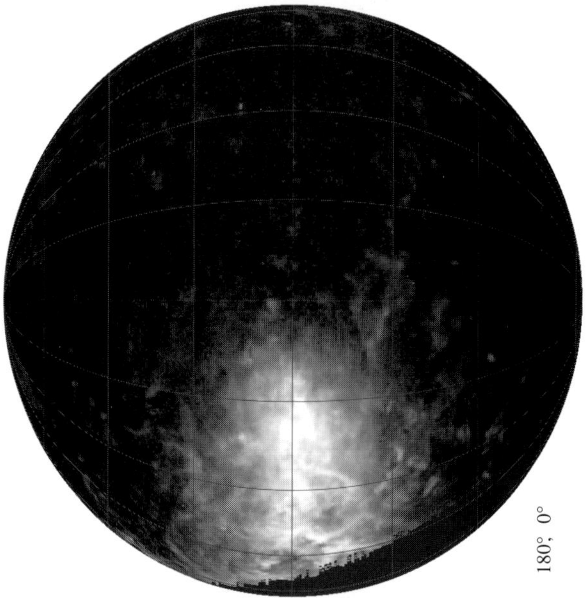

180°, 0°

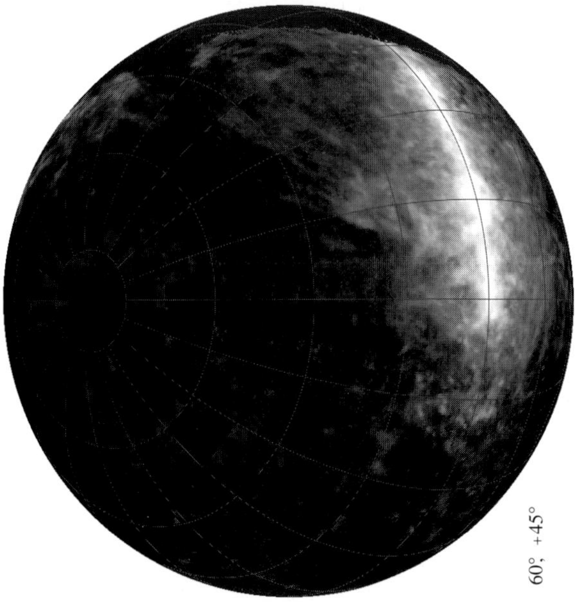

60°, +45°

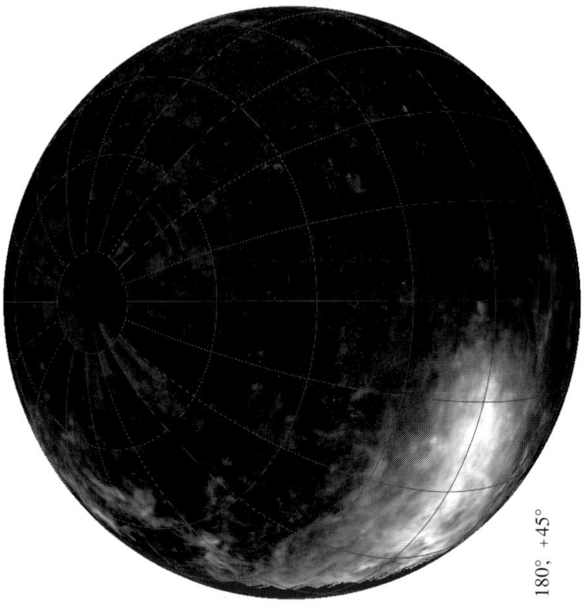

180°, +45°

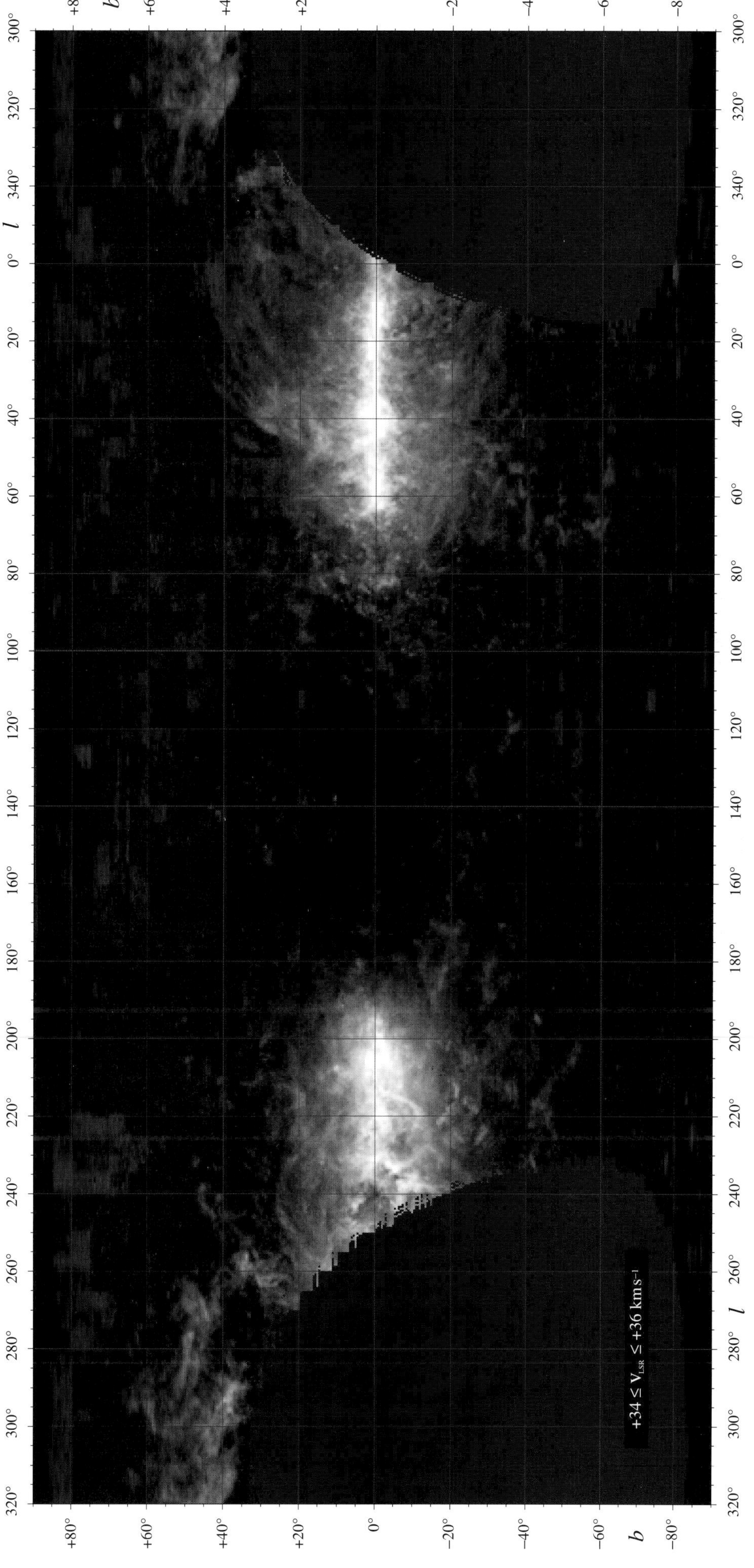
$+34 \le V_{LSR} \le +36$ km s^{-1}

$+36 \leq V_{LSR} \leq +38$ km s^{-1}

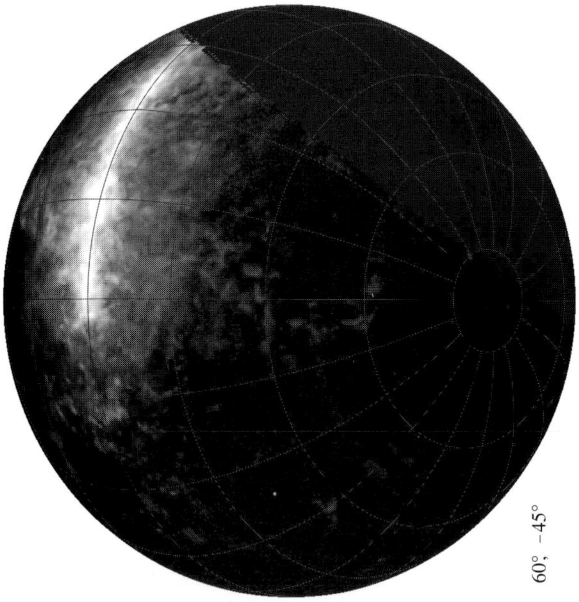

60°, −45°

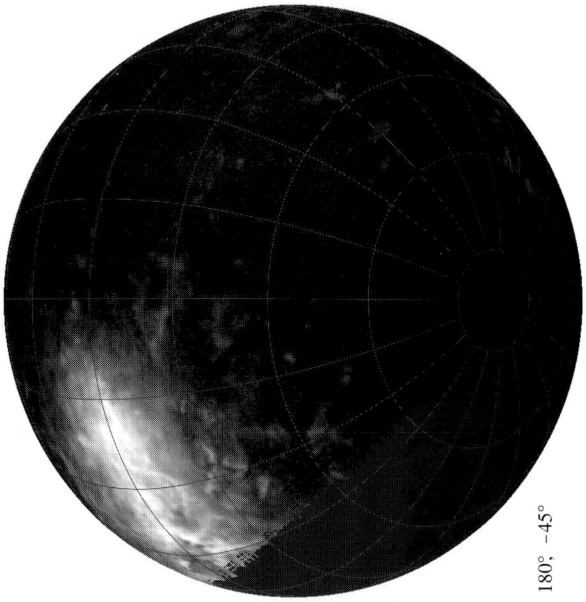

180°, −45°

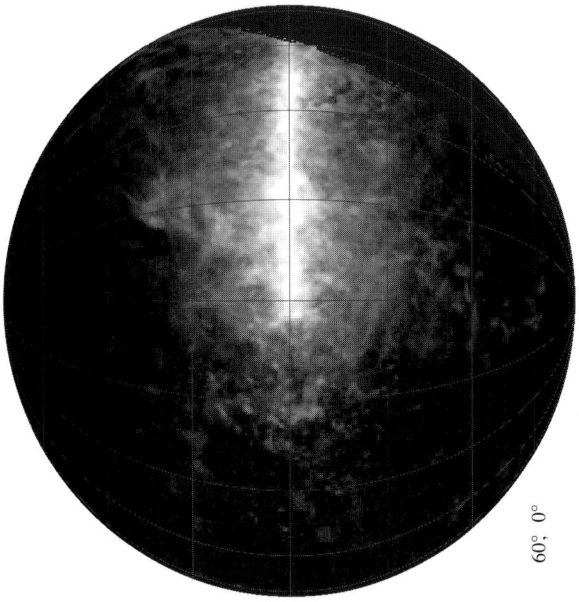

60°, 0°

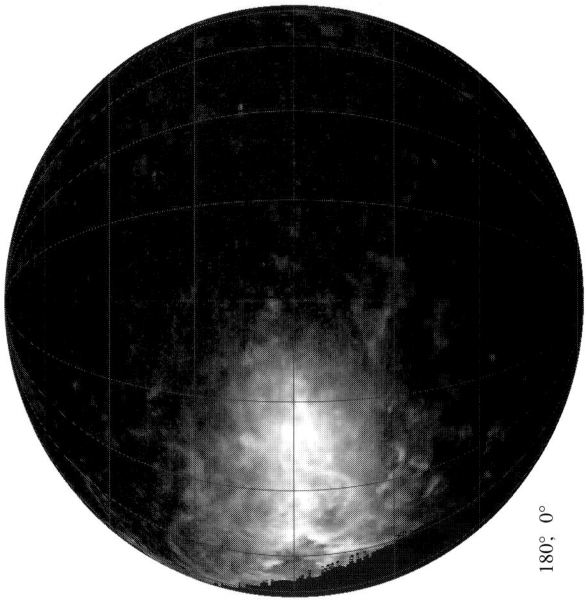

180°, 0°

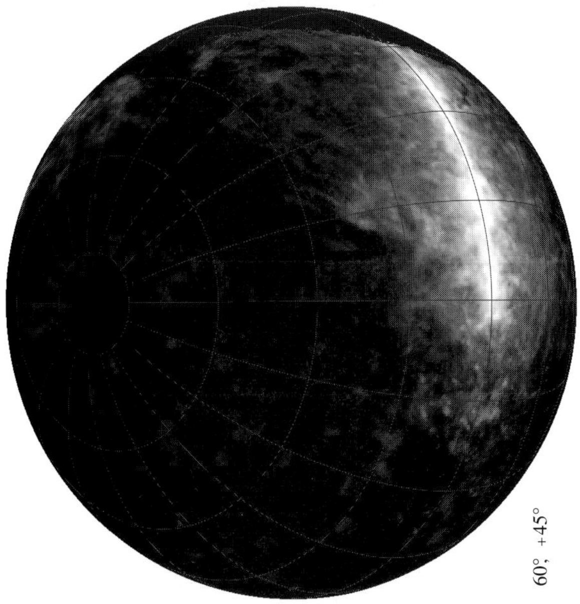

60°, +45°

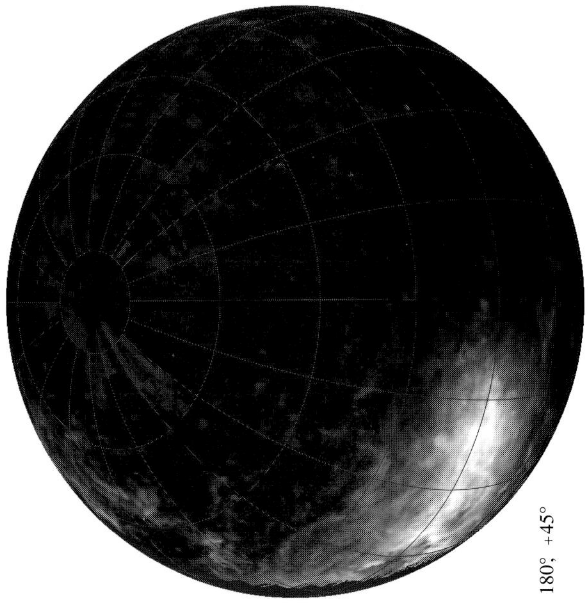

180°, +45°

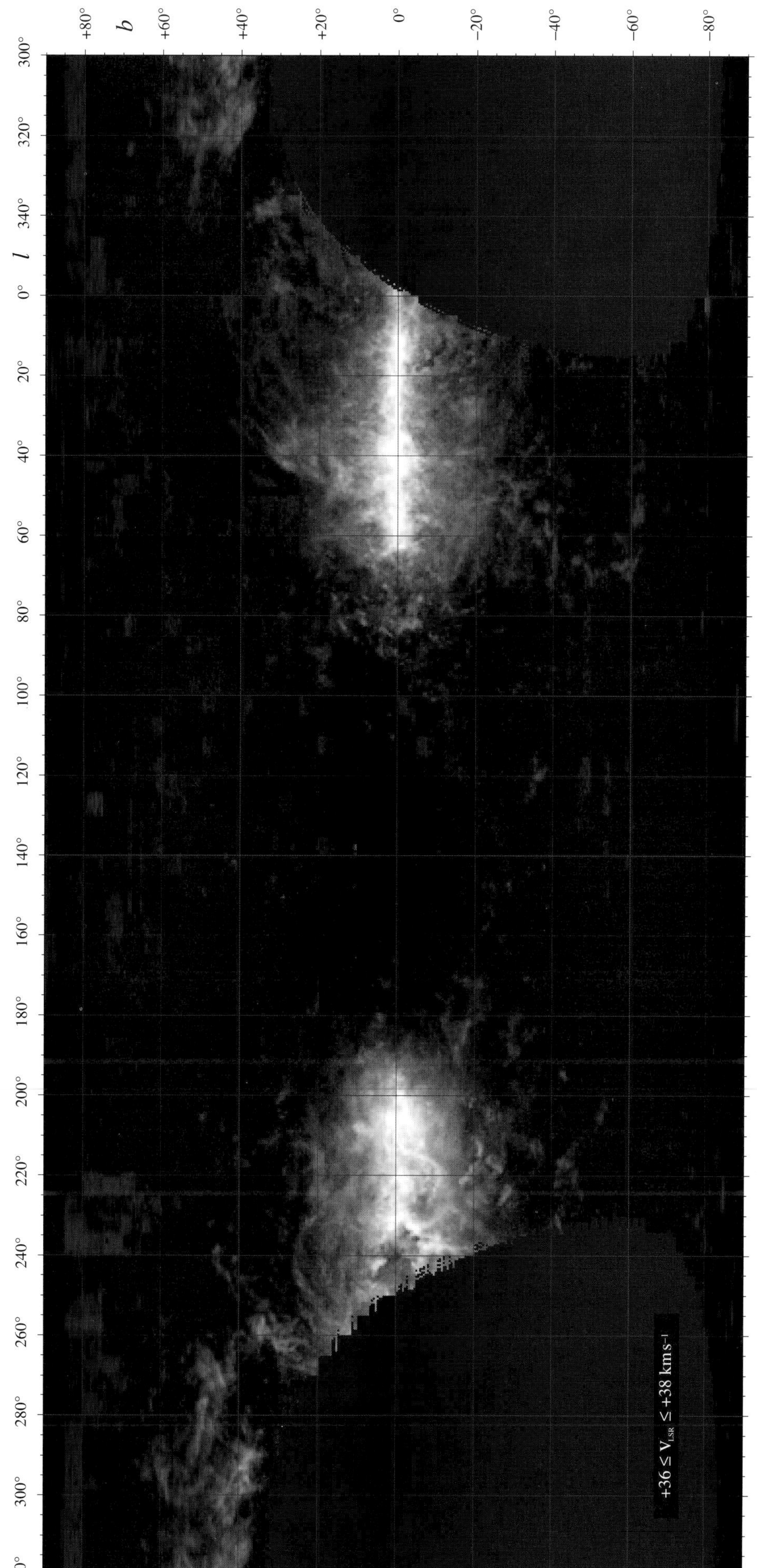

+36 ≤ V_LSR ≤ +38 km s−1
l
b
300° 320° 340° 0° 20° 40° 60° 80° 100° 120° 140° 160° 180° 200° 220° 240° 260° 280° 300° 320°
+80° +60° +40° +20° 0° −20° −40° −60° −80°

$+38 \leq V_{LSR} \leq +40$ km s^{-1}

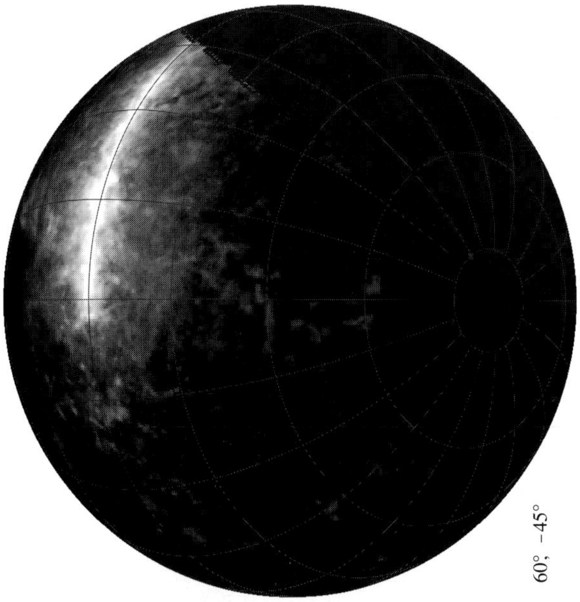

60°, −45°

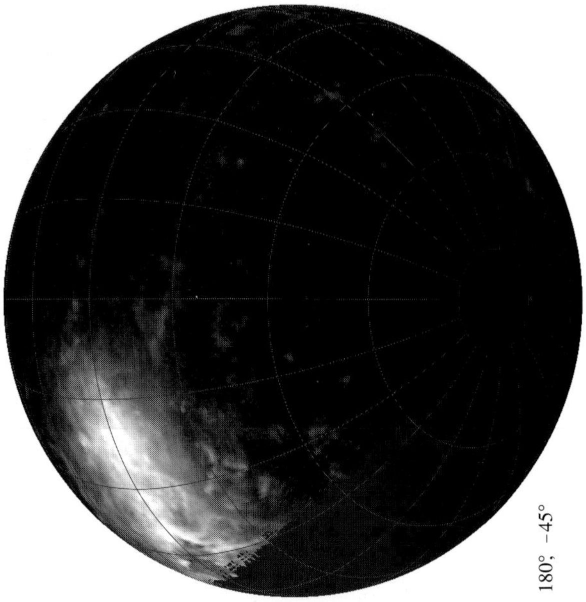

180°, −45°

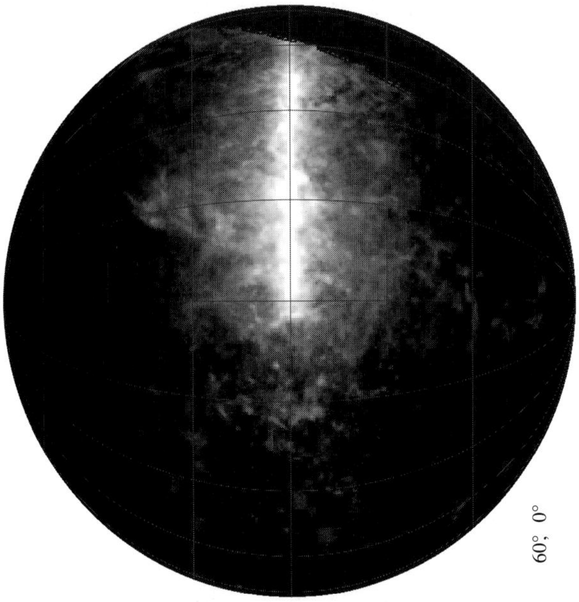

60°, 0°

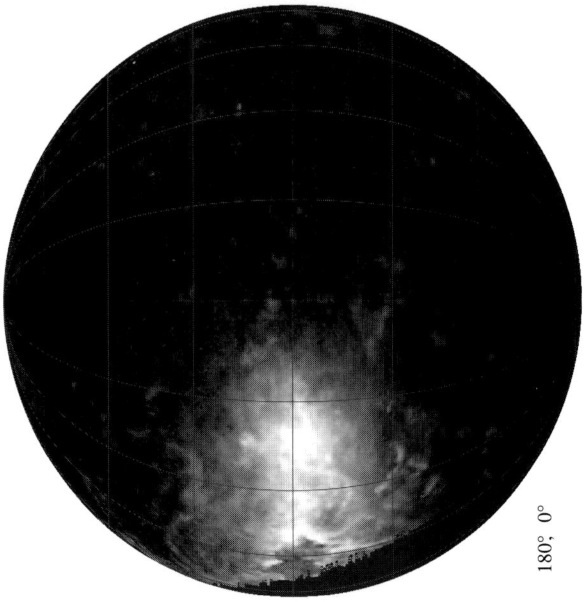

180°, 0°

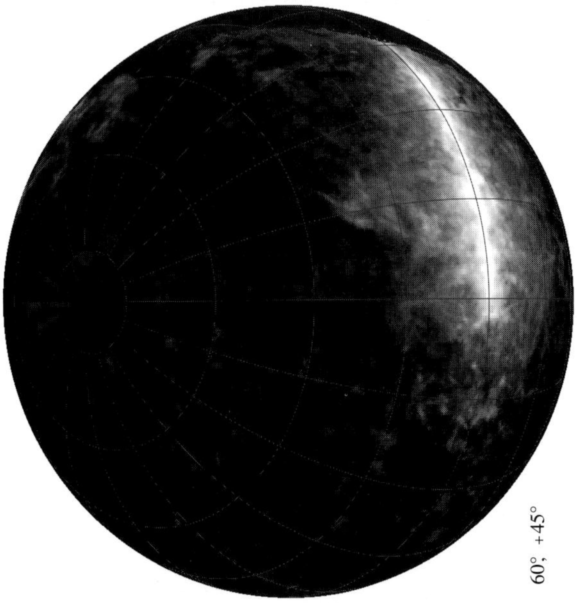

60°, +45°

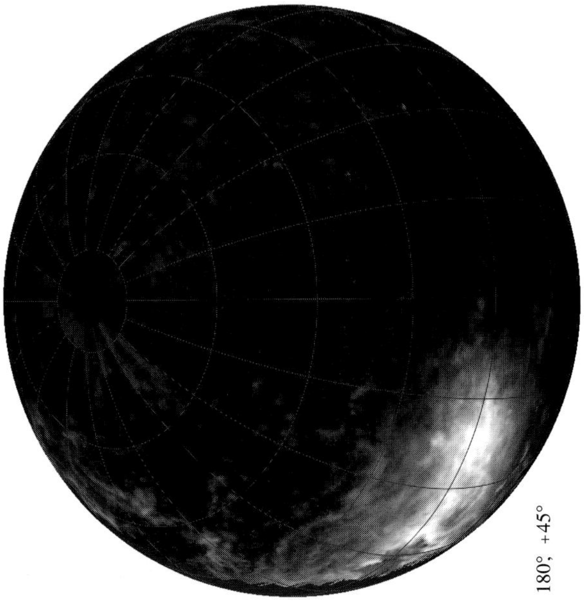

180°, +45°

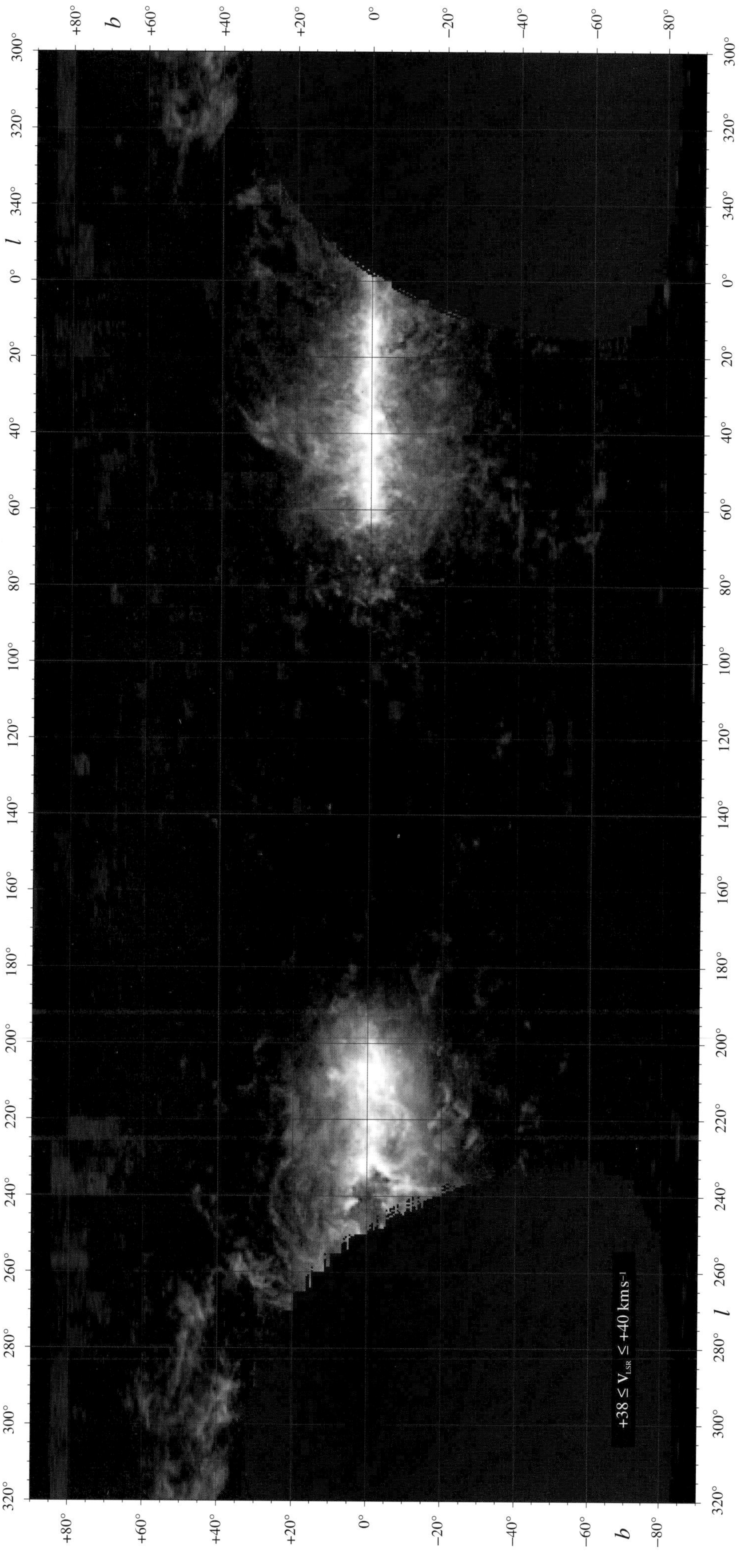

+38 ≤ V_LSR ≤ +40 km s⁻¹
l
b

$+40 \leq V_{LSR} \leq +42$ km s^{-1}

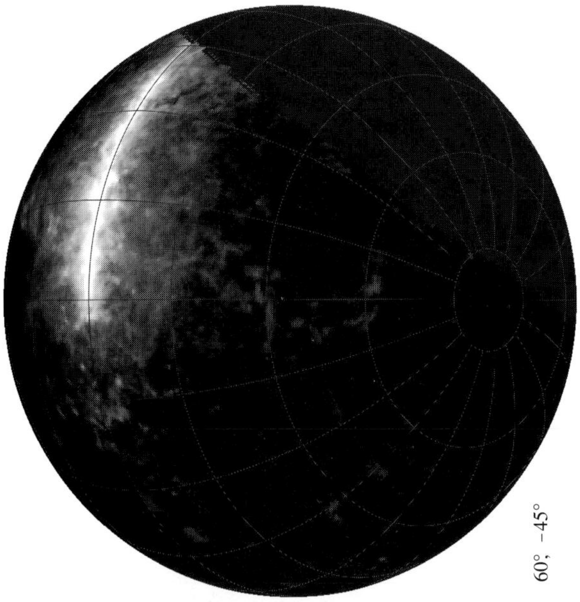

60°, −45°

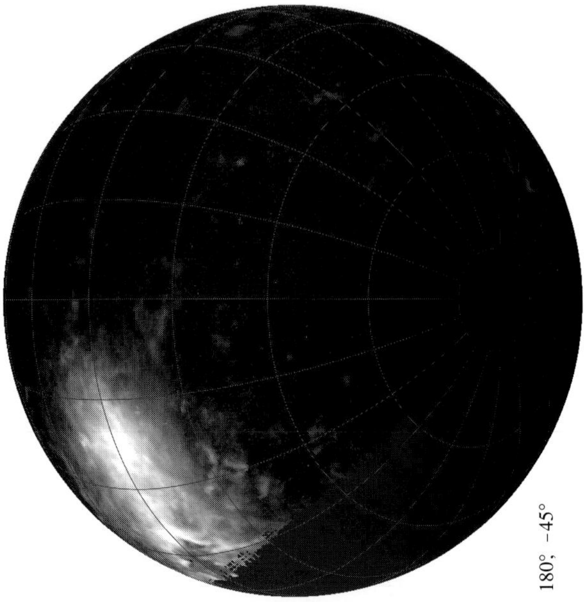

180°, −45°

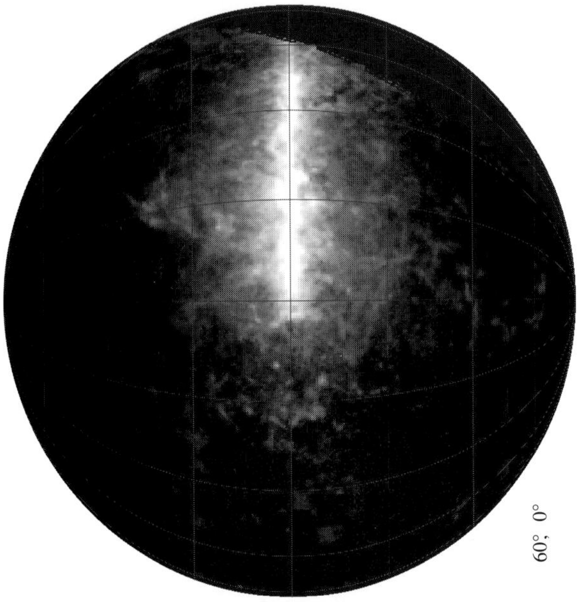

60°, 0°

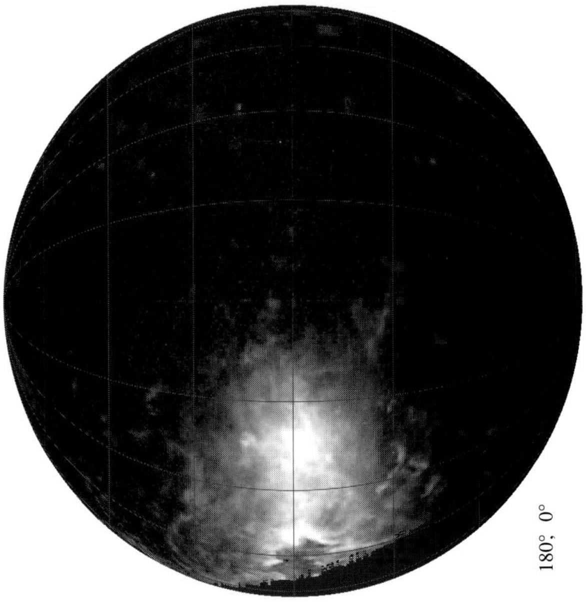

180°, 0°

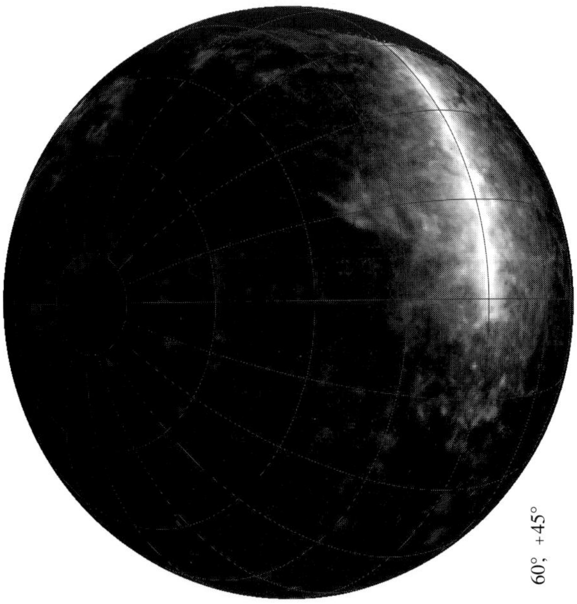

60°, +45°

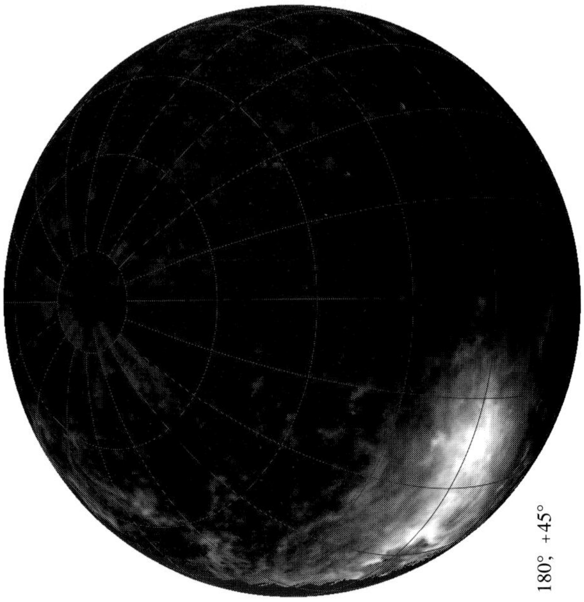

180°, +45°

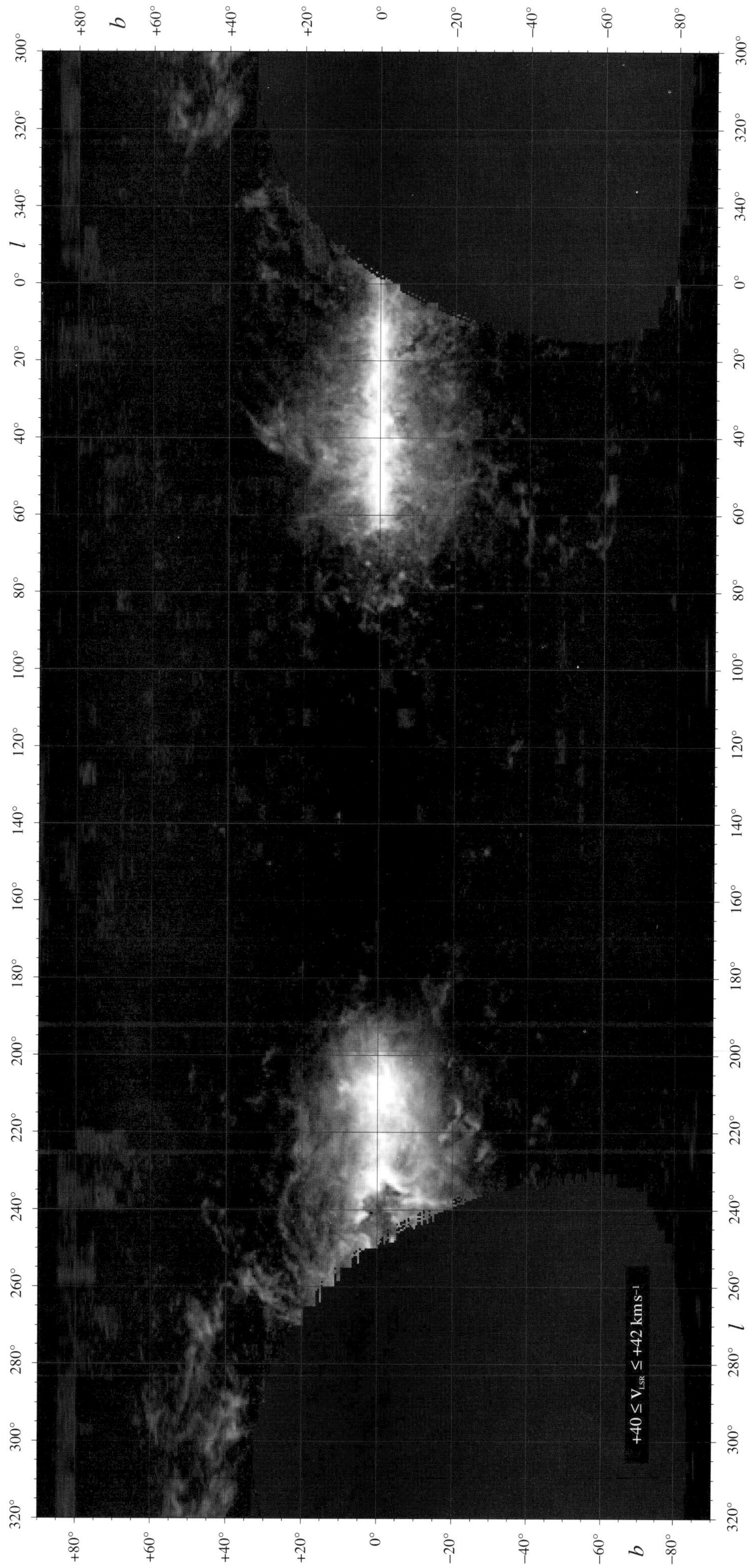
+40 ≤ V_LSR ≤ +42 km s-1
l
b

$+42 \le V_{LSR} \le +44$ km s^{-1}

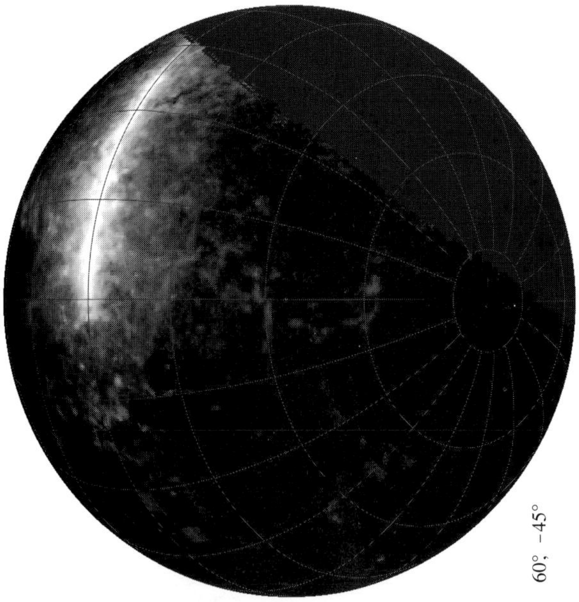

60°, −45°

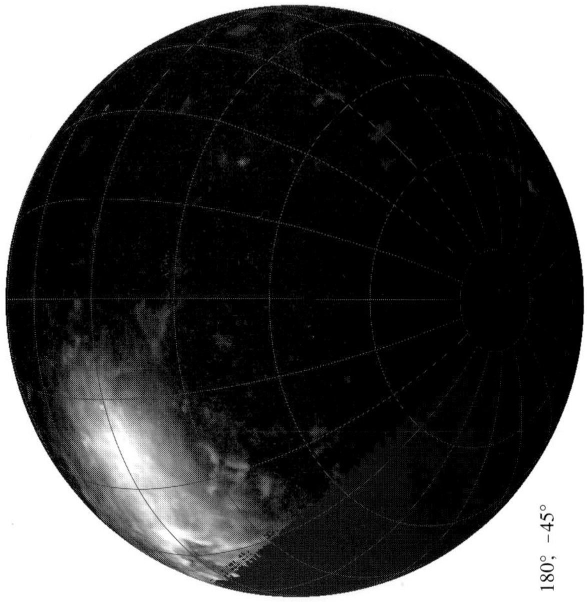

180°, −45°

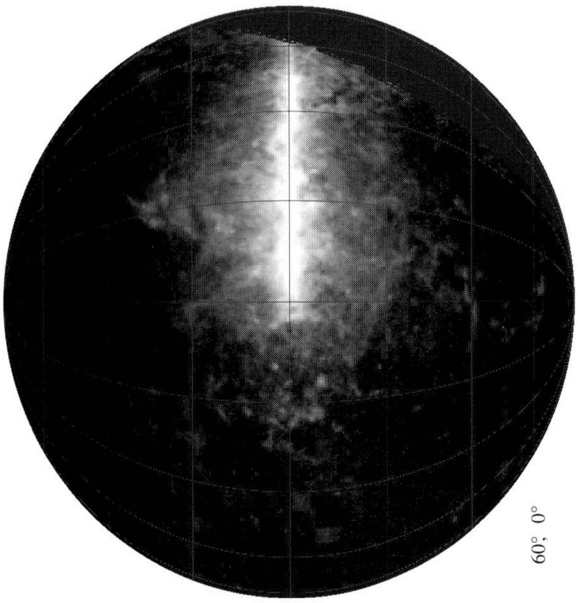

60°, 0°

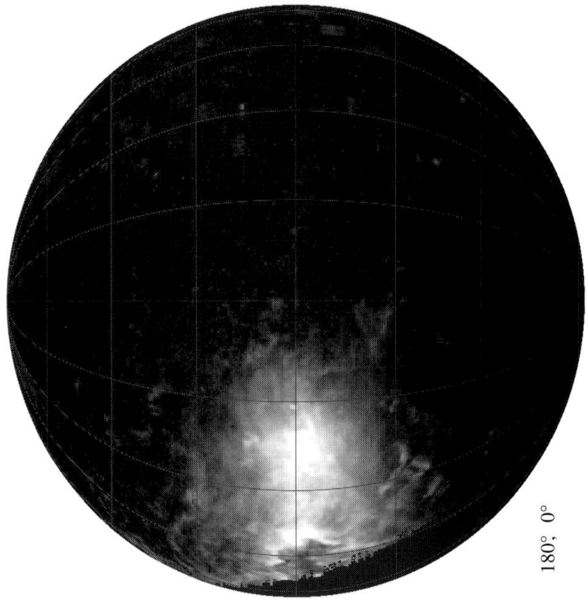

180°, 0°

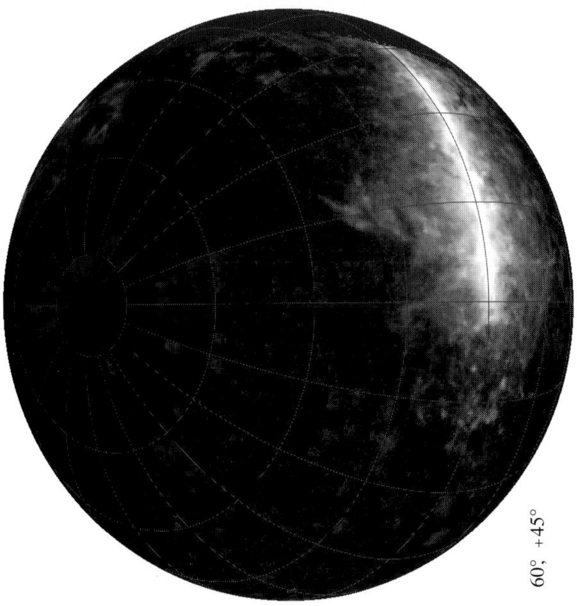

60°, +45°

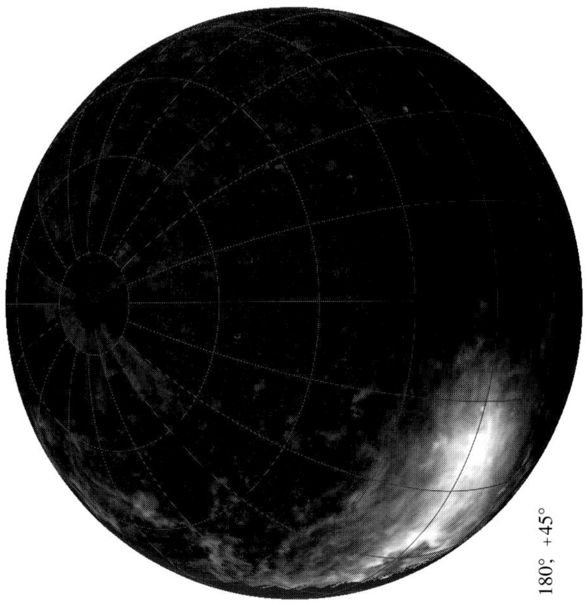

180°, +45°

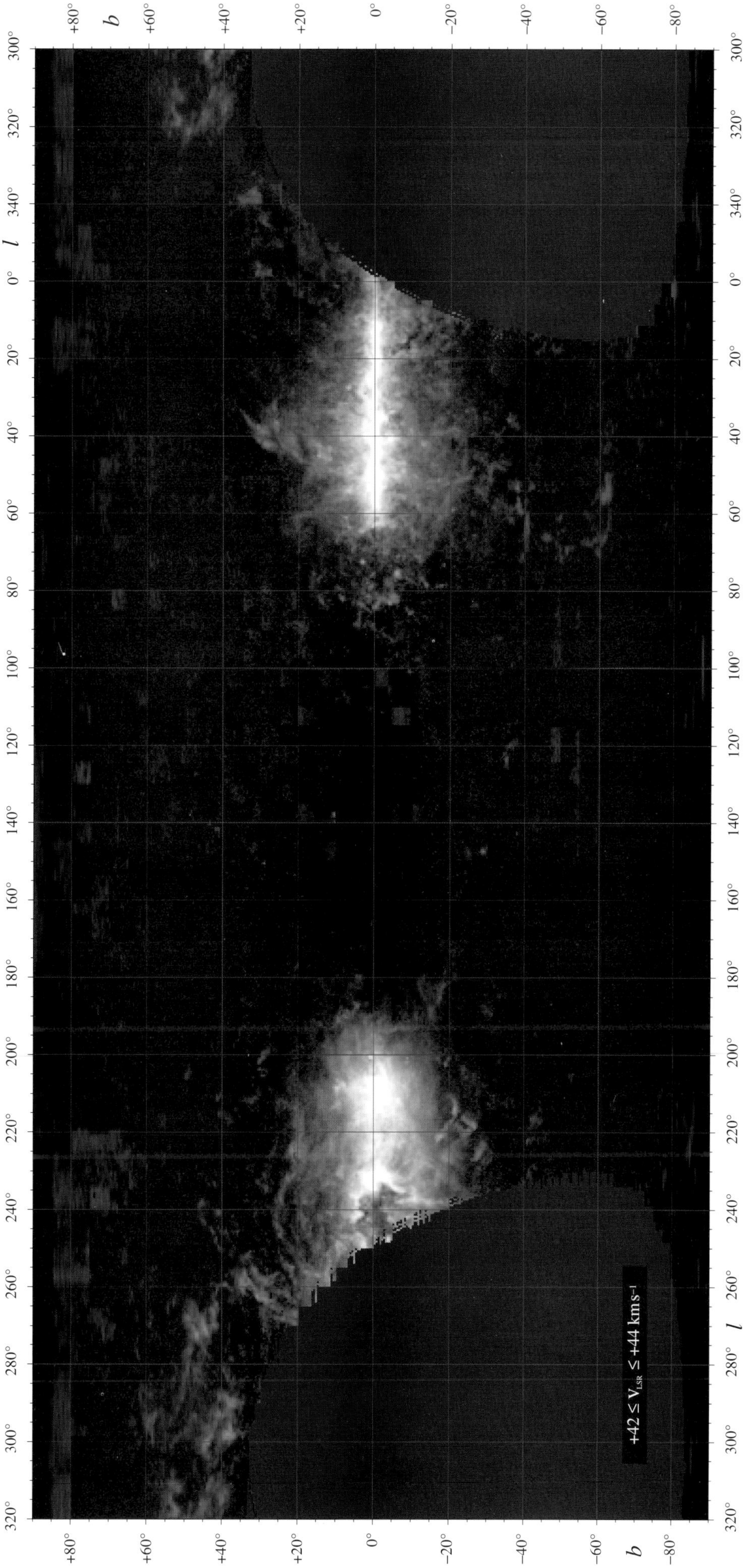
$+42 \leq V_{LSR} \leq +44$ kms^{-1}

$+44 \leq V_{LSR} \leq +46$ km s^{-1}

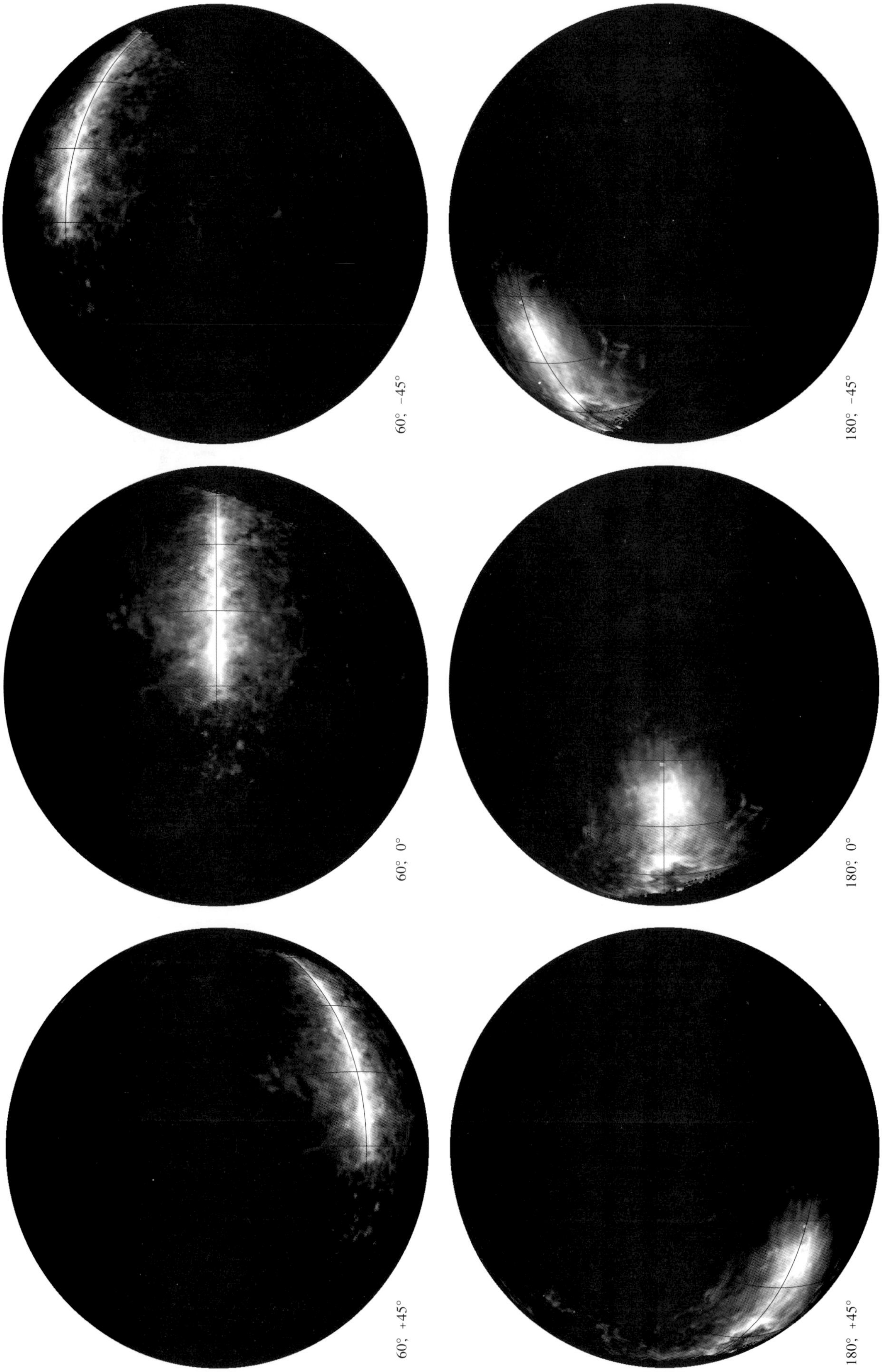

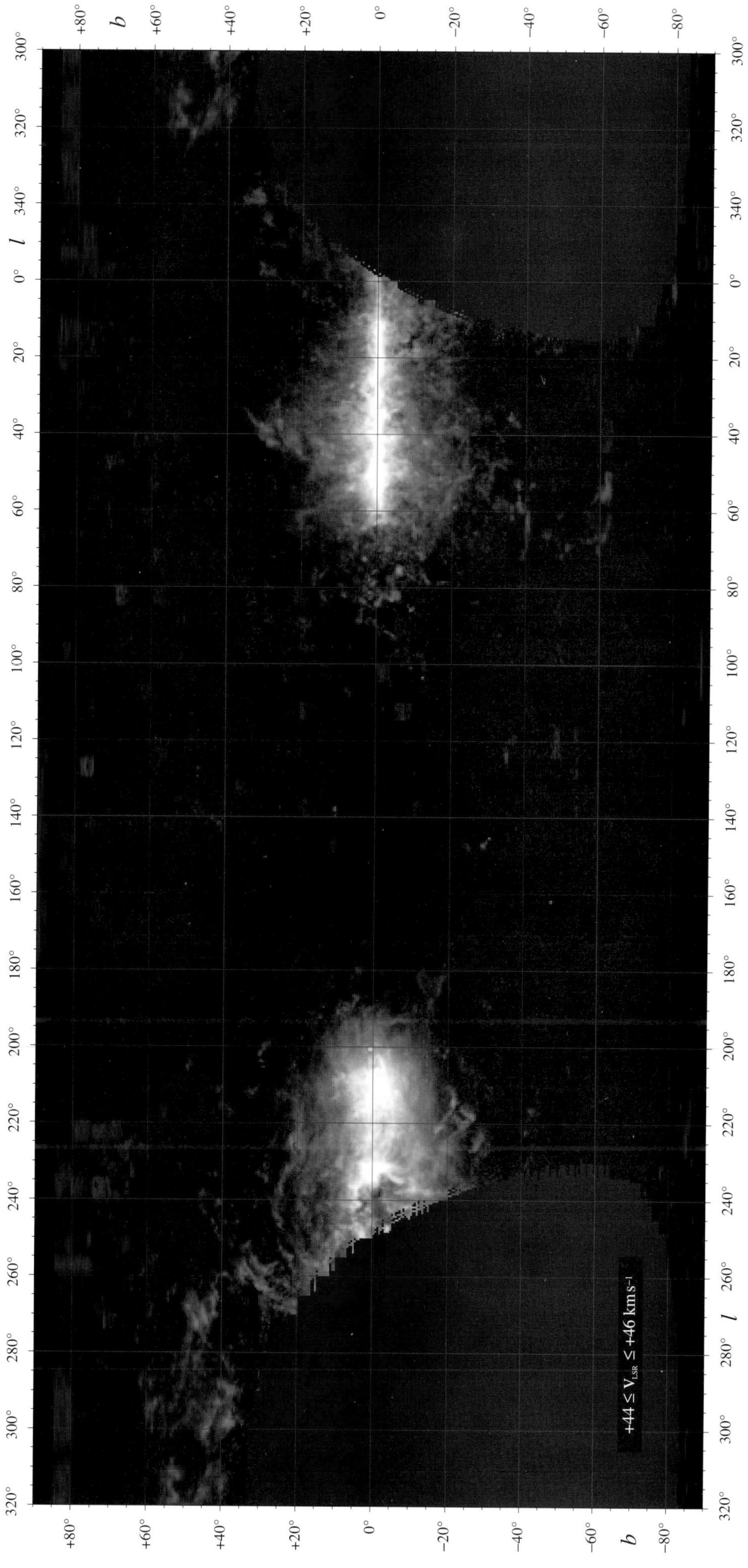
+44 ≤ V_LSR ≤ +46 km s⁻¹
l
b

$+46 \le V_{LSR} \le +48$ km s^{-1}

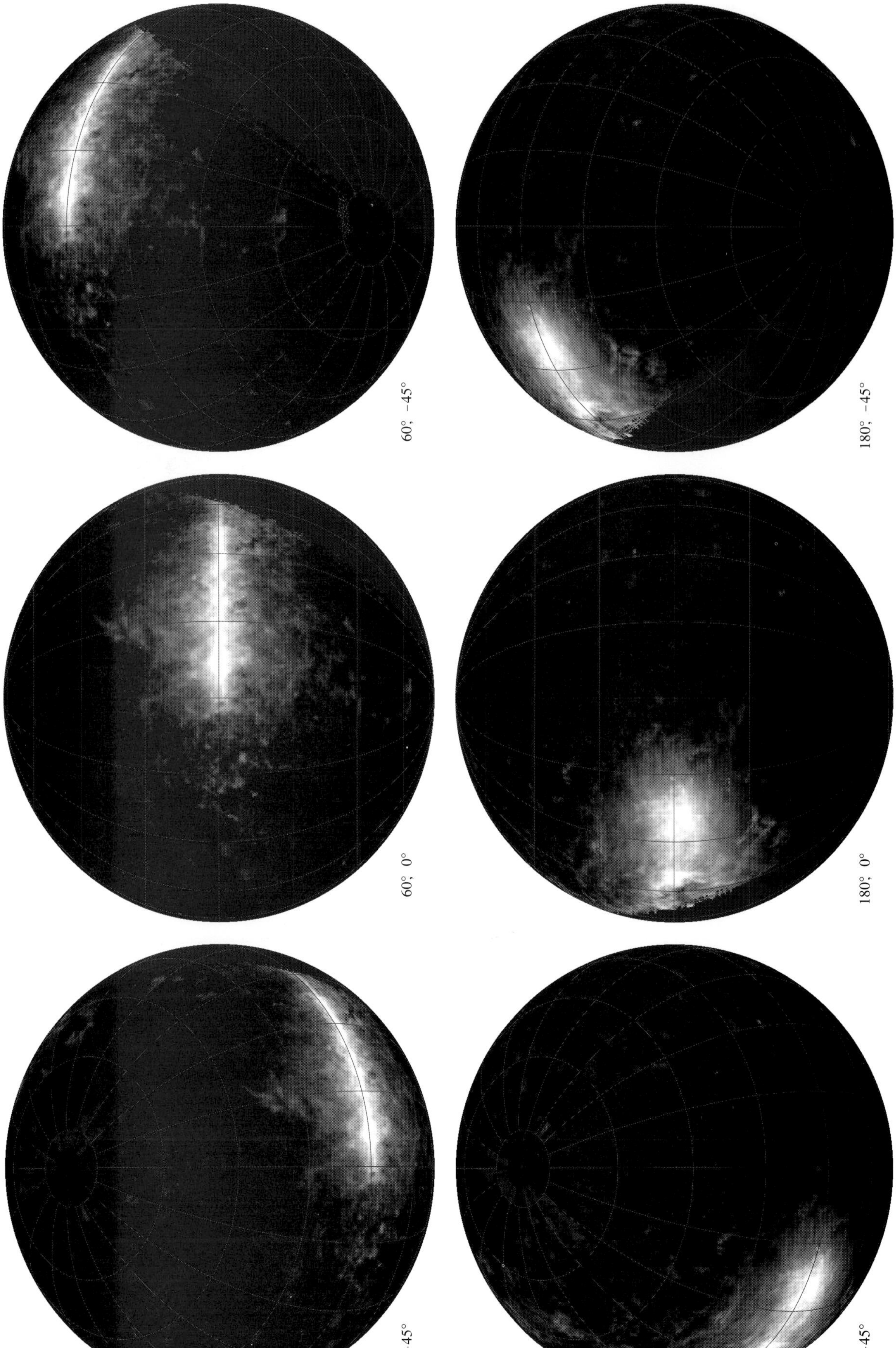

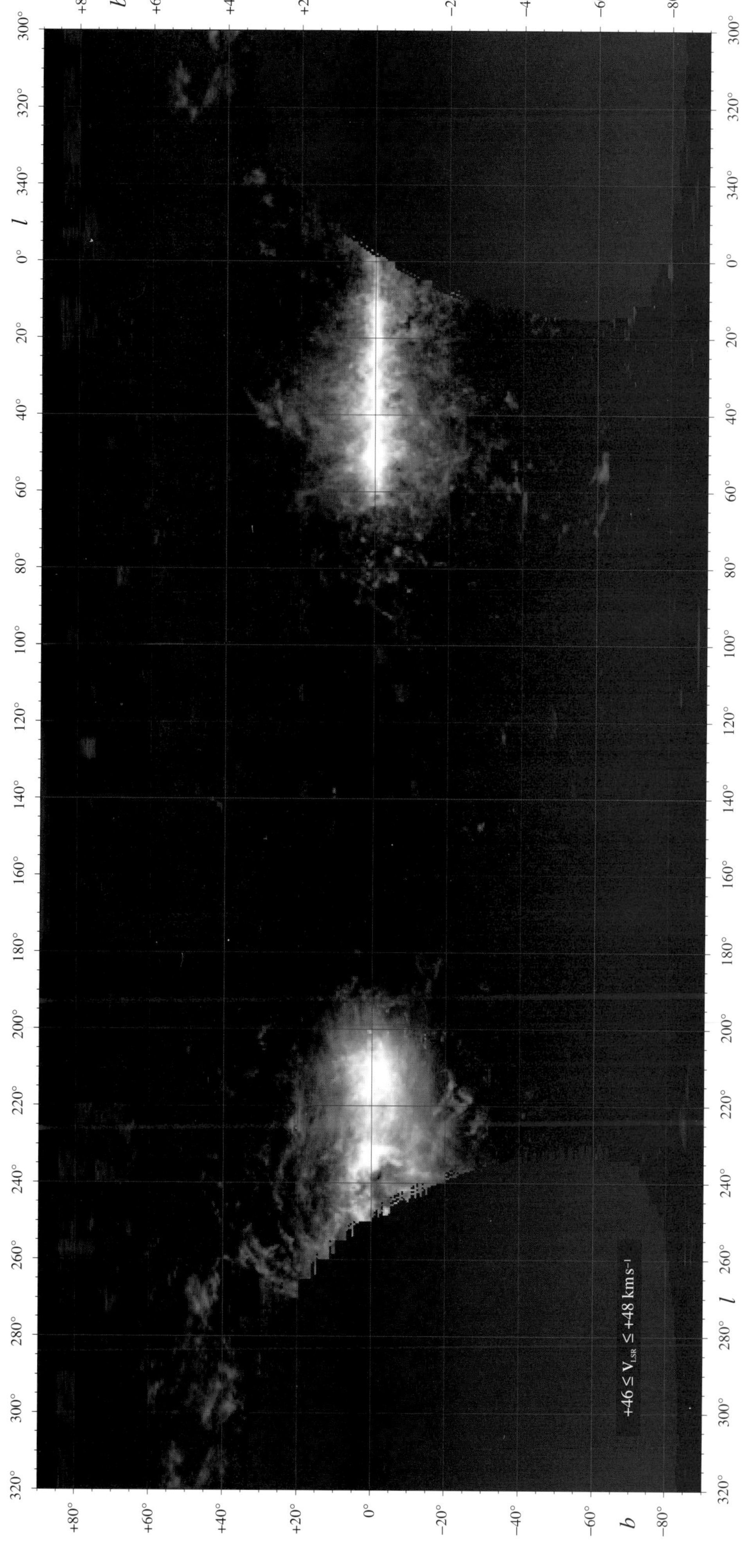

+46 ≤ V_LSR ≤ +48 km s⁻¹
l
b

$+48 \leq V_{LSR} \leq +50$ km s^{-1}

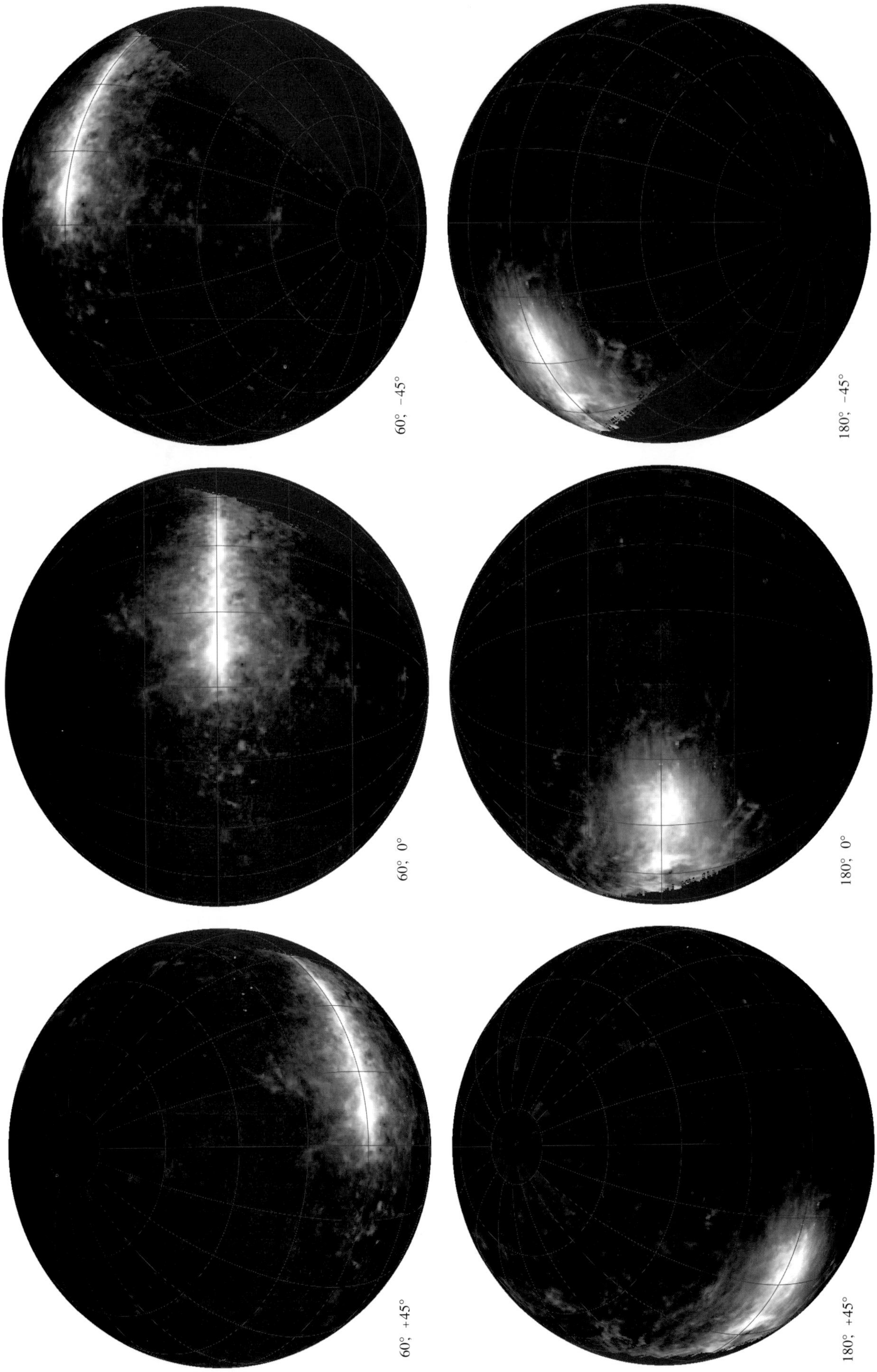

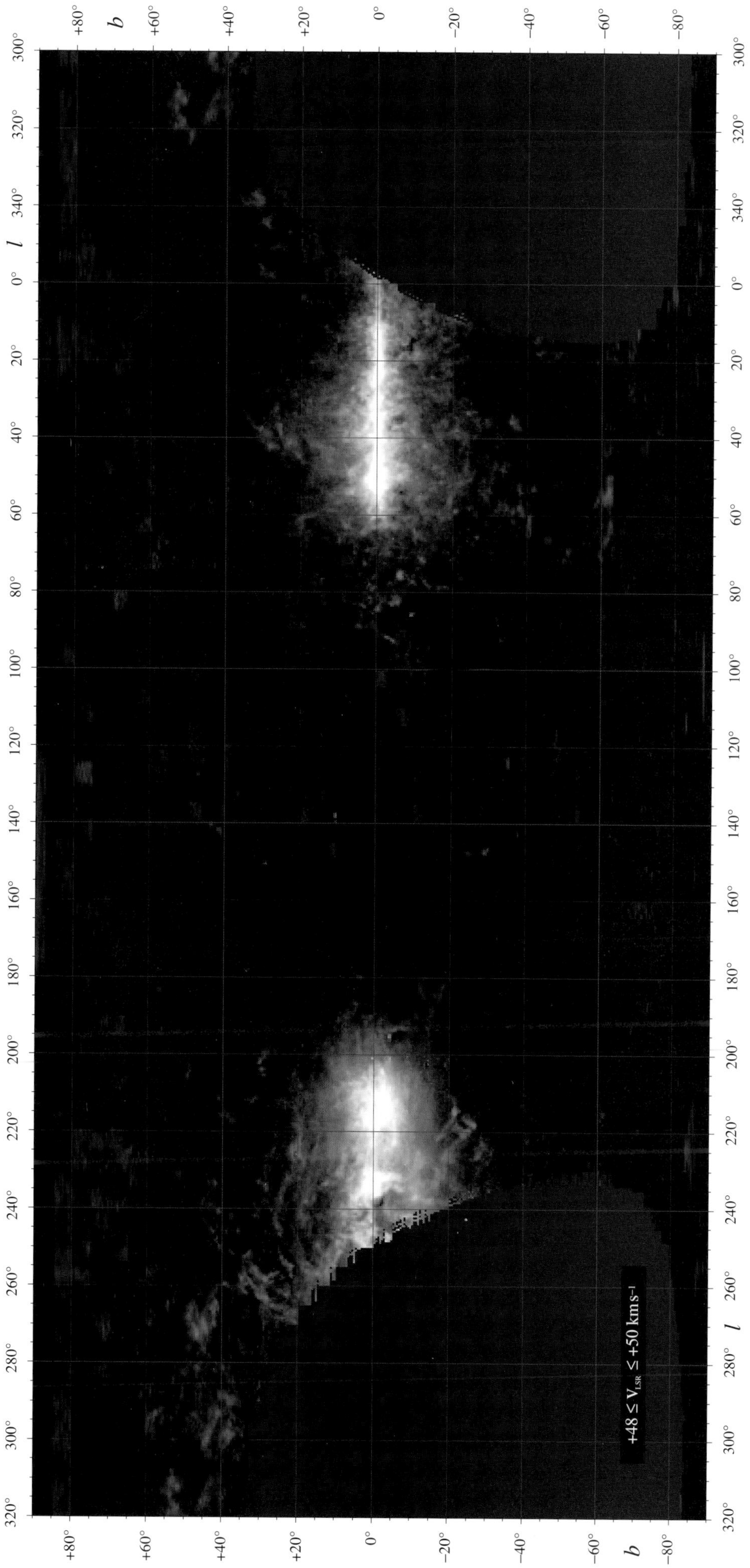
$+48 \le V_{LSR} \le +50$ kms^{-1}

$+50 \leq V_{LSR} \leq +60$ km s^{-1}

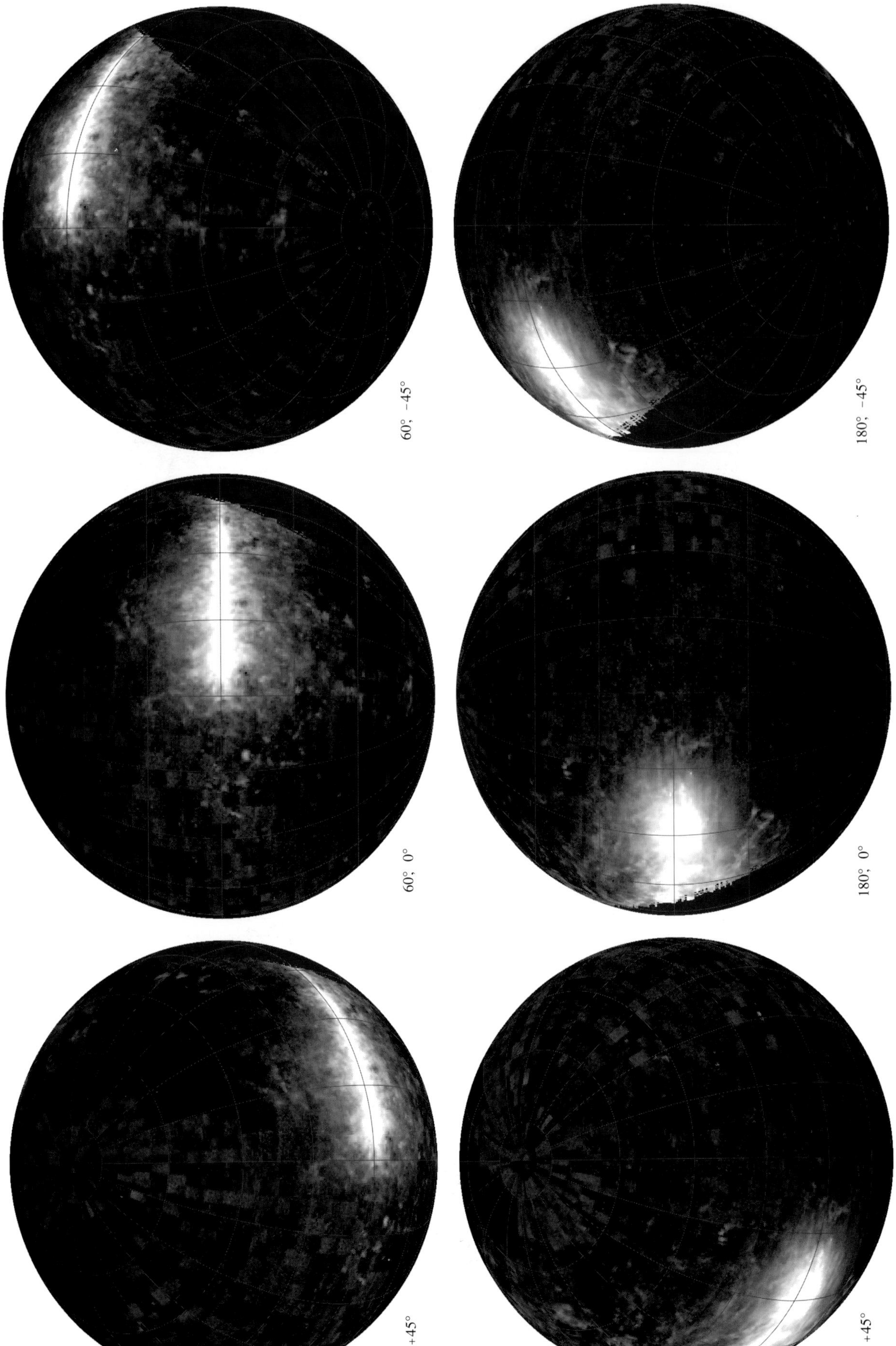

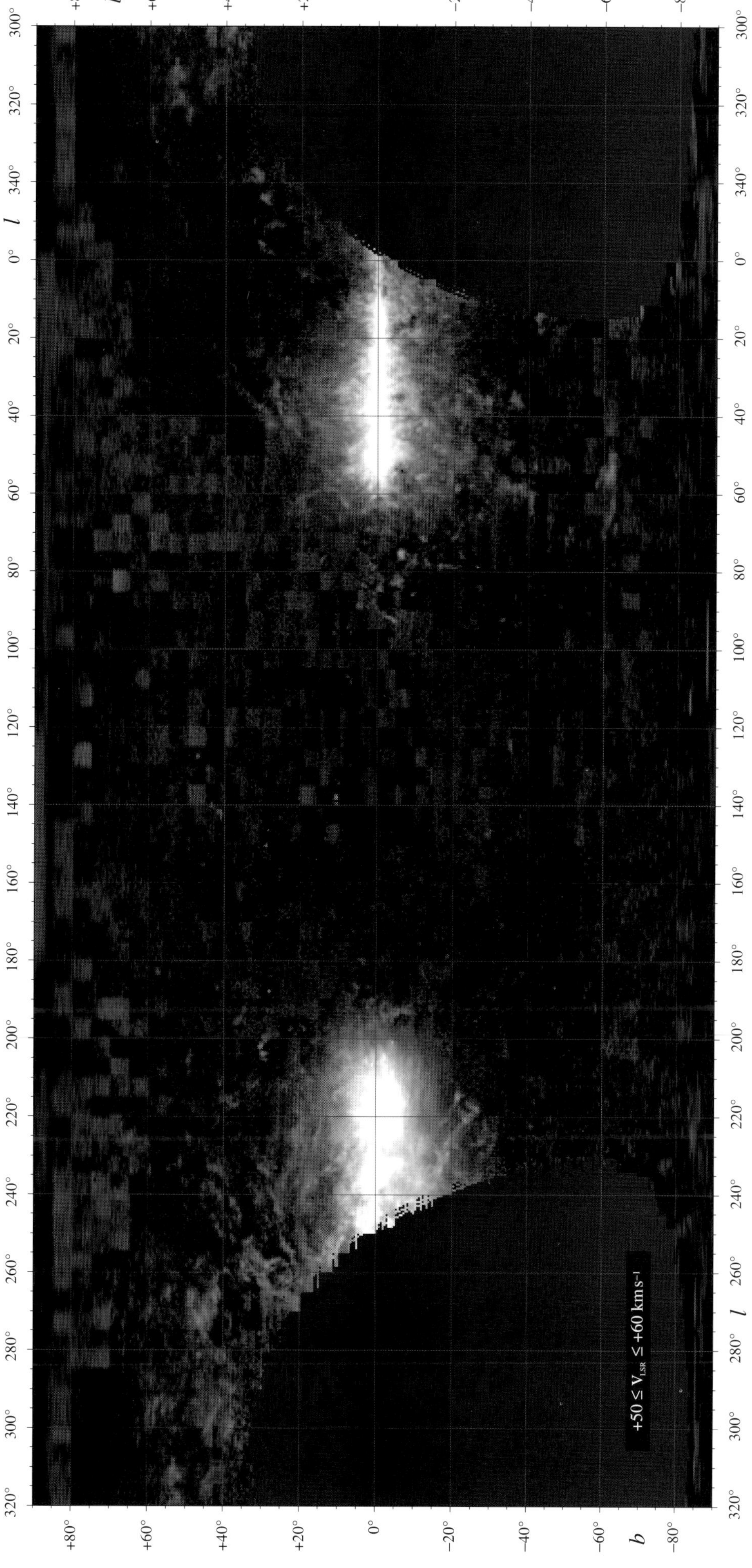
+50 ≤ V_LSR ≤ +60 km s⁻¹
l
b
300° 320° 340° 0° 20° 40° 60° 80° 100° 120° 140° 160° 180° 200° 220° 240° 260° 280° 300° 320°
+80° +60° +40° +20° 0° −20° −40° −60° −80°

$+60 \leq V_{LSR} \leq +70\ km\,s^{-1}$

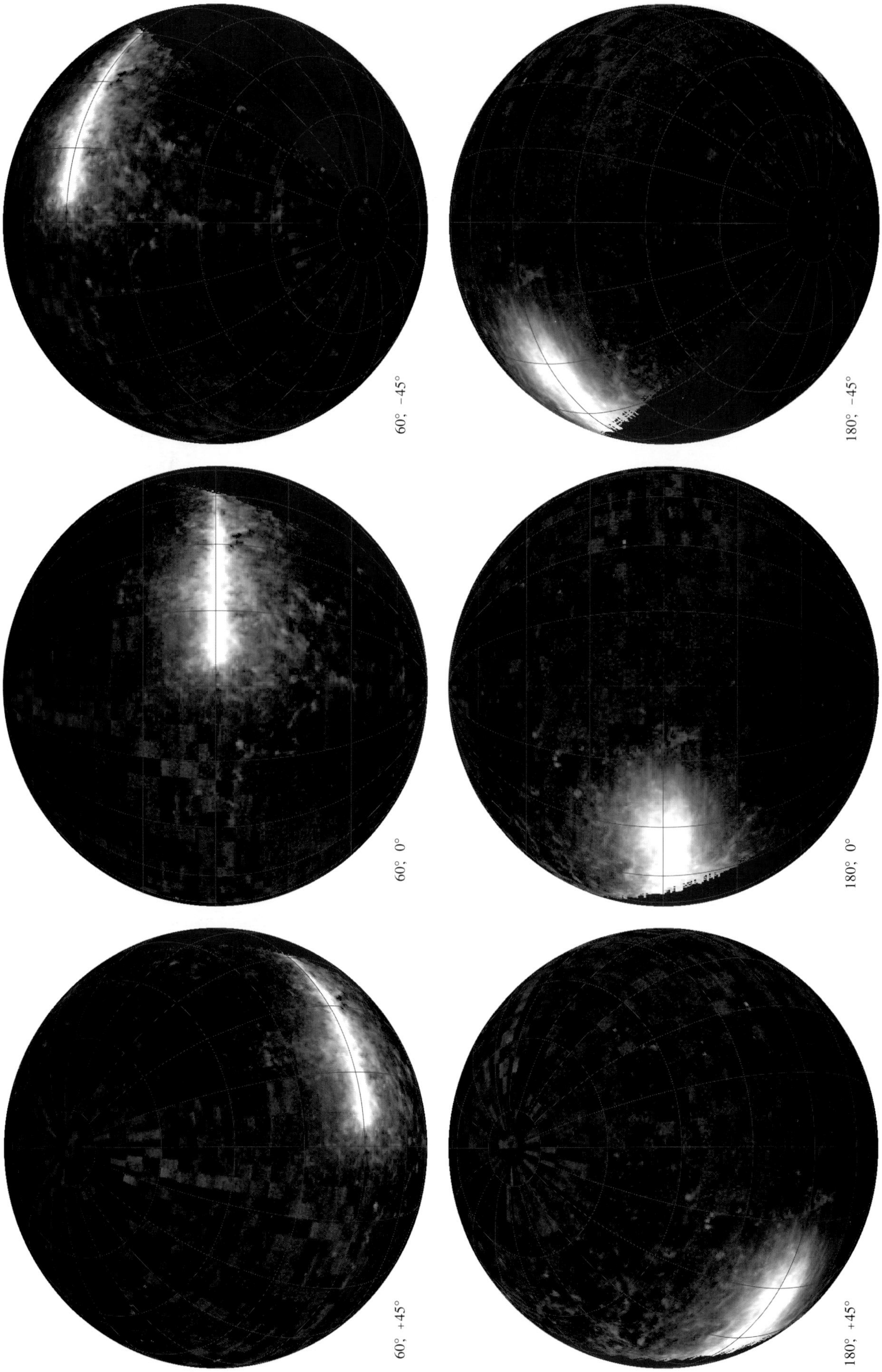

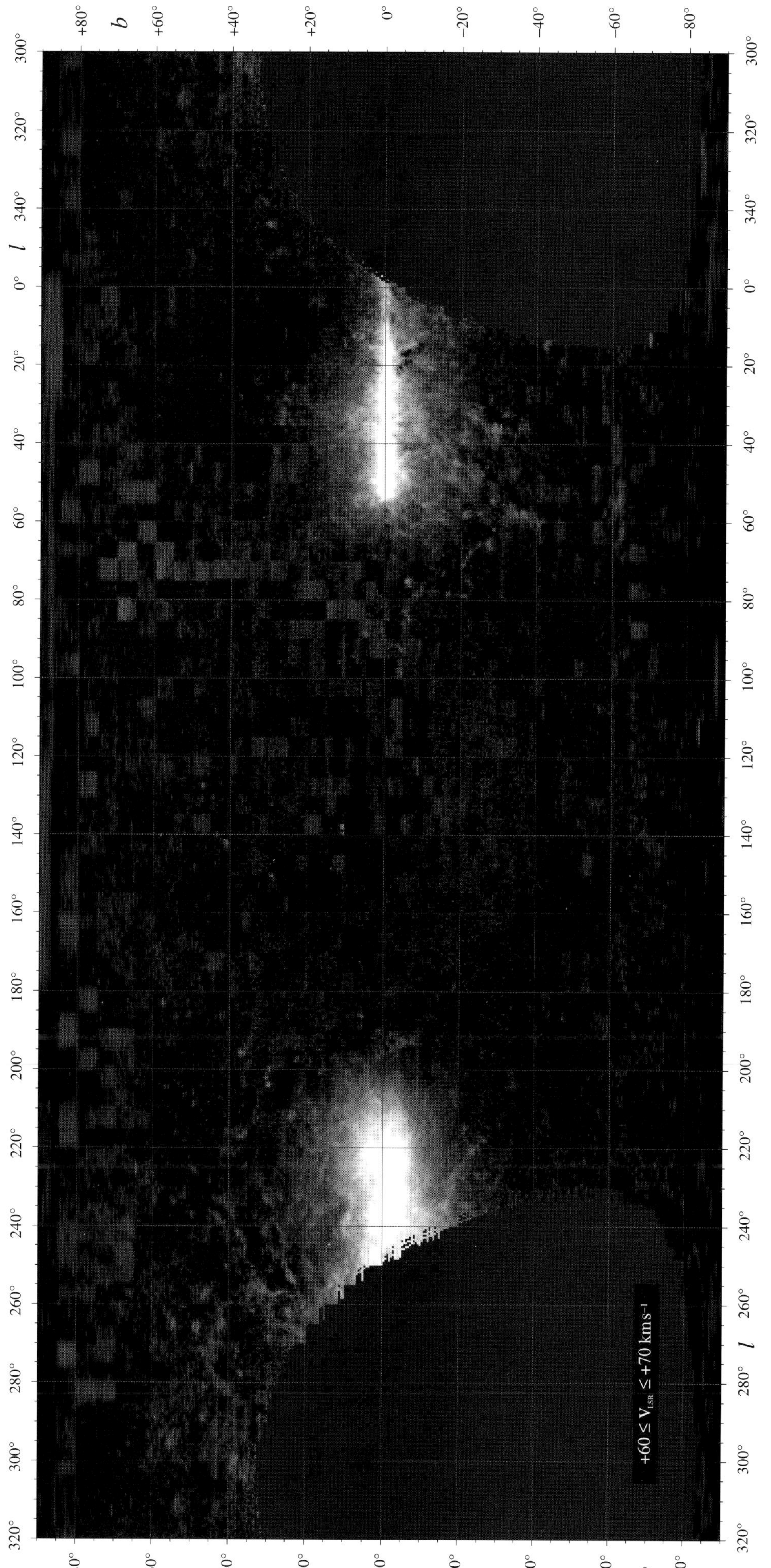
+60 ≤ V_LSR ≤ +70 km s⁻¹
l
b

$+70 \leq V_{LSR} \leq +80$ km s^{-1}

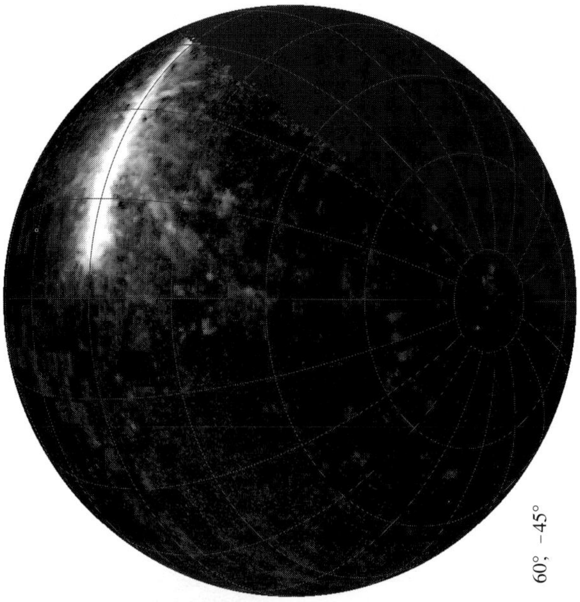

60°, −45°

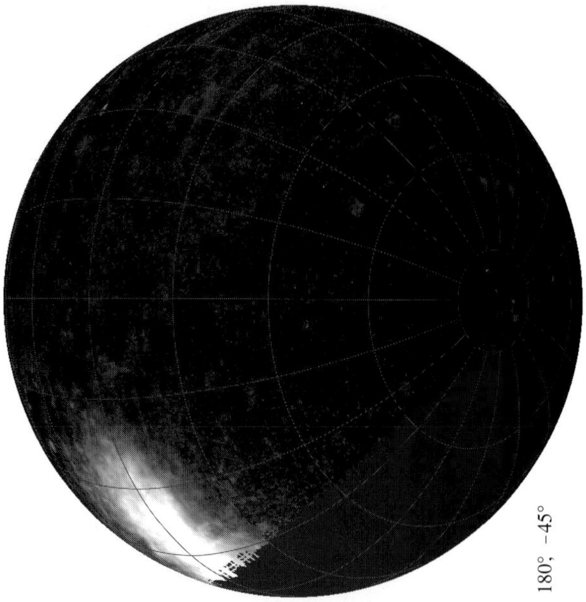

180°, −45°

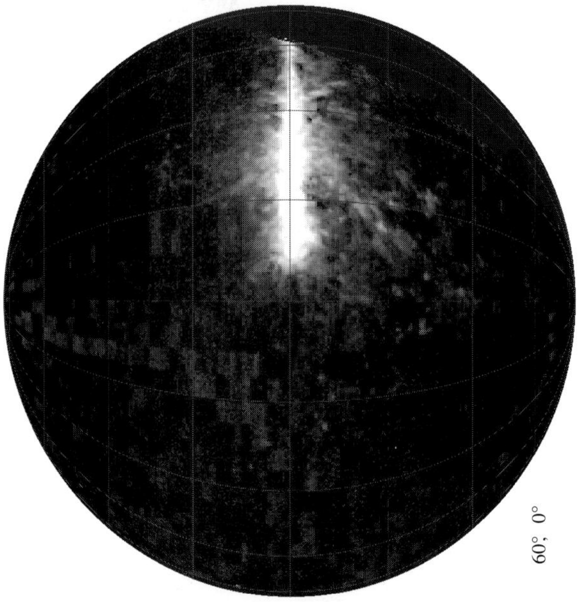

60°, 0°

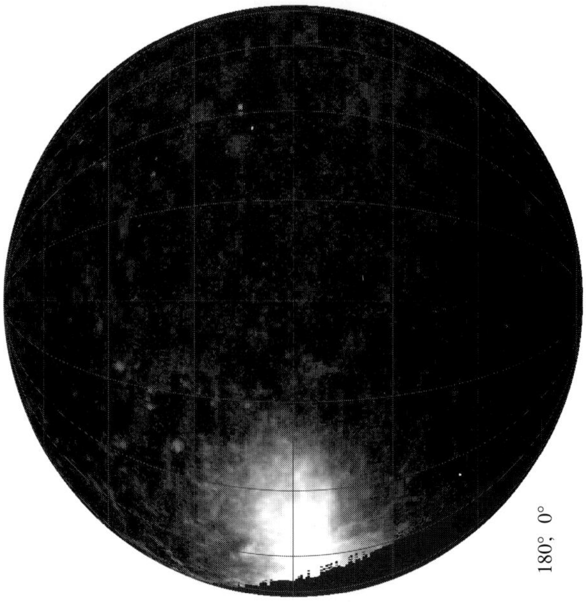

180°, 0°

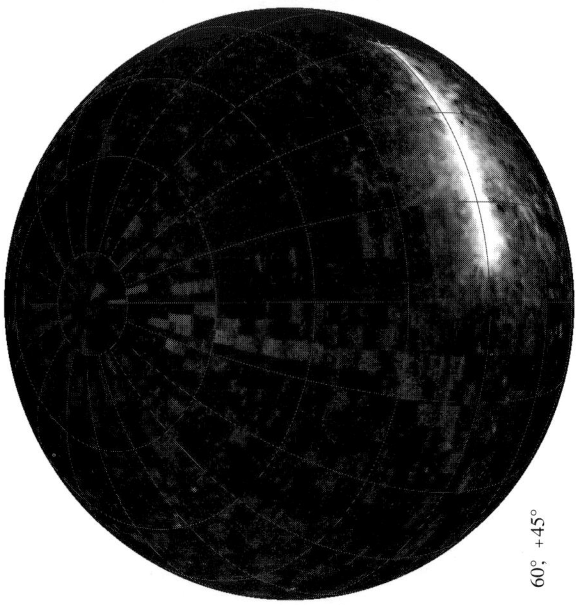

60°, +45°

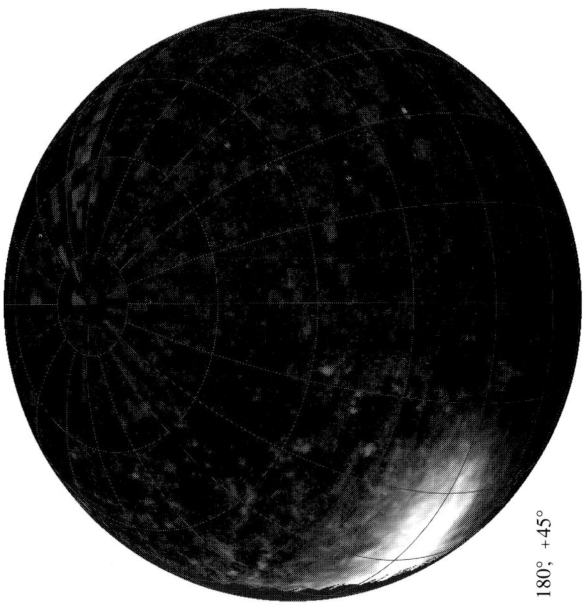

180°, +45°

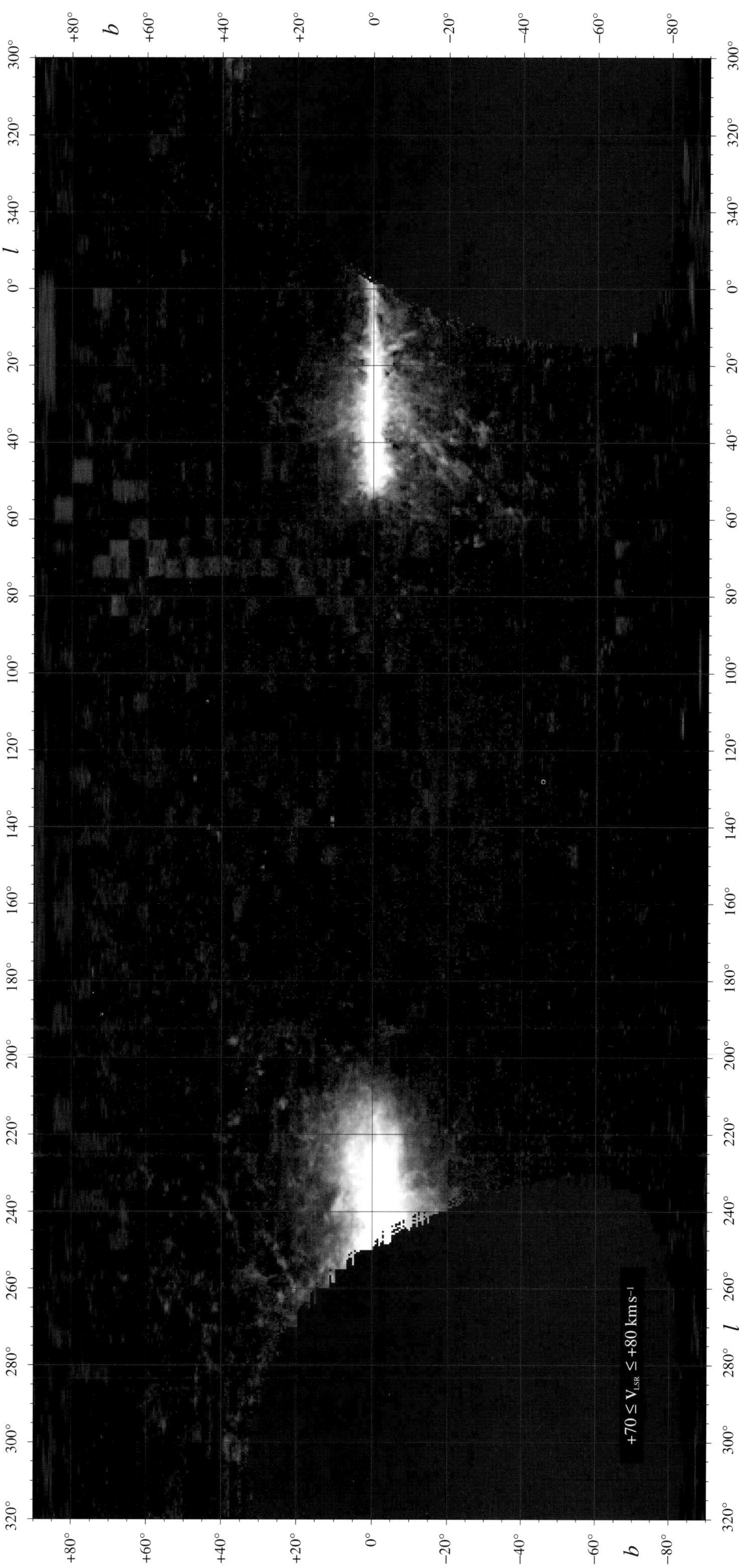

$+70 \le V_{LSR} \le +80$ kms^{-1}
l
b
300° 320° 340° 0° 20° 40° 60° 80° 100° 120° 140° 160° 180° 200° 220° 240° 260° 280° 300° 320°
+80° +60° +40° +20° 0° −20° −40° −60° −80°

$+80 \leq V_{LSR} \leq +90$ km s^{-1}

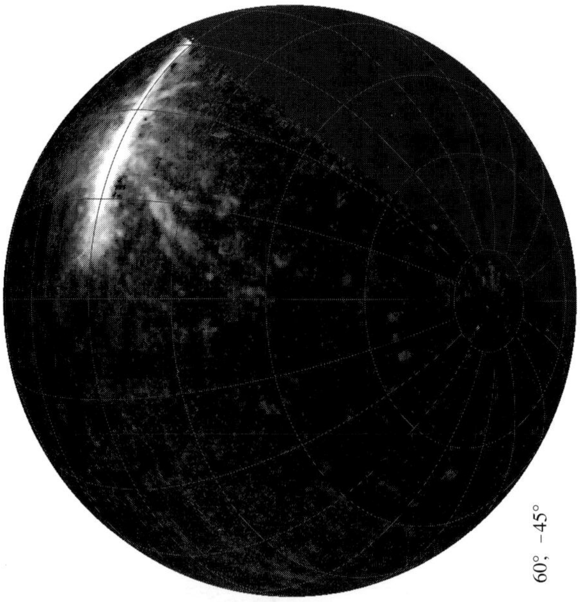

60°, −45°

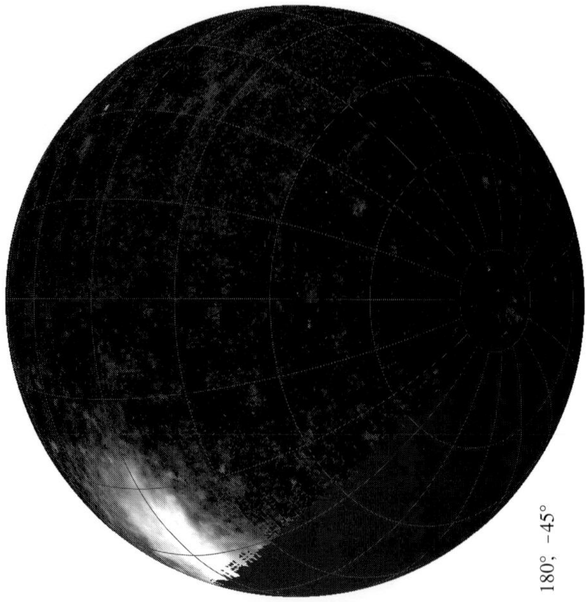

180°, −45°

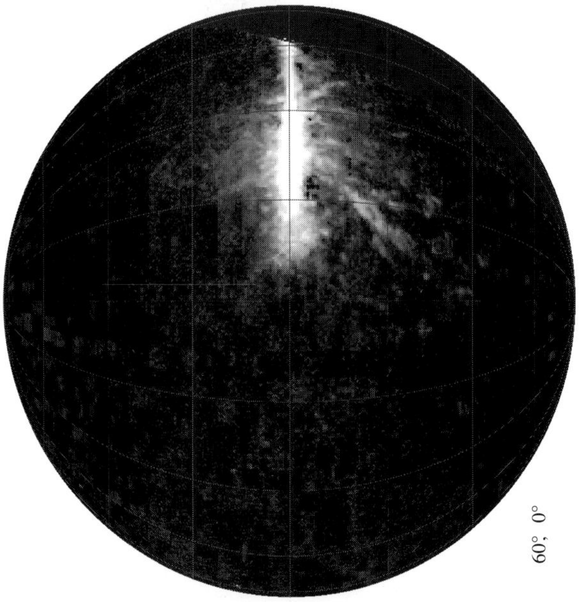

60°, 0°

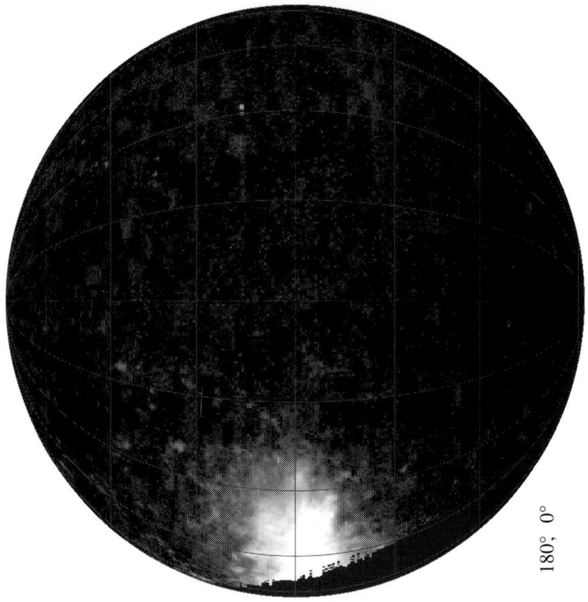

180°, 0°

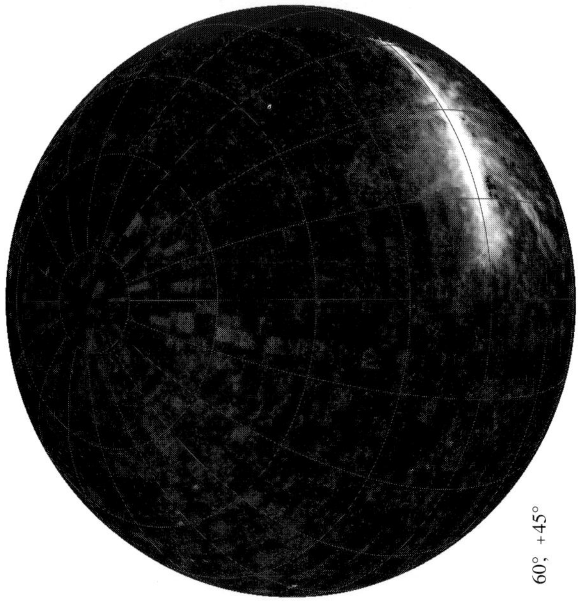

60°, +45°

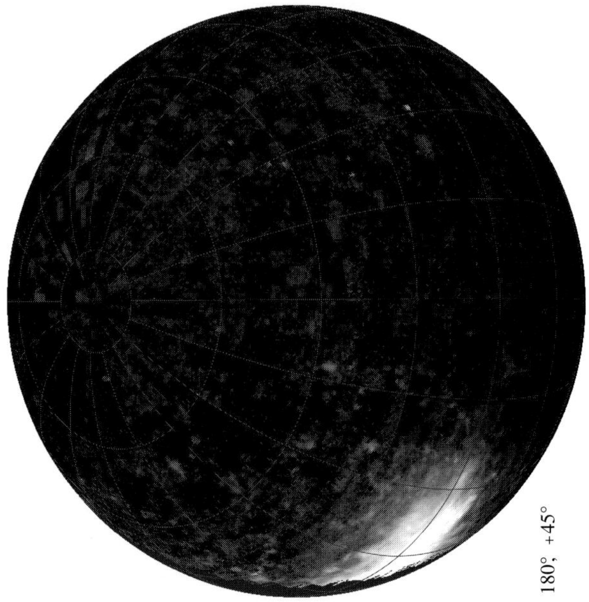

180°, +45°

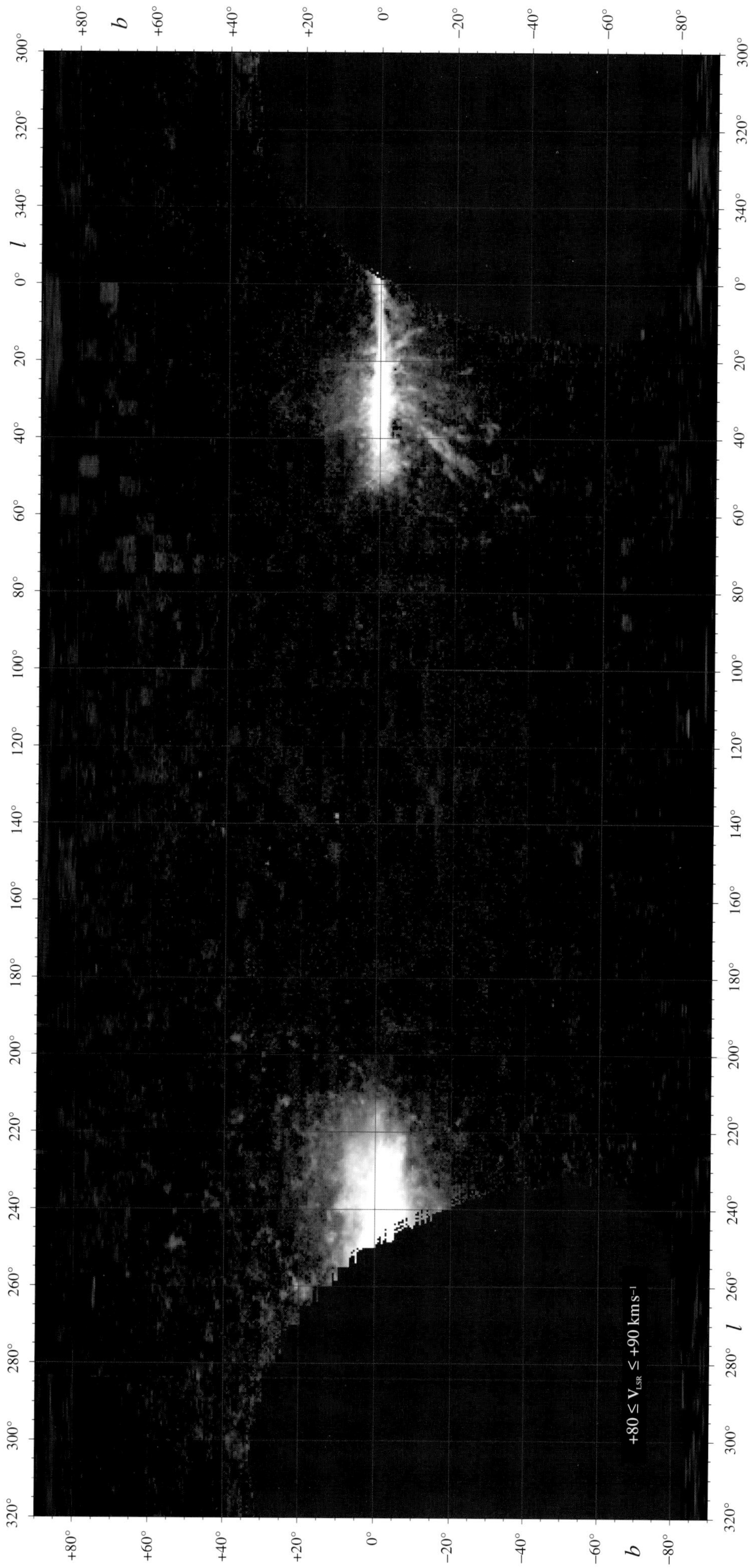
$+80 \le V_{LSR} \le +90$ km s^{-1}

$+90 \leq V_{LSR} \leq +100$ km s^{-1}

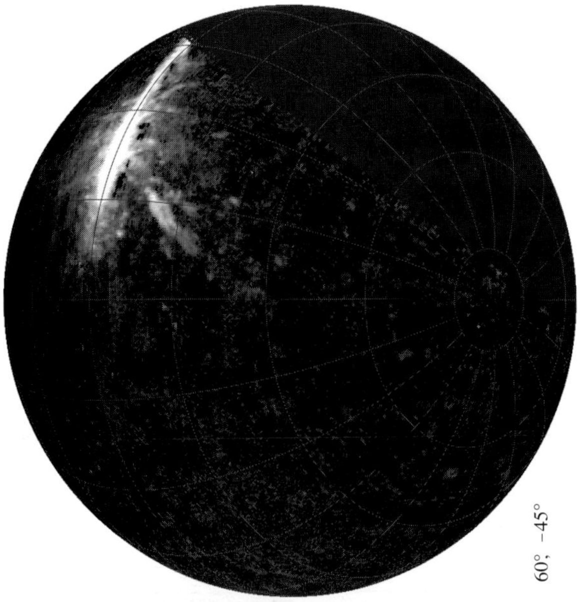

60°, −45°

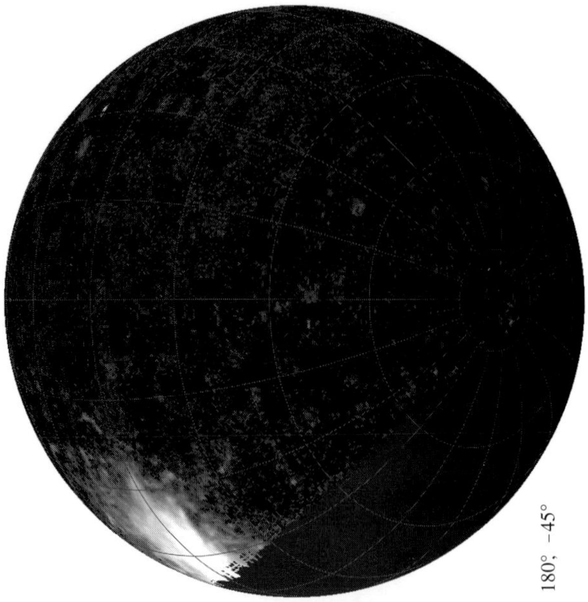

180°, −45°

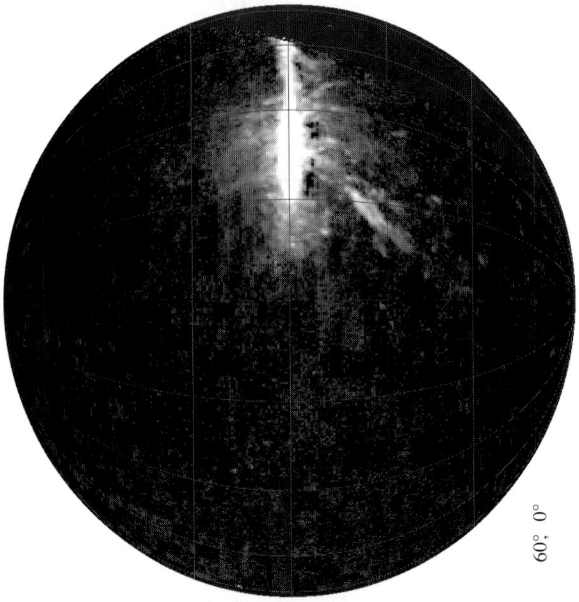

60°, 0°

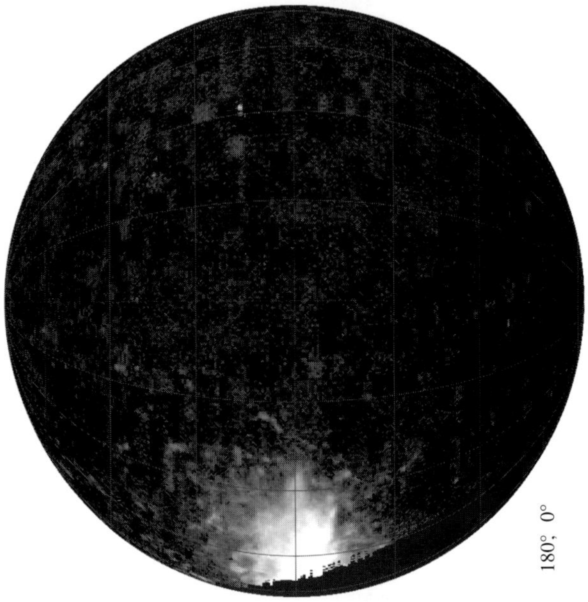

180°, 0°

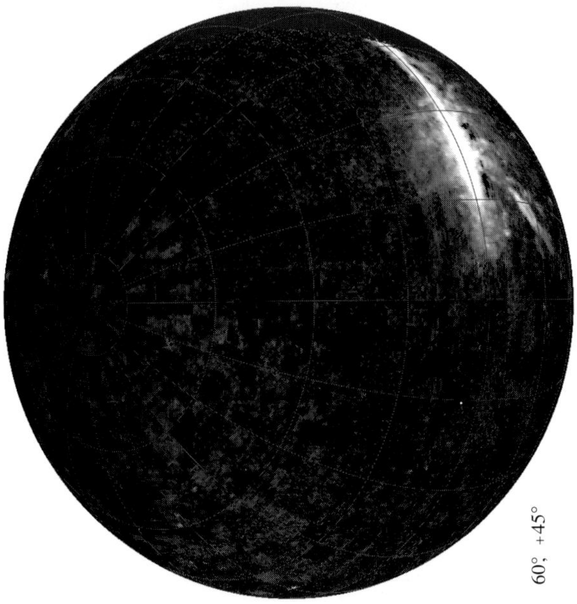

60°, +45°

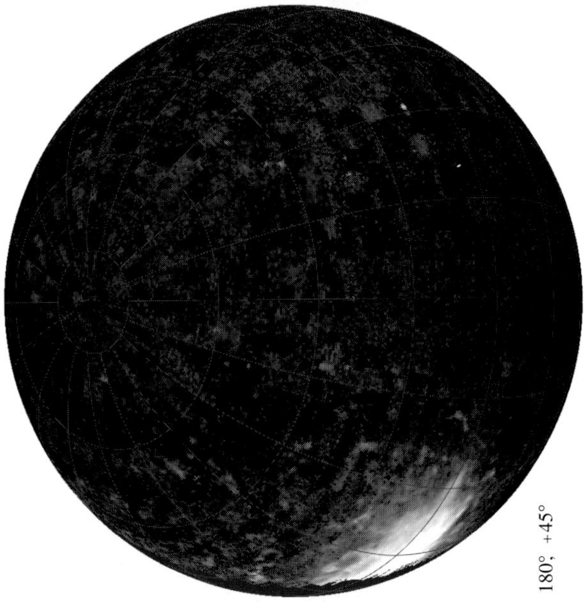

180°, +45°

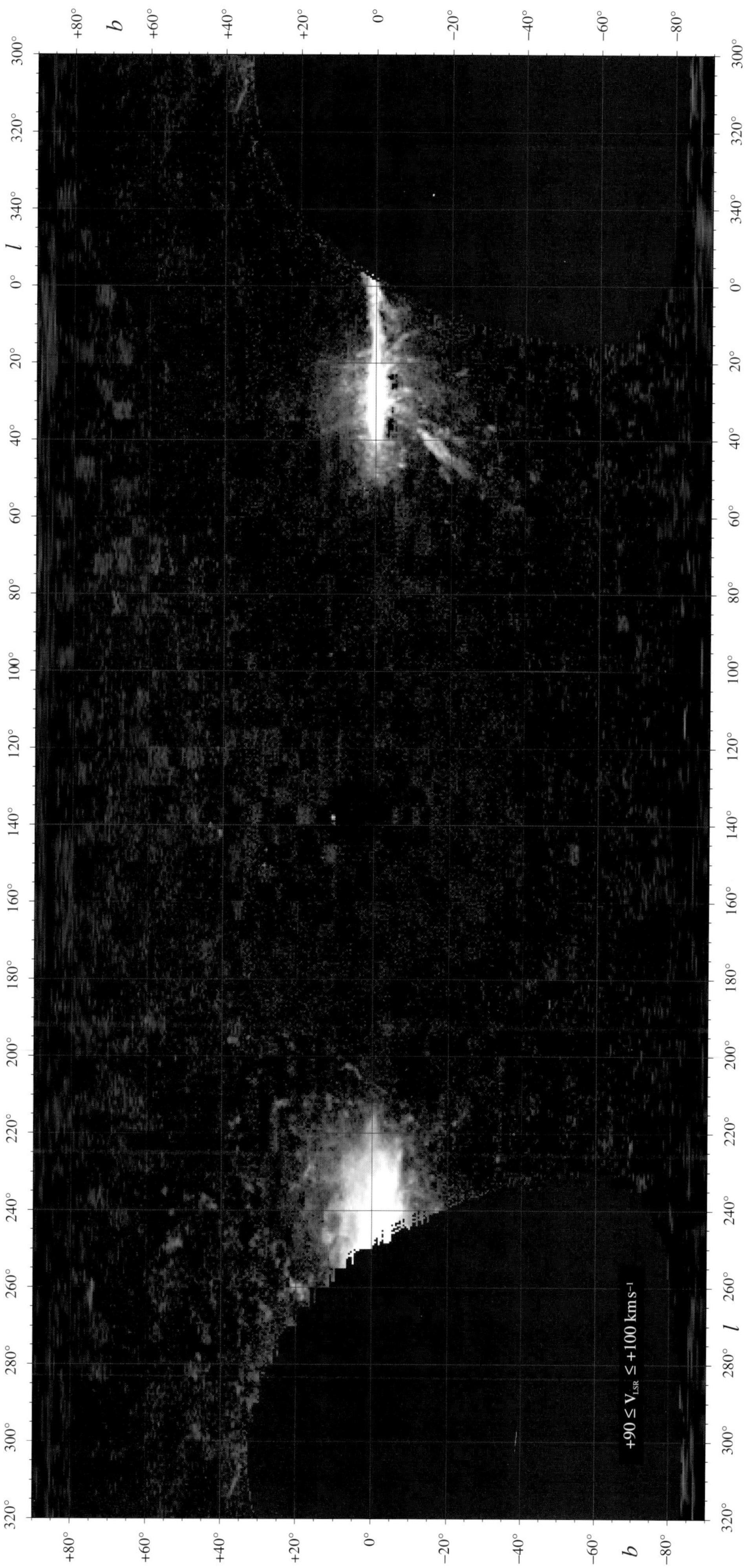
+90 ≤ V_LSR ≤ +100 kms-1

$+100 \leq V_{LSR} \leq +200$ km s^{-1}

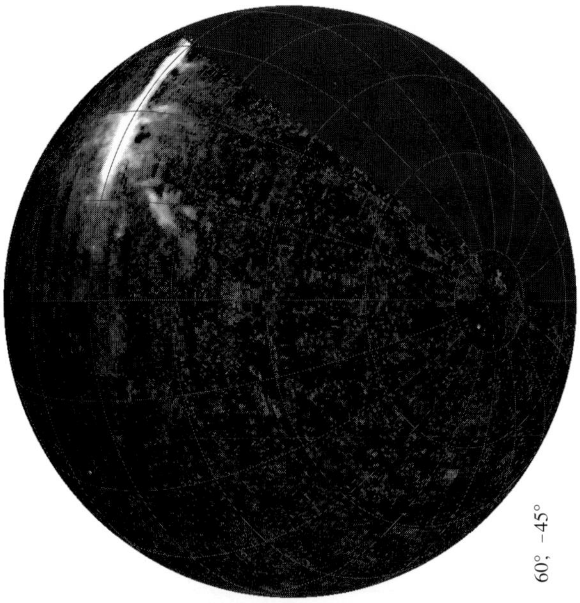

60°, −45°

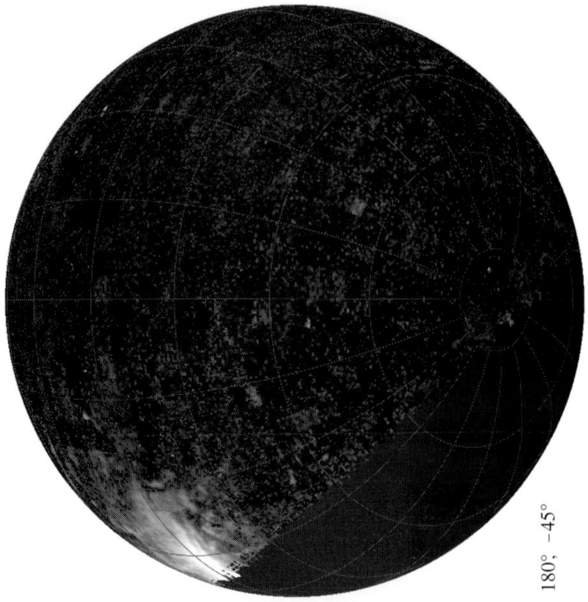

180°, −45°

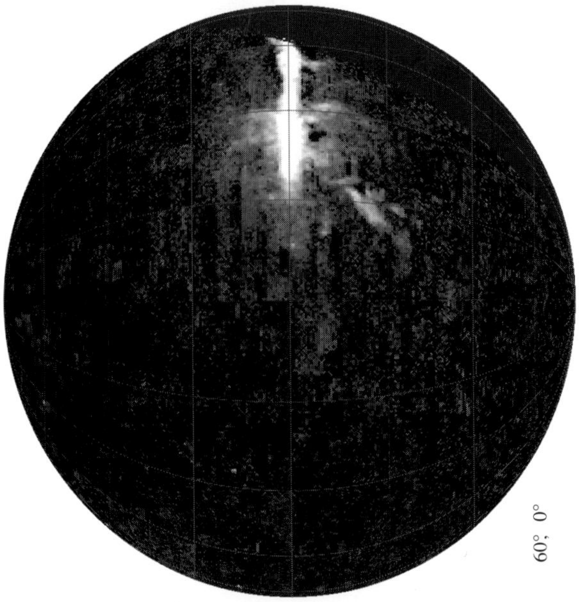

60°, 0°

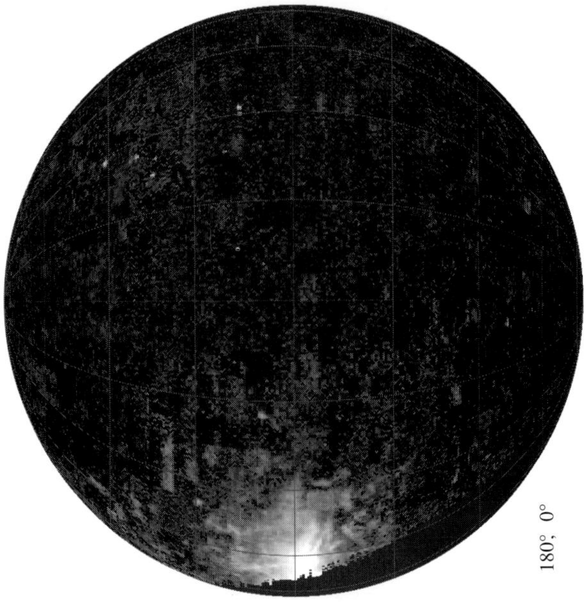

180°, 0°

60°, +45°

180°, +45°

$+100 \leq V_{LSR} \leq +200$ km s^{-1}
l
b

$+200 \le V_{LSR} \le +400$ km s^{-1}

60°, −45°

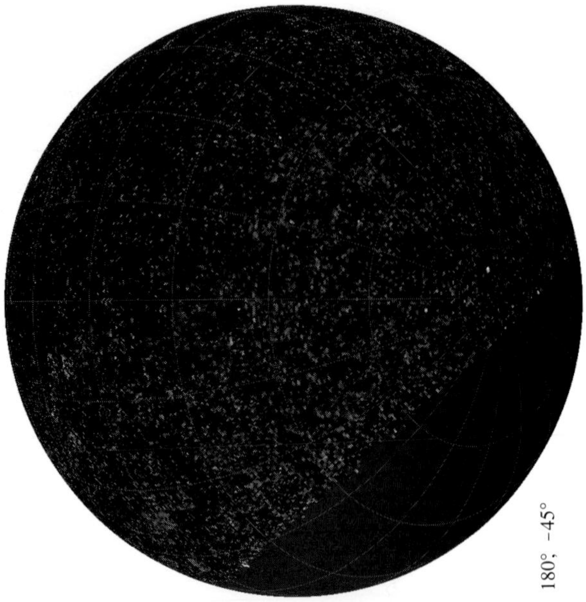

180°, −45°

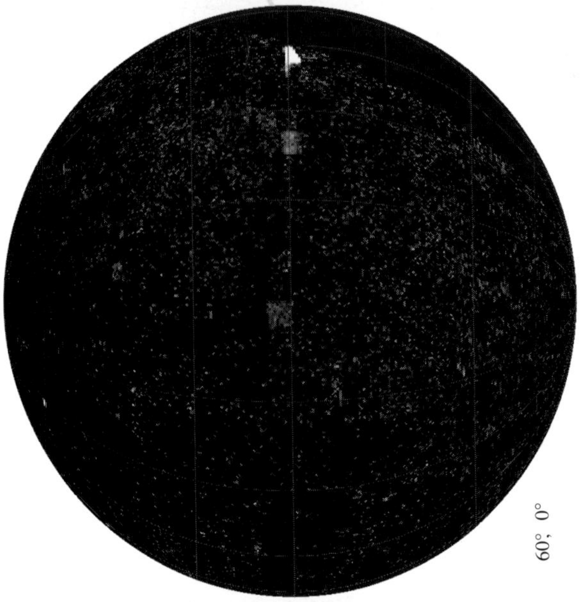

60°, 0°

180°, 0°

60°, +45°

180°, +45°

+200 ≤ V_{LSR} ≤ +400 km s^{-1}
l
b

The CD-ROM (ISO 9660) accompanies the Atlas. It contains the entire dataset of the Leiden/Dwingeloo survey in FITS format and can be used on computers with CD-ROM drives, as described in the README.TXT file on the disk. The CD-ROM also contains color images in GIF format of all the maps in the Atlas, as well as animations displaying the three-dimensional data cube.